Optimisation of Manufacturing Processes
A Response Surface Approach

Optimisation of Manufacturing Processes

A Response Surface Approach

MARK EVANS

MANEY

FOR THE INSTITUTE OF MATERIALS, MINERALS AND MINING

B0791
First published for IOM³ in 2003 by
Maney Publishing
1 Carlton House Terrace
London SW1Y 5DB

© IOM³ 2003
All rights reserved

ISBN 1-902653-86-6

Typeset in India by Emptek Inc.
Printed and bound in the UK by the Charlesworth Group

Contents

Preface   xi

I. CONCEPTS METHODS AND CASE STUDIES   1

1. PROCESS OPTIMISATION THROUGH INDUSTRIAL EXPERIMENTATION   3
    1.1   OBJECTIVES OF INDUSTRIAL EXPERIMENTATION FOR QUALITY IMPROVEMENT   3
    1.2   A SIMPLE MODEL OF A MANUFACTURING PROCESS   4
    1.3   PROCESS OPTIMISATION   6
    1.4   STATISTICALLY PLANNED (DESIGNED) EXPERIMENTS   12
    1.5   PLANNING A RESEARCH PROGRAMME   15

2. ENGINEERING CASE STUDIES   17
    2.1   THE AUSFORMING PROCESS   17
    2.2   HIGH STRENGTH STEELS   18
    2.3   HOT FORGED COPPER POWDER COMPACTS   18
    2.4   CLOSED DIE FORGING OF AERO ENGINE DISKS   21
    2.5   FRICTION WELDING   23
    2.6   THE INJECTION MOULDING EXPERIMENT   24
    2.7   THE SPRING FREE HEIGHT EXPERIMENT   25
    2.8   THE PRINTING PROCESS STUDY   26

II. LINEAR EXPERIMENTAL DESIGNS   27

3. THE 2^K EXPERIMENTAL DESIGN: FULL FACTORIALS   29
    3.1   THE 2^2 FACTORIAL DESIGN   29
         3.1.1   A Traditional Design   29
         3.1.2   A Full 2^2 Factorial Design in Standard Order Form   30
         3.1.3   A Geometric Representation of the 2^2 Experimental Design   32
    3.2   THE 2^3 FACTORIAL DESIGN   33
         3.2.1   A Geometric Representation of the 2^2 Experimental Design   33
         3.2.2   Standard Order Form for the 2^3 Factorial Experiment   33
         3.2.3   An Example of a Replicated 2^3 Factorial Experiment   35
    3.3   THE 2^k FACTORIAL DESIGN   36
         3.3.1   Standard Order Form for the 2^5 Factorial Experiment   36
         3.3.2   General Comments on the 2^k Designs   36

4. THE 2^{k-p} EXPERIMENTAL DESIGN: FRACTIONAL FACTORIALS — 39
- 4.1 BASIC CONCEPTS — 39
- 4.2 THE ONE HALF FRACTION OF THE 2^k DESIGN — 40
  - 4.2.1 Step One: Defining the Base Design — 40
  - 4.2.2 Step Two: Introduction of the Remaining Factor — 40
- 4.3 OTHER FRACTIONAL FACTORIAL DESIGNS — 43
  - 4.3.1 Step One: Defining the Base Design. — 43
  - 4.3.2 Step two: Introduction of the Remaining Factors — 43
- 4.4 A 2^{7-3}_{IV} DESIGN FOR THE AUSFORMING PROCESS — 45
- 4.5 TAGUCHI'S ORTHOGONAL ARRAYS — 46

III. OPTIMISATION OF LINEAR PROCESSES — 55

5. CONTROLLING THE MEAN : LOCATION EFFECTS IN LINEAR DESIGNS — 57
- 5.1 DEFINITION OF LOCATION EFFECTS — 57
  - 5.1.1 Control of a Mean Quality Characteristic using Main Location Effects — 57
  - 5.1.2 Control of a Mean Quality using First Order Interaction Location Effects — 58
  - 5.1.3 Higher Order Interaction Location Effects — 59
- 5.2 METHODS OF CONTROLLING THE MEAN QUALITY CHARACTERISTICS — 60
  - 5.2.1 A Control Matrix for the Mean of a Quality Characteristic — 60
  - 5.2.2 A First Order Response Surface Model for the Mean. — 62
- 5.3 THE YATES AND LEAST SQUARES PROCEDURES FOR ESTIMATING LOCATION EFFECTS — 64
  - 5.3.1 The Yates Technique — 64
  - 5.3.2 The Least Squares Technique — 66
- 5.4 LOCATION EFFECTS ESTIMATED FROM FRACTIONAL DESIGNS — 68
  - 5.4.1 Aliasing Algebra — 70
  - 5.4.2 Taguchi Designs and Aliasing — 73
  - 5.4.3 Yates Technique for Fractional Factorials — 74
- 5.5 LOCATION EFFECTS IN THE AUSFORMING PROCESS — 76
  - 5.5.1 The First Two Factors Only — 76
  - 5.5.2 The First Three Factors Only — 80
  - 5.5.3 The First Five Factors Only — 85
  - 5.5.4 All Seven Factors of the Ausforming Process — 101

6. TESTING THE IMPORTANCE OF LOCATION EFFECTS IN THE 2^K DESIGN — 109
- 6.1 A DISTRIBUTION OF EFFECT ESTIMATES — 109
- 6.2 THE STANDARD DEVIATION OF A LOCATION EFFECT ESTIMATE — 114

6.3	The *t* Test in a Replicated Design	116
	6.3.1	The Test	116
	6.3.2	Application of the *t* Test to the 2^3 Ausforming Experiment	118
6.4	The *t* Test Within the Least Squares Procedure	122
6.5	A Graphical Test for the Importance of Location Effects	125
	6.5.1	Test Derivation	125
	6.5.2	Illustration of Graphical Test Using the High Strength Steel Case Study	128
	6.5.3	Illustration of Graphical Test Using the Ausforming Process	132

7. CONTROLLING PROCESS VARIABILITY: DISPERSION EFFECTS IN LINEAR DESIGNS   137

7.1	Noise - Design Factor Interactions and Process Variability	138
7.2	A Generalised Response Surface Approach to Process Variability	141
7.3	An Application to the 2^5 Design on the Ausforming	143
7.4	Prediction Errors and Process Variability	146
	7.4.1	Estimate a Simplified Response Surface Model of the Process	146
	7.4.2	Calculate the Prediction Error Variability at Each Factor Level	146
	7.4.3	Testing the Importance of Dispersion Effects	147
7.5	The Disk Forging Operation Experiment	148
	7.5.1	Estimate a Simplified Model of the Process	149
	7.5.2	Calculate the Error Variability at Each Factor Level	149
	7.5.3	Testing the Importance of Dispersion Effects	159
7.6	A Generalised Linear Model	159
	7.6.1	Inner and Outer Arrays	159
	7.6.2	Simple Summary Statistics	160
	7.6.3	The Tendency for Process Mean and Variability to Move Together	164
	7.6.4	PerMIA Summary Statistics	166
	7.6.5	Step 1 Identify All the Control Factors	168
	7.6.6	Step 2. Obtain Reliable Estimates of the Dispersion Effects for All the Control Factors	170
7.7	The Copper Compact Experiment	171
	7.7.1	Step 1. Identify All the Control Factors for Making Copper Compacts	172
	7.2.2	Step 2. Reliable Estimates of the Dispersion Effects for Making Copper Compacts	182
7.8.	Comparing the Response Surface and Generalised Linear Models Using the Injection Moulding Experiment	184
	7.8.1	Design Problems	185
	7.8.2	Analysis of the Data	188
		7.8.2.1 The Blind Use of the $(S\text{-}N)_T$ Ratio	191
		7.8.2.2 The Lack of Analysis for Noise Factors	194

8.	**LINEAR PROCESS OPTIMISATION**	**209**
8.1	A Two Step Process Optimisation Procedure	209
	8.1.1 The Procedure	209
	8.1.2 General Techniques	212
8.2	Illustrations of Process Optimisation	214
	8.2.1 The Ausforming Process	214
	8.2.2 The Copper Compact Experiment	218
	8.2.3 The Injection Moulding Experiment	222
	8.2.4 Optimising the Disk Forging Operation	226

IV.	**NON LINEAR EXPERIMENTAL DESIGNS**	**229**
9.	**SOME NON LINEAR EXPERIMENTAL DESIGNS**	**231**
9.1	3^k Designs	231
9.2	A 3^2 Design for the Friction Welding Case Study	235
9.3	Central Composite Designs	236
9.4	A Central Composite Design for the Linear Friction Welding Case Study	242
9.5	The Box-Behnken Design	245
	9.5.1 Find all Combinations of Two	245
	9.5.2 Form 2^2 Designs for all Pairings	246
	9.5.3 Replication of Centre Points	248
9.6	Mixed Level Factorial Designs	250
	9.6.1 Factors at Two and Three Levels	251
	9.6.2 Factors at Two and Four levels	253

10.	**LINEAR AND NON LINEAR EFFECTS**	**255**
10.1	A Non Linear Effect	255
10.2	The Second Order Response Surface Model	260
	10.2.1 Structure of the Second Order Response Surface Model	260
	10.2.2 Some Models for the 3k Design	261
	10.2.3 Some Models for the Central Composite and Box Behnken Designs	263
	10.2.4 A Model for Mixed Factorial Designs	263
10.3	Estimating Response Surface Models	264
	10.3.1 Estimating a Second Order Response Surface Model	264
	10.3.2 Estimating Some Response Surface Models using Data from a 3^2 Design	266
	10.3.3 Estimating Some Response Surface Models using Data from a 3^3 Design	269
	10.3.4 Estimating Some Response Surface Models using Data from a Central Composite Design	271

	10.4 ANALYSIS OF THE FRICTION WELDING EXPERIMENT	272
	10.4.1 The First Two Process Variables	272
	10.4.2 All Three Process Variables	278

V. OPTIMISATION OF NON LINEAR PROCESSES — 283

11. SEQUENTIAL TESTING — 285

   11.1 SEQUENTIAL TESTING AND THE PATH OF STEEPEST ASCENT — 285
   11.2 SEQUENTIAL EXPERIMENTATION FOR THE AUSFORMING PROCESS — 288
       11.2.1 The First Two Factors Only — 288
       11.2.2 The First Five Factors — 294

12. DUAL RESPONSE SURFACE METHODOLOGIES — 297

   12.1 THE DUAL RESPONSE SURFACE METHODOLOGY — 297
       12.1.1 Minimise Variability Subject to a Mean Constraint — 298
       12.1.2 Minimise the Mean Square Error — 301
   12.2 THE PRINTING PROCESS CASE STUDY — 301
       12.2.1 The Experiment — 301
       12.2.2 The PerMIA — 302
       12.2.3 The Modelled Response Surface — 304
       12.2.4 Minimise Variability Subject to a Mean Constraint — 304
       12.2.5 The Mean Square Error — 312

REFERENCES — 315

INDEX — 317

Preface

This is an intermediate level textbook dealing with the optimisation of manufacturing processes using the results from statistically designed experiments. It is based on courses that I have taught at the University of Wales Swansea for over 10 years. It also reflects some of the methods that I have found useful whilst acting as a statistical consultant under the award–winning postgraduate partnership with industry run by the Materials Engineering Research Centre at Swansea University.

The book has been designed for final year undergraduate and Masters courses, as well as for researchers and practising technical professionals in engineering and the physical and chemical sciences who are new to the fields of product and process design, process improvement and quality engineering. I have used this book as the basis of an industrial short course on design and analysis of industrial experiments for practising technical professionals. There are a number of examples illustrating all the designs and analytical techniques contained within the book. These examples are drawn from a number of different fields of engineering ranging from the well-established processes in heavy manufacturing right the way through to processes that are at the forefront of modern engineering research.

The book provides a comprehensive introduction to the subjects of planning, implementation and analysis of experiments designed to both improve existing manufacturing process and to create newer and better processes and products. Such process optimisation, that results in the manufacture of products that consistently meet customer-required specifications, can be achieved in a variety of different ways and this book concentrates on the response surface approach to the problem. As such coverage includes techniques for designing linear and non-linear experiments, for identifying parsimonious models that can be used for predicting product quality and for analysing such models so that the quality required by the customer can be manufactured on a consistent basis.

Mark Evans

PART I
CONCEPTS METHODS AND CASE STUDIES

1. Process Optimisation Through Industrial Experimentation

Many engineering companies around the world have or are currently undergoing a quality control and improvement revolution that originally started in Japan many decades ago and this chapter gives a brief overview of this revolution. Robust design is a central component of the modern approach to quality improvement and is a phrase used to describe any engineering activity whose objective is to develop high quality products (and processes) at low cost. A key characteristic of robust design is the use of statistically planned (designed) experiments to identify those process variables that determine product quality. Robust design was developed in Japan by G. Taguchi[1] in the early 1950s and its wide spread use throughout Japanese industry is one of the main reasons why Japan has emerged as a major producer of relatively cheap high quality products, especially in the automobile, home electronics and microprocessing sectors. Despite its early success in Japan, robust design remained virtually untried in the United States and Europe until the early 1980s. The realisation that quality is a vital ingredient required for success in today's highly global and competitive markets has prompted Western companies to embrace the robust design concept.

1.1 OBJECTIVES OF INDUSTRIAL EXPERIMENTATION FOR QUALITY IMPROVEMENT

This book is about the planning, implementation and analysis of experiments designed to both improve existing manufacturing processes and to create newer and better processes and products. The objective of such industrial experimentation is therefore to achieve product and process optimisation through a structured manipulation of those process variables that determine the quality of the finished product. An optimised process is one that is capable of consistently producing a product to a customer required specification. **Process optimisation** is essentially a programme of continued quality improvement and a process can only be optimised once it is first understood how best to control it. Any successful experiment must therefore be capable of providing exactly the right information necessary to achieve such control. This book therefore emphasises both the planning of experiments to ensure that such information is obtained and the analysis of the resulting data to ensure the information is used correctly so that effective control and optimisation is achieved.

The focus of the book is on the engineering and physical sciences so that a number of detailed case studies are presented to help develop the ideas behind the planning and analysis of experiments. These are discussed in detail in Chapter 2 and range from relatively well-known processes, such as the ausforming process for making high strength steels, to friction welding techniques that are currently at their development stage for the joining of blades to disks in modern aero engines.

The automotive sector has numerous examples of product and process optimisation. This sector is one of the main markets for finished sheet steel where it is used to form a variety of different car body panels. The car industry is highly competitive and car manufacturers are continually looking for new materials that have higher strength to weight ratios (so that fuel economy can be improved without compromising safety), that have excellent corrosion resistance and that have low purchase cost. Aluminium has a clear natural advantage over mild steel in relation to the first two of these critical properties. To maintain market share the steel industry has therefore had to invest large sums of money into research designed to improve the corrosion resistance and strength to weight ratios of its finished sheet steels whilst at the same time keeping the cost of these new steels as low as possible.

This research has been very successful and is reflected in the fact that about 50% of the steels used in today's average family motorcar were simply not available ten years ago. Further, the use of aluminium is still confined to low volume car production. This product development has involved the use of designed experiments that have typically varied both the chemical composition of the steel (e.g. the amount of alloying elements such as chromium and nickel added to a base steel), the way the steel is treated (e.g. the type of quench and its rate) and the way that it is coated so as to find new steels that have the required strength to weight ratio or the required level of corrosion resistance. By designing the experiments correctly the steel treatments and alloying elements required to maximise (optimise) these critical material properties have be found. In this way steel manufacturers have managed to meet the ever increasing demand from car manufacturers for steels with better material properties.

1.2 A SIMPLE MODEL OF A MANUFACTURING PROCESS

To gain a fuller understanding of process optimisation it is helpful to first give a simplified representation of a manufacturing process. Figure 1.1 below gives such a simplified but very illuminating overview. Any manufacturing process can be visualised as a combination of energy, machines, labour and technological methods that turn input materials into an output or finished product. This product will have a number of quality characteristics termed responses (Y). The focus of any experiment should be on those quality characteristics that help sell the product. For example, a high strength to weight ratio is one of many quality characteristics possessed by the new high strength steels (such as

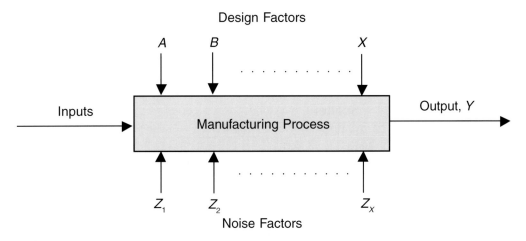

Fig. 1.1 A simple representation of a manufacturing process.

the bake hardenable steels) and maximising this ratio is critical in the fight against aluminium in the automotive market.

These quality characteristics are likely to be affected by a number of process variables, termed **factors**, at the point of manufacture. These factors can in turn be classified into two broad groups. **Design factors**, $(A, B, ..., X)$, are process variables that can be set within engineering specification. Consequently, they tend to be cheaply and easily controllable. The amount of various alloying elements added to a base steel are good examples of control factors. On the other hand **noise factors**, $(Z_1, Z_2, ..., Z_x)$, are process variables that can't be controlled in a precise way at the point of manufacture, although for the purpose of the experiment they may well be controllable within the friendly environment of a laboratory. The quench rate used in the heat treatment of a new steel alloy is a good example of a noise factor. Noise factors originate from two main sources. First, some factors are simply beyond human control. For example, the performance of a car engine will to some extent be dependant upon the air temperature through which it is travelling. Such temperatures are generally beyond human control. On the other hand, the technology required to control some factors may not yet be available or could be made available but at a tremendous cost. In fact a key characteristic of noise factors are that they are incredibly expensive to control within engineering specification at the point of production.

Some manufacturing processes are inherently **linear** in nature. For such processes a change in the amount of one factor (for given quantities of all the other factors) will always produce the same change in the level of a quality characteristic, irrespective of whether that change was made from a high or low amount of that factor. On the other

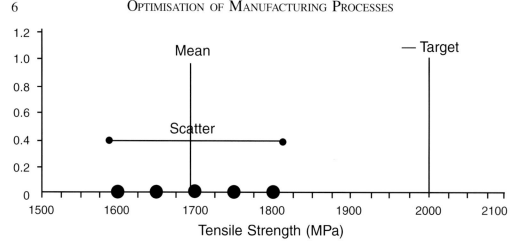

Fig. 1.2 Process variation.

hand many manufacturing processes turn out to be **non-linear** in nature. Here the change observed in a products quality characteristic following a change in the amount of a factor (all other factors remaining unchanged) depends on whether that change was made from a low or high amount of that factor. The structure of this book reflects this fundamental difference between linear and non-linear processes. Thus Part II of this book describes how to construct efficient experiments for linear processes and Part III shows how the results obtained from such experiments can be used optimise these linear processes. Then Part IV of this book describes how to construct efficient experiments for non-linear processes and Part V shows how the results obtained from such experiments can be used to optimise any non-linear process.

1.3 PROCESS OPTIMISATION

Next consider process optimisation in a little more detail. When taking repeated observations on a products quality characteristic at the point of manufacture (such as the tensile strength of a new steel alloy) under the same operating conditions two characteristics will be observed. These characteristics are critical to an understanding of how a process can be optimised to produce high quality products.

The first characteristic of many manufacturing processes is that they provide products whose quality characteristics exhibit **scatter** about the mean quality characteristics. This scatter is termed process variation. Consider, for example, measuring the tensile strength of a new steel alloy for fixed nominal values of the process variables. Figure 1.2 shows

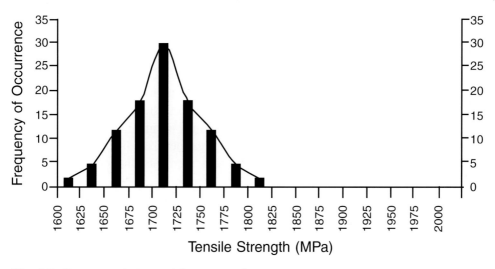

Fig. 1.3 Process variation and frequency of occurrence.

the results that could be obtained from taking five sample measurements from a process making a new steel alloy. If the target requirement set by a car manufacturer is for a strength of 2000 MPa, then this process and resulting product is clearly a long way from being optimal.

The second characteristic of many manufacturing processes is that there will be a varying **frequency of occurrence** for the measured quality characteristic. With repeated sampling it will become apparent that some values for the quality characteristic being measured will occur more frequently than others. In the steel alloy illustration above, if more than five tensile strength measurements are made, the frequency with which a range of tensile strength readings are recorded may look like that in Figure 1.3 for a suboptimal process.

In this example, the manufactured steel is again of low quality because it fails to meet the customer requirements as specified by a target tensile strength of 2000 MPa. Very few, if any, of the batches of manufactured steel meet the target requirement. The manufacturing process is therefore not optimised because it can't consistently produce the steel the customer requires. The scatter observed in the quality characteristic of a product produced by a manufacturing process is termed process variability and there are two main sources.

1. **Noise**

   Noise factors vary naturally during production and so transmit variation to the quality characteristics of a finished product. That is, the failure to achieve a desired level for a quality characteristic with high consistency can, more often than not, be attributed to excessive variability in the noise factors. Each time a response is recorded at the point of manufacture for a particular set of values for the design factors, a different set a values for the noise factors will typically occur and this will result in a variety of response values ~ even though the design factors are unchanging. In the above example, the five tensile strength measurements may have been made at different points in time. Whilst the design factors (such as the amount of each alloying element used) are the same at these points in time the values for the noise factor(s) (such as the quench rate during heat treatment) are likely to be different so giving rise to the different tensile strength measurements illustrated above.

2. **Common Cause**
Even if a process has no noise factors there will still be a small element of observable process variability. This inherent variability mainly reflects the accuracy of the instruments used to control the design factors within engineering specification, of the instruments used to measure the response, and of any microstructural variation in the material being studied from sample to sample.

This broad characterisation of a manufacturing process suggests that there are two ways to try and optimise a process. The first involves trying to obtain better control over the noise factors themselves so that they are either eliminated or bought within tight specifications. This is likely to be a very costly approach to quality improvement as noise factors- by their very nature- are expensive to control. In Figure 1.3 any reduction in the variability of the noise factors will compress the width of the frequency histogram so that the process will produce a steel with similar strengths over time. The process behaves more consistently, but still remains sub optimal because the centre of the histogram is a long way from the target.

Such suboptimal processes can also be optimised through the use of careful industrial experiments. The objectives of such experiments are to optimise through a two-step procedure. In one step an experiment is run to identify those factors or process variables that will move the mean quality characteristic closer to the target. This is referred to as location experimental design. In the steel alloy example above this might involve the identification of those design factors (such as the number and the amount of each alloy to add to a base steel) that shift the peak of the histogram in Figure 1.3 to the right so that it corresponds to a tensile strength equal to the target of 2000 MPa. Such an experiment might find that this can best be achieved by increasing the amount of chrome added to a base steel.

Chapters 3 and 9 describe how to construct some efficient experiments for linear and non-linear processes respectively when the number of factors to be studied is quite small (five or less for linear processes, three or less for non linear processes). Chapter 4 describes how to construct some efficient experiments for linear processes when the number of factors to be studied is quite large. Then Chapters 5 and 6 illustrate how the results from linear experiments can be used to put the mean quality characteristic of a manufactured product on target, and Chapter 10 looks at this topic within the framework of a non-linear design.

In the second step an addition to the same or a new experiment is run to identify those factors or process variables that will reduce the scatter in the data around a mean that is now on target. This can be done using either **robust experimental design** or **dispersion experimental design**. The former designs are used predominantly for linear processes whilst the latter can be used for linear and non-linear processes. **Interactions** between design and noise factors are critical in reducing the process variability that stems from the existence of noise factors. Robust experimental designs are experiments that allow for the identification of such interactions and the first half of Chapter 7 shows how data from such robust designs can be used to minimise process variability. Then Chapter 8 brings together the content of Chapters 5, 6 and 7 to show how linear manufacturing processes can be optimised. Before outlining such detail, Figure 1.4 gives an overview on how the identification and use of such interactions can reduce process variability.

Figure 1.4 can be thought of as a graphical summary of the results obtained from a robust experimental design where information is sought on how two process variables determine the quality characteristic of a manufactured product (e.g. the tensile strength of a new steel alloy). One process variable can be controlled and so is a design factor (e.g. the quantity of chrome added to a base steel in the steel alloy example above). In the experiment this factor is varied between two different amounts (e.g. a high and a low amount of chrome). The other factor is a noise factor and so can't be controlled (e.g. the quench rate of a heat treatment in the steel alloy example above). This factor varies naturally within the limits of the distribution shown on the horizontal axis of each graph. The height of this distribution shows the frequency with which each value of the noise factor occurs over time.

The top graph illustrates a situation in which the experiment finds no interaction between the noise and design factor and this reflects itself as two parallel lines corresponding to the high and low values for the design factor. As a result of this, the variability in the quality characteristic, shown by the width of the two distributions on the vertical axis, resulting from the uncontrollable variability in the noise factor is not affected by the amount of the design factor used. Manipulating the design factor can't reduce variability. On the other hand the bottom graph shows the results of an experiment that finds a strong interaction between the noise and design factor and this reflects itself as two non parallel lines corresponding to a high and low value for the design factor. As a result of this, the variability in the quality characteristic resulting from the uncontrollable variability in the noise factor is affected by the amount of the design factor used. Lowering the amount of the design factor used will now reduce the amount of process variation (e.g. reducing the chrome content of the base steel).

Once such noise-design factor interactions are found a simple two-step optimisation procedure exists. First manipulate all those design factors (in the way illustrated above) that interact with a noise factor to minimise the variability in the products quality characteristics. Then, if the mean quality characteristic under these design factors settings is off target, manipulate those deign factors that do not interact with noise factors, and which influence the mean response, until the mean is on target. This is discussed in more detail in the first half of Chapter 8.

10 OPTIMISATION OF MANUFACTURING PROCESSES

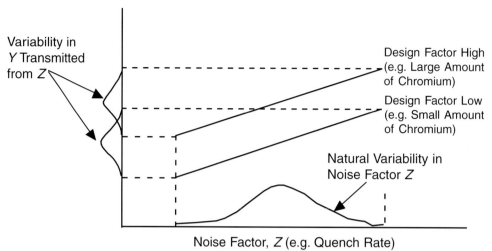

No Interaction Between Noise and Design Factors

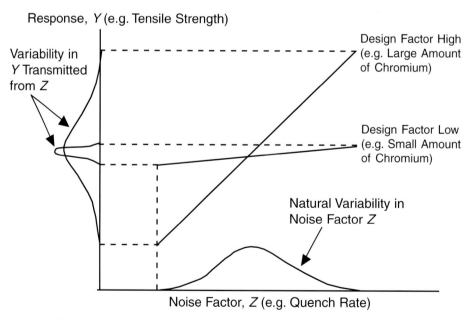

Interaction Between Noise and Design Factors

Fig. 1.4 The role of design – noise factor interactions in controlling process variation.

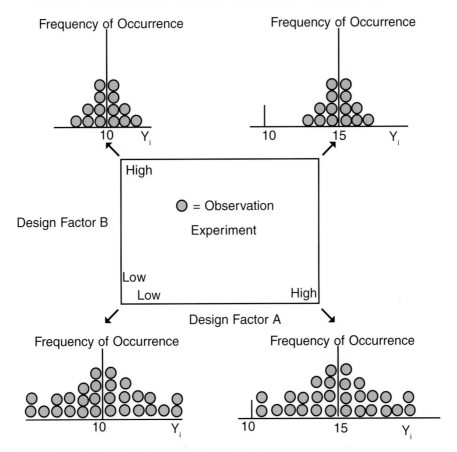

Fig. 1.5 Illustration of dispersion experimental design.

Dispersion experimental design attempts to reduce variability through a replication of tests within an experiment. Such designs are particularly well suited to a process where the process variation is mainly common cause. Through such replications, variability can be directly measured and a two-step procedure to achieve optimisation can once again be used. First, the settings of the design factors that minimise the direct measure of variability can be found. Then the design factors that do not influence this direct measure of variability but do influence the mean quality characteristic can be manipulated to put the mean on target. This is discussed in more detail in the second half of Chapters 7 and 8, but Figure 1.5 gives an overview.

Figure 1.5 shows a hypothetical experiment in which the effect of two design factors, A and B, on a critical quality characteristic, Y, are being analysed. Each factor is set at two different settings, (low and high), giving four different test conditions in total. At each test condition a number of measurements are made of the quality characteristic and these are summarised in a frequency histogram. In this figure, factor A alters the mean response whilst factor B alters the variability in the recorded responses. It is of course quite conceivable that a single factor could affect both the mean and variability in recorded responses. However, in this simple example the objective of maximising the mean response, whilst minimising the variability around that mean, could be achieved by increasing the level for each of the design factors. Notice that all the variability in this process is common cause because there are no noise factors being considered.

Finally, Chapters 11 and 12 offer alternatives to the two-step optimisation procedures discussed above. However, they can only be used for non-linear processes and replications must be carried out. These techniques require sequential testing to find the approximate optimal conditions (discussed in Chapter 11) and once this is completed the exact optimal conditions are found through a processes of **constrained optimisation** (discussed in Chapter 12). Here the process variability is minimised subject to the constraint that the mean quality characteristic is on target.

This book therefore goes into some detail on the different approaches to process optimisation for both linear and non-linear manufacturing processes. Whatever approach is used, the optimisation procedure always involves putting the mean on target with minimum process variation. This is summarised graphically in Figure 1.6 opposite. In this figure, distribution A summarises the current operating conditions for the process being illustrated (this may be the steel alloying example above where the quality characteristic being monitored is the tensile strength). The product being produced is therefore of poor quality because none of the items produced meets the target requirement. To optimise the process, those process variables that influence the variability but not the mean quality characteristic are identified from a suitable experiment and manipulated to minimise the variability. Such an action may lead to distribution B being achieved. This distribution is much narrower. Then those process variables that influence the mean but not the variability in the quality characteristic are identified from the same experiment and manipulated to increase the mean towards the target. Such an action may lead to distribution C being achieved. This distribution is shifted to the right.

1.4 STATISTICALLY PLANNED (DESIGNED) EXPERIMENTS

It was mentioned above that Chapters 3 and 4 will discuss efficient experimental designs for linear processes and Chapter 9 efficient experimental designs for non-linear processes. Before going into the details of these designs, it is useful to give a broad review of the different approaches to experimentation together with their relative advantages and the common terminologies used.

There is a tendency amongst engineers to spend much of their research time thinking about how to analyse results rather than about how they are going to generate them. Yet

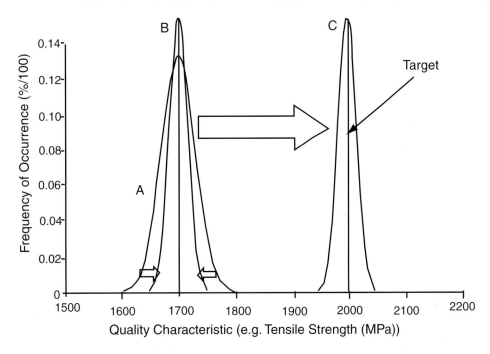

Fig. 1.6 Process optimisation: Mean on target with minimum variation.

the conclusions drawn can only be as good as the data they are based on and so it is important to spend some time designing the experiment itself. It was mentioned above that the key technique used in a robust design is a statistically designed experiment. A **designed experiment** usually involves a series of tests in which useful and structured changes are made to the settings of the design and noise factors that make up the manufacturing process under investigation. There are a vast variety of such designs currently available but the terminology used is common to many of them. It is very important for the engineer to become familiar with this terminology so that discussions between engineers and statisticians can take place without too much confusion being generated.

To start off then, the **experimental unit** refers to the item actually being tested. For example, a steel tube in a creep test, sheet steel in an ausforming test or an aluminium alloy in a corrosion - fatigue test.

The term **factor** was briefly introduced in Section 1.2 above and refers to a specific test condition. For example, stress and temperature in a creep test, temperature and the amount of deformation in an ausforming test and humidity in a corrosion test. Conventionally, factors are represented in shorthand by capital letters.

Level refers to the values associated with the test conditions. For example, stresses equal to 90 and 150 MPa and temperatures equal to 500 and 550°C represents a creep test at two levels for the factors stress and temperature respectively. In shorthand, levels are represented by lowercase letters.

Again the term **response** was mentioned in section 1.2 above and it refers to what is actually being measured during the experiment. For example, life in hours for a creep test, tensile strength in an ausforming experiment or cycles to failure in a corrosion ~ fatigue test. The symbol Y will be used to represent the response.

The experimental unit, factors, levels and responses then define the complete experiment. There is available a vast library of statistically designed experiments. All engineers will be familiar with the traditional approach to experimentation that involves changing the level of one factor at a time. A factorial design on the other hand is a specific type of experiment in which all the combinations of the levels of all the factors are investigated during the trial or experiment. This is clearly distinct from the traditional design. Factorial designs have many advantages to offer over the more traditional approach. They offer time saving economies in the laboratory and yield more reliable results because a greater number of tests can be averaged when working out the impact of changes in the levels of each factor on the response. Then when it comes to analysing the data such designs have a number of useful statistical properties that simplify the calculations required. Perhaps most importantly these designs allow noise variation to be built directly into the experiment so that robustness can be easily achieved.

On this last point consider again Figure 1.4. This figure summarised an experiment where each factor was varied between two different levels. The end points of each line in the figure come from an individual test and the four end points define all the combinations of two levels for two factors. It is thus a factorial experiment and from it an interaction (or not as the case may be) can be identified by two non-parallel lines. This interaction then enables the process variability to be reduced. Yet in a traditional experiment (that changes one factor at a time) involving two factors at two different levels, only three of the four tests needed to identify this interaction would be carried out and the information required to minimise process variation would not be available.

The 2^k factorial design is a design containing k factors each at two levels. Such designs are capable of identifying linear relationships only (see Chapters 3 and 4). Carrying out a fraction of a full factorial design can often save time and money. These are the so called 2^{k-p} fractional factorial designs (Chapter 4). On top of this, Taguchi[1] has published many two level orthogonal designs capable of analysing just a few or alternatively very many factors. At the start of any research programme it is recommended that experimentation be confined to these two level designs. Then if it is felt necessary to look at non-linear relationships, a higher-level design can be used that involves only that subset of factors identified as being important from the two level designs. This procedure often results in fewer tests.

If non-linear relationships need to be studied a number of experimental designs can be used. One possibility is to set up a factorial involving k factors each at three levels, i.e. a 3^k design (see Chapter 9). Such a design is capable of identifying quadratic relationships.

However, these designs are not the best way to analyse non-linear effects because they are large even for a relatively small number of factors, (over eighty tests when looking at just four factors). Better designs include factorials with mixed levels, the most popular of which are designs having factors at two and three levels or factors and two and four levels, central composite designs and Box–Behnken Designs (see Chapter 9).

1.5 PLANNING A RESEARCH PROGRAMME

A successful and effective research programme can often be achieved by splitting it up into two distinct phases. The first phase involves progressing sequentially through the following steps.

i. Clearly define the problem at hand.
ii. Define the objective(s) of the research programme.
iii. Studying the manufacturing process in detail so that all the factors that may be of relevance can be identified.

Any industrial based research programme will require technically sound engineering input if it is to succeed. The choice of what quality characteristics to model, what design and possibly noise factors to vary and over what ranges is engineering rather than a statistical judgement. A poor choice of factors and their ranges, resulting from incomplete engineering knowledge of the manufacturing process, is likely to result in an unsatisfactory solution – no matter how good the experimental plan and ingenuity in data analysis. A team approach to this phase of the programme is highly recommended and the interested engineer is referred to the work of Box and Jones[2] and Coleman and Montgomery.[3]

It is only once this stage of the programme has been successfully completed that phase two of the research programme can begin. This phase is statistically based and involves progressing through all or some of the following sequential steps (depending on the nature and depth of the research programme).

iv. Decide on a statistically planned experiment from those discussed briefly in Section 1.4 above.
v. Carry out the experiment.
vi. Analyse the experiment, interpret the results and quantify a model for the manufacturing process.
vii. Run a confirmation test (s) to assess the validity of the model.
viii. If the confirmation test(s) fail to support the model, identify reasons for this and seek solutions.
ix. Having identified an adequate model, use it to **optimise** the process. This may involve tests that are additional to those used in stage iv.

2. Engineering Case Studies

It is beneficial to discuss the subject of statistically designed experiments with reference to real manufacturing processes and products. Consequently, the following case studies will be used in the subsequent chapters to illustrate different experimental designs and methods of analysis. Some of the case studies relate to well established manufacturing processes (such as steel making via the ausforming process) whilst some relate to processes which are still very much at the development stage (such as the friction welding of blades to compressor disks in commercial and military aero engines).

2.1 THE AUSFORMING PROCESS

The ausforming process[4] involves taking alloyed steel and heating it to a temperature somewhere between 850 and 1200°C. Within this temperature range, known as the austenitising temperature, the steels have an allotropic form known as austenite. The steel is then kept within this temperature range for a specified length of time after which it is cooled to a temperature between 300 and 600°C. At this temperature, called the deformation temperature, the austenite is no longer in a stable form but remains unchanged for a considerable length of time. During this time, and at the deformation temperature, the steel is deformed by either extrusion, rolling or forging. Finally, the deformed steel is cooled to room temperature through a blast of air or by quenching it in a bath of oil or water. The result is a steel of very high strength.

This brief description of the ausforming process suggests that responses of interest might include the strength of the finished steel as measured by its tensile strength or the percentage reduction area. Process factors which could influence these responses are the austenitising temperature (factor A), the deformation temperature (factor B), the amount of deformation (factor C) and the quench rate (factor D). Other important factors could include the rate of deformation (factor E), the time at which the steels remain at the austenitising temperature (factor F, or the austenitising time) and the time at which the steels remain in a furnace at the deformation temperature before deformation takes place (factor G, the isothermal incubation time).

Some of these process factors, such as factors A, B, C, F and G are easy to control at the plant at precisely defined levels, whilst the remainder can only be controlled within a range defined by a tolerance band. However, it is the case that in a laboratory some of these noise factors can be more accurately controlled. Laboratory studies of the ausforming process using statistically planned experiments have been published in the engineering literature and this book draws on some of the results presented by Duckworth and Taylor[5] for a 3Cr1Ni1Si steel.

2.2 HIGH STRENGTH STEELS

The last decade or so has seen the competition between steel, aluminium and plastics for use in the manufacture of passenger car bodies intensify. For example, in 1975 about 59% of a typical European car consisted of steel (by weight) with a 4% contribution coming from aluminium.[6] By 1990 these contributions had changed to 55 and 12% for steel and aluminium respectively. To a large extent steel is managing to mitigate the threat from aluminium by developing new low weight high strength steels and corrosion resistant coated steels. Indeed 50% of the steels used in today's cars were just not available 10 years ago.[6] Many of these new high strength steels now used to manufacture the panels of a passenger car are obtained by adding alloying elements to low carbon precipitation harden able ferretic steels.

An understanding of the contributions made by the addition of alloying elements to the mechanical properties of steel is essential for the manufacture of high strength steels that can meet the stringent specifications set by car manufactures. By considering the quantity of an alloying element added to base steel as a factor, statistically planned experiments can provide a very useful framework for addressing such problems. DePaul and Kitchen[7] have carried out such a study using nickel, copper and niobium as the alloying elements.

In their study each steel alloy was manufactured in a laboratory using a magnesia crucible by induction melting in air an Armco iron. All melts were deoxidised with silicon ~ manganese and killed with aluminium prior to pouring into 102 by 102 mm ingots. These ingots were then soaked and forged into 25.4 mm slabs at a temperature of 1232°C. A two pass rolling operation was then used which consisted of a 6.35 mm reduction at 1232°C and a 12.5 mm reduction at 816°C. Both rolling passes were at 90 degrees to the forging axis. This was then followed by an air cool to room temperature and a subsequent roll at room temperature to achieve a further 33.3% reduction. Tensile specimens and full size Charpy V notch specimens were then cut from the plates parallel to the rolling direction. Each specimen was then aged for 1 hour at 566°C.

The chemical composition of the base steel was 0.03% carbon, 0.44% magnesium, 0.15% silicon, 0.11% aluminium, 0.004% phosphorous and 0.0185% sulphur. Factor A was taken to be the percentage addition of the nickel alloying element, factor B the percentage addition of the copper alloying element and factor C the percentage addition of the niobium alloying element to this base steel. The mechanical properties studied included the yield point, the yield stress, ultimate tensile strength and the percentage elongation.

2.3 HOT FORGED COPPER POWDER COMPACTS

Powder forging is a hybrid process in which preforms made using conventional PM techniques are hot forged in fully closed impression dies. The process involves several

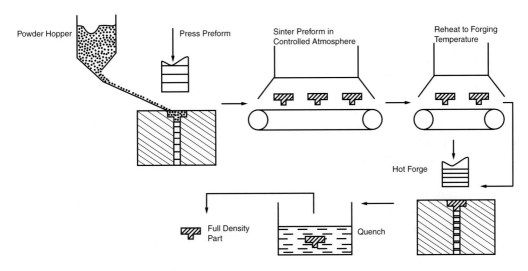

Fig. 2.1 Schematic route for powder forging.

stages and at each stage changes can be made to both the material and process factors which could have an effect on the properties of the final forging. Figure 2.1 shows a schematic representation of the powder forging process. At the first stage of the process high purity irregular copper powder is sieved to a specified powder size. Next the powder is compacted into a preform and sintered at a temperature of 1050°C in a protective atmosphere of 96%N – 4%H. The preheating furnace used by Evans and McColvin[8] to achieve this is illustrated in Figure 2.2. The sintered preforms or compacts are then coated with a suspension of powdered graphite in methanol and allowed to dry.

Conventionally, the preform would then be placed in a furnace until the preform reached the required density. Such a process usually takes between 24 and 36 hours. Powder forging technology removes this time consuming process by placing the compact, i.e. preform, into a forge and hitting it until the required density is reached. That is, the compacts are preheated after sintering and placed into a forging die such as that shown in Figure 2.3, which itself is preheated close to the forging press in a wide mouthed muffle furnace. The die and compacts are then transferred to the forging machine so that forging can take place. The forging process is then terminated when the load on the compacts reaches a critical value. This critical value is given by the ratio of forging load to shear modulus so as to allow for variations of shear modulus with temperature. Finally, the specimens are removed from the forge and after delays of 5 seconds are water quenched to room temperature.

Evans and McColvin[8] have studied this forging process in a laboratory using the equipment shown above and, using a statistically planned experiment, analysed the effects

Fig. 2.2 Sintering and preheat furnaces with controlled atmosphere apparatus.

Fig. 2.3 Forging die for copper compacts.

Fig. 2.4 A typical aero engine disk (reproduced with permission from UEF Aerospace).

of various process factors on the ultimate tensile strength, 0.1% proof stress, percent elongation and the percent reduction in area to fracture of these hot forged copper powder compacts. From the above description of the manufacturing process several design factors might influence such responses. These include the powder size (factor A), the density of the compacted powder (factor B), the length of time over which sintering takes place (factor C), the preheat temperature of the sintered preforms (factor E), the preheat temperature of the forging die (factor F), the speed at which forging takes place (factor D) and the length of time over which forging takes place as determined by the forging load/shear modulus (factor G).

2.4 CLOSED DIE FORGING OF AERO ENGINE DISKS

As Figure 2.4 illustrates, aero engine disks have a complex geometry that requires the use of a two stage forging operation. Nickel and titanium materials are often purchased in the form of cylindrical shapes weighing between 18 and 200 kg. This raw material is then prepared and taken to a vacuum load lock chamber where it is preheated to a temperature of around 1090°C. Once heated an automated arm moves the raw material to the forging chamber of a hydraulic 8000 ton single action press where it is forged into

Fig. 2.5 Forging hammer for aero engine disks (reproduced with kind permission from UEF Aerospace).

a preform. The preform is then cooled using argon and nitrogen gases in an exit lock chamber. The thickness of this preform is determined by the dimensions of the final disk to be produced. A particular thickness is suitable for a range of final disk sizes.

The preform next goes to a vertical turret lathe where a centring dimple is machined in to the preform to ensure the preform is centred in the die during final forging. Next the preform is reheated and lubricated ready for final forging. With the final shape die in place the preform undergoes a second round of forging in a press similar to that shown in Figure 2.5 After final forging the disks are within 0.15 cm of the required dimensions for the final disk. After cooling the disks go off for heat treatment.

Large sums of money have been invested into the development of computer models that can predict the energy requirements of this second stage of forging. Evans[9] has developed one such model using the finite element technique. The energy requirement is thought to be a function of four boundary conditions that can be considered as design factors for this forging process. These are the surface area of the aero engine disk (factor A), the coefficient of friction (factor B), the heat transfer coefficient (factor C) and the emissive power (factor D). The computer model can be used to predict the energy required to forge an aero engine disk under various values for these boundary conditions. If the levels for these conditions are set up as a statistically planned experiment the values for the boundary conditions that minimise the energy requirements can be found. This is a good example of a virtual experiment.

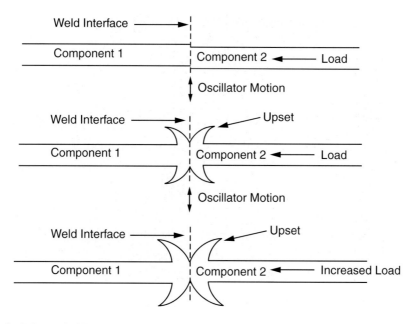

Fig. 2.6 Schematic illustration of linear friction welding.

2.5 FRICTION WELDING

Friction welding belongs to a family of techniques including flash welding and traditional forge welding. Linear friction welding offers many advantages over the more traditional approaches. For example, dissimilar metals can be joined using friction welding, no filler metal is required so that no compositional change takes place and none of the problems associated with liquid weld metal joints, such as porosity and shrinkage cracking, apply to friction welding. For these reasons Rolls Royce have built pilot equipment to friction weld blades onto the compressor disks of their military and commercial aircraft engines. Factorial designs are currently being used at Rolls Royce to identify the levels at which each process variable must operate in order to achieve a weld with the required weld property on a consistent basis.

With linear friction welding two components are vibrated together along a common axis under an applied load. The friction generated causes the interface to heat up and when a predetermined time has been reached the rubbing action is halted and the load increased so that the two components become forged together. This process is illustrated schematically in Figure 2.6.

The friction welding process is a complex one, but the quality of the welded interface seems to be dependent in some way on the local strain that occurs. It is this strain that measures the deformation of surface asperities and the break up of contaminant oxides and allows new atomic bonds to form across the interface. The amount of flow that occurs during welding is a function of the local temperatures as well as the applied forces and these heating and deformation process are intimately connected. The temperature gradients with respect to distance and time are very steep and hence difficult to measure experimentally, but it is possible to set up a physical model based on viscoplastic finite element procedures which allow an investigation of this process to be made (Evans).[10] This is another example of a virtual experiment.

The major factors which influence the heating are the weld frequency (factor B), axial force (factor A) and the weld amplitude (factor C) and other factors could also be important. The ideal response to measure would be the quality of the weld itself but this cannot be measured directly. However, it is known that the weld quality is directly related to the degree of upset (i.e. overall shortening of the weld piece close to the interface) or the shear strains at the interface itself. Both of the quantities can be taken as the relevant response of the system to the above mentioned design factors.

2.6 THE INJECTION MOULDING EXPERIMENT

For high volume production, injection moulding is the most widely used and economic of all the thermoplastic processes. Engel[11] has described and analysed an experiment that was used to improve the injection moulding of plastic pipes. A major problem in plastic pipe manufacturing is pipe shrinkage during the cooling off period. Hence the response variable of most interest is the percentage shrinkage. Figure 2.7 below shows the basic elements of an injection moulding machine used in the manufacture of plastic pipes. The manufacturing process itself involves the loading of plastic in powder or granular form into a feed hopper located at the top of the machine. This loading process takes place at regular intervals during a typical manufacturing day with the number of loadings per day being governed by the cycle time (factor C). The plastic then falls from the hopper into the machine barrel that contains a rotating a reciprocating screw. The rotating action of the screw compresses and moves the plastic into a chamber located immediately in front of the screw head. By the time the plastic reaches this chamber it is in a molten state partly as a result of the shearing action of the screw but also through the controlled use of the heater bands around the barrel (factor B).

As soon as the chamber is full a control system activates a hydraulic system that plunges the screw forward thereby injecting the molten plastic into the cavity of the mould. This control system can be used to manipulate the speed with which the plastic is injected into the mould (factor E), the pressure exerted on the plastic (factor A) and the length of time the screw is held in the forward position (factor D). On top of these design factors the engineers conducting the experiment believed that the die cavity thickness (factor F) and the size of the safety gate (factor G) might be important in determining pipe shrinkage.

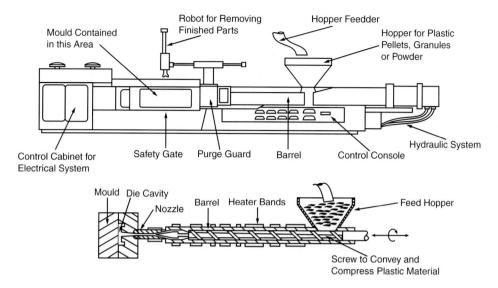

Fig. 2.7 An injection-moulding machine.

It was also realised that percentage regrind (factor Z_1), air moisture content (factor Z_2) and ambient temperature (factor Z_3) could well be important in determining percentage shrinkage. Yet the engineers realised that these three factors were not being controlled at the point of manufacture partly because it would be too costly to do so. These three factors were therefore treated as noise factors in the experiment that was carried out in a laboratory-controlled environment where a small scale model of the injection moulding process was located. These noise factors could be controlled within engineering specification in the laboratory because the technology was available to do so.

2.7 THE SPRING FREE HEIGHT EXPERIMENT

Pignatiello and Ramberg[12] have studied a manufacturing process that heat treats leaf springs used in truck suspension systems. Such heat treatment involves the transportation of the leaf spring assembly through a high temperature furnace using a conveyor belt. Next the part is transferred to a forming machine where the curvature of the spring is induced by holding the spring in a high pressure press for a short length of time. Finally, the spring is placed into an oil quench. An important quality characteristic of the finished product is the free height of the spring, and an ideal heat treatment process is considered to be one that produces a spring free height of exactly twenty centimetres. Deviations above or below this target are therefore considered undesirable.

This description of the manufacturing process suggests there are five process factors to consider. Four of these are design factors and one is a noise factor. The design factors include the furnace temperature, (factor A), heating time, (factor B), transfer time (time taken for the part to reach the forming machine from the furnace), (factor C) and the hold down time (the length of time the camber former is held down), (factor D). The oil quench temperature, (factor E), on the other hand is the noise factor.

2.8 THE PRINTING PROCESS STUDY

The printing of coloured inks onto packaging labels can be achieved in a variety of ways. The following experiment was first described in Box and Draper[13] and involves the application of two layers of paint applied using paint spray guns. A thin coating of paint is applied by automated spray guns at each coating to ensure a high quality finish. These industrial robots are taught by skilled human painters to ensure efficient paint utilisation and a good quality finish. The package labels are passed through fixed spray heads using a conveyor belt system. The quality of the finished surface was thought to be determined by the speed of this conveyor belt (factor X_1), the pressure of the spray from the gun (factor X_2) and the spray distance of the labels from the spray gun (X_3). The quality of the finished surface was measured as the number of printing defects over the surface of the label, Y. The target specification being aimed for by the designers of this experiment was no more than 500 surface defects.

PART II
LINEAR EXPERIMENTAL DESIGNS

3. The 2^k Experimental Design: Full Factorials

This chapter deals with designs that are suitable for studying manufacturing processes that are inherently linear in nature and where the number of process variables is quite small (five or less). In such a process, the relationship between the response being studied and a factor (for given levels of all the other factors) can be represented by a series of straight lines and so each and every factor only needs to have two different levels. This chapter starts off with a comparison between a traditional linear design and the simplest full factorial design for a linear process. This is the 2^2 factorial design involving two factors, each set at two different levels. These designs are illustrated using the ausforming process described in Section 2.1. The 2^2 design is then generalised to the 2^k design so that linear processes with more than two factors (in fact k factors) can be studied. All these designs are illustrated using the ausforming process. Geometric representations of these 2^k designs will be given together with the important tabular standard order form representation.

3.1 THE 2^2 FACTORIAL DESIGN

To illustrate the 2^2 factorial design, consider the ausforming process of Section 2.1. Suppose a group of research engineers would like to quantify that way in which the process variables of the ausforming process determine the strength of the manufactured steel and as a first step towards this they confine themselves to an analysis of the first two factors of this process. These two factors are design factors in that they can be controlled within specification. Suppose it is already known from past studies that the effect of austenitising temperature (factor A) on strength is approximately linear over the temperature range 930 to 1040°C and the effect of deformation temperature (factor B) on strength is nearly linear over the range 400 to 550°C. Given that the process is believed to be linear over these ranges each factor only needs to be set at two different levels and the above temperatures can therefore be chosen as the high and low levels to be used in the experiment.

3.1.1 A Traditional Design

The above is an illustration of the sort of information that should be available to a research team once steps (i) to (iii) of Section 1.5 have been carried out. The next step in a well

Table 3.1 The traditional approach to experimentation.

Test	Factor A	Factor B	Response Y
	Austenitising Temperature, °C	Deformation Temperature, °C	Tensile Strength, MPa
1	930	400	$Y_1 = 2331$
2	1040	400	$Y_2 = 2100$
3	1040	550	$Y_3 = 1668$

planned research program would be to select the type of experimental design to use. One possibility is to carry out a **traditional experiment** where just one factor at a time is changed. Table 3.1 shows what such an experiment would look like when considering the austenitising temperature (factor A) and deformation temperature (factor B) at the two different levels stated above. Three tests would be carried out with only one of the factors changing level from one test to the next. The first test has both factors set are their low levels and then in the next test the first factor is increased to its high level with the second factor remaining unchanged. Then the final test involves changing the second factor whilst leaving the first factor unchanged. The final column of Table 3.1 shows the tensile strengths recorded from the ausforming process under these test conditions, where Y_1 is the tensile strength recorded for test 1 and Y_3 the tensile strength recorded in test 3. (All the other process variables of the ausforming process are held fixed at their low levels during this traditional experiment).

3.1.2 A Full 2^2 Factorial Design in Standard Order Form

An alternative to this design is to set up and implement a **full factorial** design. In such an experiment the levels of each factor are varied together. Remember from Section 1.3 that a full factorial experiment involves carrying out at least one test at each possible combination of the levels of each factor. The number of different combinations is found by raising the number of levels for each factor to the power of the number of factors to be studied. So when studying two factors, each at two different levels, there are $2^2 = 4$ such test combinations to consider. This is the so called 2^2 **factorial design**. Table 3.2 shows a 2^2 experimental design for studying the first two factors of the ausforming process. The ausforming temperature is represented by the letter A and the deformation temperature by the letter B. The high and low values for each factor are the same as above.

Each test in a full factorial design is given a shorthand descriptor. (1) is used to symbolise a test carried out when both factors are set at their low levels. The absence of a letter is then used to indicate when a factor is set at its low level. Thus a indicates a test

Table 3.2 A 2^2 design for part of the ausforming process in standard order form.

Test	Actual Test Conditions		Coded Test Conditions			Tensile Strength
	Ausforming Temp., °C	Deformation Temp., °C	A	B	Test Order	MPa
(1)	930	400	-1	-1	3	2331
a	1040	400	1	-1	1	2100
b	930	550	-1	1	2	2146
ab	1040	550	1	1	4	1668

carried out when factor A is set high and factor B low, whilst b represents a test in which factor B is high and factor A low. Finally, *ab* represents a test carried out when both factors A and B are at their high levels. Notice that a full factorial involves carrying out one additional test to those tests making up the traditional experiment, namely test *b*.

To help identify all the tests that need to be carried out in a full factorial experiment, test conditions and results are usually written out in a **standard order form**. For the 2^2 design this is made up of test (1) followed by test *a*, then test *b* and finally test *ab*. The rule used for the derivation of this standard order form is quite straightforward. It involves the creation of a vertical column of descriptors with test condition (1) at the top. The rest of the vertical column is then created by introducing each factor in alphabetical order and multiplying it by each and every term above it. For example, after test (1), introduce the next factor in alphabetical order and multiply it by the previously identified test, i.e. (1) × a = test a. Then (1) × b or test b and finally $a \times b$ or test ab. The resulting column is the first column of Table 3.2 and this is the 2^2 experiment written out in standard order form. There are no other combinations of two factors at two different levels to derive.

In Table 3.2 each test condition is given a **coded value** such that -1 is used to indicate when a factor is at its low level and +1 is used when a factor is at its high value. Such coded levels can be explicitly derived from the formula

$$\text{Coded test condition} = \frac{\text{Actual test condition} - \text{mean test condition}}{\text{Range of test condition}/2} \quad (3.1)$$

As an illustration, consider the ausforming temperatures shown in Table 3.2. The range of the test condition is simply the difference between the high and low temperatures, i.e. $1040 - 930 = 110°C$ for the ausforming temperature. The mean test condition is calculated as $(1040 + 930)/2 = 985°C$. Thus when the ausforming temperature is at 1040°C, the corresponding coded test condition is, from equation (3.1),

$$\text{Coded test condition} = \frac{1040 - 985}{110/2} = +1$$

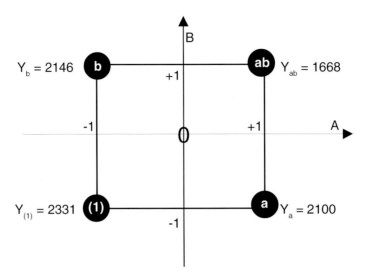

Fig. 3.1 A geometric representation of the 2^2 design for the ausforming process.

This equation can also be rearranged to give the actual test condition from a coded test condition

Actual test condition = mean test condition +
(coded test condition × range of test condition/2) (3.2)

It is important to point out, however, that the actual tests should never be carried out in this standard order. Tests must always be done in a random order such as the one shown in the test order column of Table 3.2. The tensile strength measurements obtained from each of these four test conditions are those shown in the last column of Table 3.2. This is an example of an unreplicated 2^2 experiment in that only one measure of the response variable (in this example tensile strength) is made under each test condition. The symbol N is used to represent the number of tests carried out at each of the 2^2 test conditions. So for an unreplicated experiment $N = 1$.

3.1.3 A GEOMETRIC REPRESENTATION OF THE 2^2 EXPERIMENTAL DESIGN

The 2^2 factorial experiment can be given a simple geometric representation that takes the form of a square. The coded level for each factor is represented along the horizontal and vertical axis as in Figure 3.1. Each corner of the square then represents one of the four unique tests defining the 2^2 factorial design. Thus the top right hand corner of the square represents test ab, where both factors are set high. In coded units both factors are set at the +1 level. In turn $Y_{(1)}$ represents the value for the recorded response when all the factors are at their low level, Y_a represents the value for the recorded response obtained when factor A alone is set at its high level, Y_b reads the value for the recorded response obtained when factor B alone is set at its high level and Y_{ab} represents the value for the recorded response when both factors are set high. The responses in Figure 3.1 are the tensile strengths recorded from the ausforming process.

3.2 THE 2^3 FACTORIAL DESIGN

A more realistic understanding of the ausforming process can be obtained by introducing a third factor into the previous experiment – the amount of deformation. Suppose it is already known that the effect of deformation on strength is linear over the range 50 to 80% deformation so that these can be chosen as the high and low levels for the amount of deformation. A linear process with three factors can be studied using the 2^3 factorial experiment where each factor has two different levels.

3.2.1 A GEOMETRIC REPRESENTATION OF THE 2^3 EXPERIMENTAL DESIGN

In the **2^3 design** there are $2^3 = 8$ different test combinations. This 2^3 design can therefore be visualised as a cube as in Figure 3.2 below where factor C is the amount of deformation. The level for each factor is again shown along each of the three axis so that each corner of the cube defines one of the eight tests making up the 2^3 factorial experiment. Figure 3.2 also shows the results obtained from carrying out this factorial experiment.

Notice that the square on the front of the cube is nothing more than the above 2^2 factorial design. To run a 2^3 factorial experiment four additional tests need to be added to the 2^2 design. These are shown at the corners of the square at the back of the cube. The first of these is test c where only factor C is set at its high level, the second is test ac where only factors A and C are set high, the third is test bc where factors B and C are set high and the fourth is test abc in which all three factors are set at their high levels. The response measured at each of these eight test conditions is again the ultimate tensile strength measured in MPa.

3.2.2 STANDARD ORDER FORM FOR THE 2^3 FACTORIAL EXPERIMENT

When considering more than three factors it is not possible to identify all the tests conditions making up a full factorial experiment graphically in the way shown above. The best way to identify all the tests of a full factorial is to write them out in standard order form using the rule described in Section 3.1.2. The standard order form for the 2^3 design is shown in the test column of Table 3.3 and is found as follows. For a single

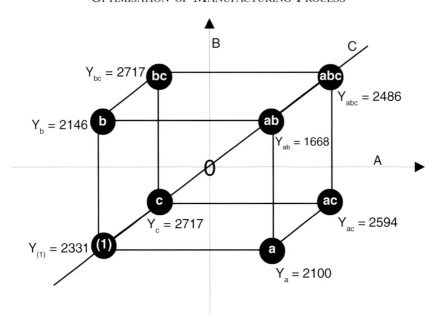

Fig. 3.2 A 2^3 factorial design for the ausforming process.

Table 3.3 A 2^3 design for part of the ausforming process in standard order form.

Test	Actual Test Conditions			Coded Test Conditions				Tensile Strength
	Ausforming Temp., °C	Deformation Temp., °C	Deformation %	A	B	C	Test Order	MPa
(1)	930	400	50	-1	-1	-1	5	2331
a	1040	400	50	1	-1	-1	8	2100
b	930	550	50	-1	1	-1	1	2146
ab	1040	550	50	1	1	-1	2	1668
c	930	400	80	-1	-1	1	6	2717
ac	1040	400	80	1	-1	1	3	2594
bc	930	550	80	-1	1	1	7	2717
abc	1040	550	80	1	1	1	4	2486

Table 3.4 A replicated 2^3 design for part of the ausforming process.

Test	Coded Test Conditions			Replicate 1	Replicate 2
	A	B	C	Strength, MPa	Strength, MPa
(1)	-1	-1	-1	2331	2331
a	1	-1	-1	2100	2270
b	-1	1	-1	2146	2270
ab	1	1	-1	1668	1559
c	-1	-1	1	2717	2810
ac	1	-1	1	2594	2455
bc	-1	1	1	2717	2671
abc	1	1	1	2486	2440

factor the standard order is (1) followed by a. For two factors we add to this list b followed by ab, derived by multiplying the first two factor combinations by the letter b. ((1) × b = b and a × b = ab). For three factors we add c followed by ac then bc then abc, derived by multiplying the first four factor combinations by the additional letter c. ((1) × c = c, a × c = ac, c × b = bc and c × ab = abc).

Note, however, that each test is not carried out in the standard order form. As usual the tests are carried out in the random order shown in the test order column of Table 3.3. Notice (by comparing Tables 3.2 with 3.3) that the results obtained in the 2^2 experiment above were all obtained when the amount of deformation was fixed at its low level of 50%.

3.2.3 AN EXAMPLE OF A REPLICATED 2^3 FACTORIAL EXPERIMENT

The above experiment is again an unreplicated 2^3 design where eight tests were conducted under eight different test conditions. If this experiment of the ausforming process were to be repeated, the likelihood is that the same strength values would not be recorded under the same eight test conditions – the explanation for this being process variability. Further, because the three factors considered so far are design factors this process variability would be solely attributable to common cause variation. To confirm this, the unreplicated 2^3 design for the ausforming process above is repeated and the results from this replicate are shown, (together with the first replicate), in Table 3.4. Notice that the strength measurement is not always the same at each test condition. This experiment is an example of a replicated 2^3 design where $N = 2$.

3.3 THE 2^k FACTORIAL DESIGN

2^k full factorials can be used to study linear manufacturing processes that have k process variables. A higher order design of this nature can be used to obtain further insights into the nature of the ausforming process and its ability to produce a very high strength steel. The description of the ausforming process given in section 2.1 suggested that the strength of the manufactured steel is also likely to be dependant upon the rate of quenching (factor D) and the rate of deformation (factor E). Both of these are likely to be noise factors at the point of manufacture but not in the laboratory where the following experiment was carried out. Again it is believed from past studies that strength is linearly related to the rate of deformation over the range 0.30 to 5 mm per minute and linearly related to the quench rate over the range 1K per second to 7 K per second. These can therefore be chosen as the low and high values for these two new factors in a full 2^5 factorial design, ($k = 5$).

3.3.1 STANDARD ORDER FORM FOR THE 2^5 FACTORIAL EXPERIMENT

In the **2^5 design** there are $2^5 = 32$ different test combinations. It is impossible to visualise this design geometrically and so the thirty two tests making up the 2^5 experiment can best be identified by writing them out in standard order form. Table 3.5 shows the 2^5 experiment in standard order form together with the results obtained from carrying out this experiment. (The remaining two factors are held fixed at their low levels). The first eight tests in standard order form are those associated with the 2^3 design. Alphabetically, the next factor to introduce is factor D and when d is multiplied into the test descriptors of the 2^3 design the next eight tests are obtained, namely tests $(1) \times d = d$ to test $abc \times d = abcd$. Alphabetically, the next factor to introduce is factor E and when e is multiplied into the test descriptors of the 2^4 design the final sixteen tests are obtained, namely tests $(1) \times e = e$ to test $abcd \times e = abcde$. Again this experiment is an unreplicated design because one test result is obtained at each of the thirty two test conditions.

For factor D (the quench rate) the coded test condition -1 corresponds to 1K per second and the coded test condition $+1$ corresponds to 7K per second. For factor E (the deformation rate) the coded test condition -1 corresponds to 0.30 mm per minute and the coded test condition $+1$ corresponds to 5 mm per minute. The coded test values for factors A to C remain as in the previous sections.

3.3.2 GENERAL COMMENTS ON THE 2^k DESIGNS

From the discussion above it is clear that the 2^3 and 2^4 designs are contained within the 2^5 design. This is a general characteristic of full factorial experiments, so that all lower factorials go into making up a full 2^k experiment.

A repeating pattern is clearly present in Tables 3.3 and 3.5. In the first coded test condition column, the numbers change sign alternately. In the second coded test condition

Table 3.5 A 2^5 design for part of the ausforming process in standard order form.

Test	Coded Test Conditions					Test Order	Tensile Strength MPa
	Factor A	Factor B	Factor C	Factor D	Factor E		
(1)	-1	-1	-1	-1	-1	7	2331
a	1	-1	-1	-1	-1	16	2100
b	-1	1	-1	-1	-1	27	2146
ab	1	1	-1	-1	-1	6	1668
c	-1	-1	1	-1	-1	30	2717
ac	1	-1	1	-1	-1	17	2594
bc	-1	1	1	-1	-1	26	2717
abc	1	1	1	-1	-1	28	2486
d	-1	-1	-1	1	-1	4	2038
ad	1	-1	-1	1	-1	18	1498
bd	-1	1	-1	1	-1	32	1930
abd	1	1	-1	1	-1	8	1529
cd	-1	-1	1	1	-1	22	2779
acd	1	-1	1	1	-1	15	1899
bcd	-1	1	1	1	-1	31	2594
abcd	1	1	1	1	-1	1	1853
e	-1	-1	-1	-1	1	19	2517
ae	1	-1	-1	-1	1	11	2084
be	-1	1	-1	-1	1	23	2285
abe	1	1	-1	-1	1	3	1992
ce	-1	-1	1	-1	1	29	2609
ace	1	-1	1	-1	1	20	1837
bce	-1	1	1	-1	1	9	2285
abce	1	1	1	-1	1	12	1652
de	-1	-1	-1	1	1	5	2887
ade	1	-1	-1	1	1	21	2455
bde	-1	1	-1	1	1	24	2656
abde	1	1	-1	1	1	10	2362
cde	-1	-1	1	1	1	13	2980
acde	1	-1	1	1	1	2	2208
bcde	-1	1	1	1	1	25	2656
abcde	1	1	1	1	1	14	2023

column, the numbers change sign in pairs (double that in the previous column). In the third coded test condition column, the sign changes in blocks of four (double that in the previous column). In the fourth coded test condition column, the sign changes in blocks of eight (double that in the previous column) and in the fifth coded test condition column the sign changes in blocks of 16 (double that in the previous column). This characteristic of a full factorial design can be used to identify all the tests conditions of any 2^k design. For example, the 2^6 design involves 64 tests. The first 32 tests are those shown in Table 3.5 but with the sixth factor always set low. The next 32 tests are those shown in Table 3.5 but with the sixth factor always set high. In the sixth coded test condition column, the sign changes in blocks of 32 (double that in the previous column).

In a competitive business environment, time and money impose constraints on the number of tests that can be carried out. Without replication, these designs are often unsuitable for processes with more than five factors. Six factors would require $2^6 = 64$ tests! So if replication is required, time and money will ensure that even fewer factors can be studied. A solution to this problem is dealt with in the next chapter. Finally, these 2^k factorial designs can only be used when the manufacturing process is inherently linear. The identification of non-linear relationship will require designs with factors at three or more levels, i.e. non-linear designs. Non-linear designs are discussed later on in Part IV of the book.

4. The 2^{k-p} Experimental Design: Fractional Factorials

As the name suggests, fractional factorial designs use only a fraction of the experimental tests required to cover all factor level combinations. Consequently, some information has to be sacrificed and the idea is to maximise the amount of information that can be gleamed from the minimum number of tests. This chapter starts by introducing the basic concepts of fractional factorial designs. Then a more detailed look at the one-half fractional factorial design is given and a two-step procedure for designing such an experiment is also outlined. This is illustrated using the first five factors of the ausforming process described in Section 2.1. This two-step procedure is then generalised to other fractional factorial designs. Next a one eighth fractional factorial experiment is carried out using all of the seven factors of the ausforming process outlined in Section 2.1. Finally, an alternative approach to designing fractional factorial experiments suggested by Taguchi is described.

4.1 BASIC CONCEPTS

To illustrate some of the basics, return to the unreplicated 2^2 design of Chapter 3 with responses $Y_{(1)}$, Y_a, Y_b and Y_{ab}. Now suppose that only half of these tests are actually carried out. This would give a one-half fraction of the 2^2 design or simply a 2^{2-1} design. Consequently, it would require only half the number of tests, $2^{2-1} = 2$. Suppose it is decided to undertake the (1), ab half of the full 2^2 design. This half of the 2^2 design is called the **principle fraction**. The remaining tests define the **alternate fraction** and the sum of the principle and alternate fractions gives the full 2^2 design. The one-half fraction of the 2^k design is usually written as the 2^{k-1} design and requires only half the number of tests in the full 2^k design, i.e. 2^{k-1} test instead of the full 2^k tests. Further the principle and alternate fractions of the 2^{k-1} together define the full 2^k design.

Further savings on the number of tests required for a particular experiment can be obtained by setting up smaller fractions. For example, there is the one-quarter fraction of the 2^k design. This is usually written as the 2^{k-2} design and requires 2^{k-2} test instead of the full 2^k tests. Another important fractional factorial design is the one eight fraction of the 2^k design. This is written as the 2^{k-3} design and requires 2^{k-3} tests. In general, it possible to construct a 2^{k-p} design where p is the size of the fraction used in the fractional design. $p = 4$ would imply a one sixteenth fractional factorial design which is very popular with Taguchi advocates as will be seen in Section 4.5 below.

Because only a fraction of the tests defining a full 2^k design are carried out in a fractional factorial experiment, it follows that only a fraction of the information on how to optimise a process can be obtained from this design compared with the information that can be obtained from a full 2^k experiment. The trick is to minimise the loss of such information. The following two-step procedure does exactly this. The rest of this chapter describes how to build a fractional factorial design and the next chapter explains why this two-step procedure minimises the information loss.

4.2 THE ONE-HALF FRACTION OF THE 2^K DESIGN

The **one-half fractional factorial** experiment, often referred to as the 2^{k-1} fractional factorial design, can be constructed for any 2^k experiment by following two sequential steps. These steps will be illustrated using a 2^{5-1} fractional factorial design of the ausforming process described in Section 2.1.

4.2.1 STEP ONE: DEFINING THE BASE DESIGN

The one-half fractional factorial requires 2^{k-1} tests to be carried out. The **base design** for any 2^{k-1} fractional factorial is the full 2^{k-1} factorial design. In the ausforming process studied in Section 3.3.1, $k = 5$ factors were looked at so the base design is a full $2^{5-1} = 2^4$ factorial. Using the standard order for test runs this can be written out as in Table 4.1.

Factor A is austenitising temperature, factor B deformation temperature, factor C the amount of deformation and factor D the quench rate. The high and low values for each of these factors are as in Section 3.3.1 and are in coded units.

4.2.2 STEP TWO: INTRODUCTION OF THE REMAINING FACTOR

The final factor considered in Section 3.3.1 was factor E, the deformation rate. This is introduced into the fractional design by setting its high and low levels equal to the elements, (i.e. the minus and plus ones), given in the columns defining the base design. That is, the test levels for factor E are found by multiplying together the numbers in the columns of the base design, $E = ABCD$. $E = ABCD$ is called the **design generator** for this fractional factorial design. For the alternate fraction the design generator is $E = -ABCD$. The actual 2^{k-1} tests making up the one half fractional factorial design are then defined by the coded test conditions of this amended base design. Table 4.2 illustrates how the principle one-half fractional factorial design is constructed for $k = 5$.

The first five columns of Table 4.2 is the full 2^4 design. The 2^{5-1} design in shown in the remaining columns of Table 4.2 and these are derived as follows. Columns A to D in the second half of Table 4.2 are identical to columns A to D in the first half of the Table.

The 2^{K-P} Experimental Design: Fractional Factorials

Table 4.1 Base design for the 2^{5-1} fractional factorial.

Tests	Coded Test Conditions			
	A	B	C	D
(1)	−1	−1	−1	−1
a	1	−1	−1	−1
b	−1	1	−1	−1
ab	1	1	−1	−1
c	−1	−1	1	−1
ac	1	−1	1	−1
bc	−1	1	1	−1
abc	1	1	1	−1
d	−1	−1	−1	1
ad	1	−1	−1	1
bd	−1	1	−1	1
abd	1	1	−1	1
cd	−1	−1	1	1
acd	1	−1	1	1
bcd	−1	1	1	1
abcd	1	1	1	1

Table 4.2 The principle 2^{5-1} fractional factorial design.

	Principle Fraction for the 2^{5-1} Design									
	Full 2^4 Design				2^{5-1}					
	A	B	C	D	A	B	C	D	E = ABCD	Test
(1)	−1	−1	−1	−1	−1	−1	−1	−1	1	e
a	1	−1	−1	−1	1	−1	−1	−1	−1	a
b	−1	1	−1	−1	−1	1	−1	−1	−1	b
ab	1	1	−1	−1	1	1	−1	−1	1	abe
c	−1	−1	1	−1	−1	−1	1	−1	−1	c
ac	1	−1	1	−1	1	−1	1	−1	1	ace
bc	−1	1	1	−1	−1	1	1	−1	1	bce
abc	1	1	1	−1	1	1	1	−1	−1	abc
d	−1	−1	−1	1	−1	−1	−1	1	−1	d
ad	1	−1	−1	1	1	−1	−1	1	1	ade
bd	−1	1	−1	1	−1	1	−1	1	1	bde
abd	1	1	−1	1	1	1	−1	1	−1	abd
cd	−1	−1	1	1	−1	−1	1	1	1	cde
acd	1	−1	1	1	1	−1	1	1	−1	acd
bcd	−1	1	1	1	−1	1	1	1	−1	bcd
abcd	1	1	1	1	1	1	1	1	1	abcde

Table 4.3 The alternate 2^{5-1} fractional factorial design.

	\multicolumn{9}{c}{Alternate Fraction for the 2^{5-1} Design}	
	Full 2^4 Design				\multicolumn{5}{c}{2^{5-1}}	
	A	B	C	D	A	B	C	D	E = –ABCD	Test
(1)	–1	–1	–1	–1	–1	–1	–1	–1	–1	(1)
a	1	–1	–1	–1	1	–1	–1	–1	1	ac
b	–1	1	–1	–1	–1	1	–1	–1	1	bc
ab	1	1	–1	–1	1	1	–1	–1	–1	ab
c	–1	–1	1	–1	–1	–1	1	–1	1	ce
ac	1	–1	1	–1	1	–1	1	–1	–1	ac
bc	–1	1	1	–1	–1	1	1	–1	–1	bc
abc	1	1	1	–1	1	1	1	–1	1	abce
d	–1	–1	–1	1	–1	–1	–1	1	1	de
ad	1	–1	–1	1	1	–1	–1	1	–1	ad
bd	–1	1	–1	1	–1	1	–1	1	–1	bd
abd	1	1	–1	1	1	1	–1	1	1	abde
cd	–1	–1	1	1	–1	–1	1	1	–1	cd
acd	1	–1	1	1	1	–1	1	1	1	acde
bcd	–1	1	1	1	–1	1	1	1	1	bcde
abcd	1	1	1	1	1	1	1	1	–1	abcd

The first number in column E of Table 4.2 is then found by using the design generator $E = ABCD$, i.e., multiplying together the numbers in the first row of columns A, B, C and D to give +1. The minus and plus ones in columns A to E, (first row), then define the first test for the 2^{5-1} design. Note that only column E has a positive number so the first test to be carried out involves setting factors A to D at their low levels and factor E at its high level. That is, test e. The second number in column E is found by multiplying together the numbers in the second row of columns A, B, C and D to give –1. The minus and plus ones in columns A to E, (second row), then define the second test of the 2^{5-1} design. Note that only column A has a positive number so the second test to be carried out involves setting factors B to E at their low levels with factor A at its high level. That is, test a. Carrying on in this way gives the sixteen tests shown in the last column of Table 4.2 that are required for the 2^{5-1} design, i.e. tests e, a, b, abe, c, ace, bce, abc, d, ade, bde, abd, *cde*, *acd*, *bcd* and *abcd*. Note that this is half the number of tests required for the full 2^5 design shown in Section 3.3. The final and not first column defines the test that need to be carried out for a 2^{5-1} design of the ausforming process.

The alternate fraction is found in a similar way except now the numbers in column E are found using the design generator $E = -ABCD$, i.e. by multiplying together the numbers in columns A to D and then reversing the sign. This design is shown in Table 4.3. Note the test combinations for each of these two fractional designs are unique and that the two fractions together define the full 2^5 experiment shown in Section 3.3.

4.3 OTHER FRACTIONAL FACTORIAL DESIGNS

Further savings on the number of tests required for a particular experiment can be obtained by setting up smaller fractions. For example, there is the **one-quarter fraction** of the 2^k design. This is usually written as the 2^{k-2} design and requires 2^{k-2} test instead of the full 2^k tests. Another important fractional factorial design is the **one-eighth fraction** of the 2^k design. This is written as the 2^{k-3} design and requires 2^{k-3} tests. In general, it possible to construct a 2^{k-p} design where p is the size of the fraction used in the fractional design. $p = 4$ would imply a one sixteenth fractional factorial design which is very popular with Taguchi advocates as will be seen in Section 4.5 below. Any pth fractional factorial design can be constructed by following the two steps outlined in Section 4.2. These steps need to be generalised as follows:

4.3.1 STEP ONE: DEFINING THE BASE DESIGN

The base design for any 2^{k-p} fractional factorial is the full 2^{k-p} factorial design.

4.3.2 STEP TWO: INTRODUCTION OF THE REMAINING FACTORS

p additional factors or columns must next be added to the base design using p design generators. The test levels for the p additional factors are set as the product of the elements in the columns given by the p design generators. These design generators must be chosen so as to obtain the maximum amount of information on how to optimise a process from the minimum number of tests. Such design generators have been worked out by Montgomery[14] and are shown in Table 4.4.

This table is very easy to use. For example, suppose you wanted to study a linear manufacturing process that had seven process variables. You have a choice of carrying out one of four possible experiments. You could carry out a one-half fractional factorial of the full 2^7 experiment. This would involve carrying out 64 tests if the experiment is not to be replicated. The one remaining factor left out of the base design would then be

Table 4.4 Selected 2^{k-p} fractional factorial designs.

Number of Factors, k	Fraction	Number of Tests, M	Design Generators
3	2_{III}^{3-1}	4	$C = \pm AB$
4	2_{IV}^{4-1}	8	$D = \pm ABC$
5	2_{V}^{5-1}	16	$E = \pm ABCD$
5	2_{III}^{5-2}	8	$D = \pm AB$ $E = \pm AC$
6	2_{VI}^{6-1}	32	$F = \pm ABCDE$
6	2_{IV}^{6-2}	16	$E = \pm ABC$ $F = \pm BCD$
6	2_{III}^{6-3}	8	$D = \pm AB$ $E = \pm AC$ $F = \pm BC$
7	2_{VII}^{7-1}	64	$G = \pm ABCDEF$
7	2_{IV}^{7-2}	32	$F = \pm ABCD$ $G = \pm ABDE$
7	2_{IV}^{7-3}	16	$E = \pm ABC$ $F = \pm BCD$ $G = \pm ACD$
7	2_{III}^{7-4}	8	$D = \pm AB$ $E = \pm AC$ $F = \pm BC$ $G = \pm ABC$
8	2_{V}^{8-2}	64	$G = \pm ABCD$ $H = \pm ABEF$
8	2_{IV}^{8-3}	32	$F = \pm ABC$ $G = \pm ABD$ $H = \pm BCDE$
8	2_{IV}^{8-4}	16	$E = \pm BCD$ $F = \pm ACD$ $G = \pm ABC$ $H = \pm ABD$
9	2_{VI}^{9-2}	128	$H = \pm ACDFG$ $J = \pm BCEF$
9	2_{IV}^{9-3}	64	$G = \pm ABCD$ $H = \pm ACEF$ $J = \pm CDEF$
9	2_{IV}^{9-4}	32	$F = \pm BCDE$ $G = \pm ACDE$ $H = \pm ABDE$ $J = \pm ABCE$
9	2_{III}^{9-5}	16	$E = \pm ABC$ $F = \pm BCD$ $G = \pm ACD$ $H = \pm ABD$ $J = \pm ABCD$
10	2_{IV}^{10-5}	32	$F = \pm ABCD$ $G = \pm ABCE$ $H = \pm ABDE$ $I = \pm ACDE$ $J = \pm BCDE$

reintroduced using either the design generator $G = +ABCDEF$ for the principle fraction or using the design generator $G = -ABCDEF$ for the alternate fraction.

Alternatively, you could carry out a one-quarter fractional factorial of the full 2^7 experiment. This would involve carrying out 32 tests if the experiment is not to be replicated. The two remaining factors left out of the base design would then be reintroduced using either the design generators $F = +ABCD$ and $G = +ABDE$ for the principle fraction or using the design generators $F = -ABCD$ and $G = -ABDE$ for the alternate fraction.

Or you could carry out a one-eighth fractional factorial of the full 2^7 experiment. This would involve carrying out 16 tests if the experiment is not to be replicated. The three remaining factors left out of the base design would then be reintroduced using either the design generators $E = +ABC$, $F = +BCD$ and $G = +ACD$ for the principle fraction or using the design generators $E = -ABC$, $F = -BCD$ and $G = -ACD$ for the alternate fraction.

Finally, you could carry out a one-sixteenth fractional factorial of the full 2^7 experiment. This would involve carrying out just eight tests if the experiment is not to be replicated. The four remaining factors left out of the base design would then be reintroduced using either the design generators $D = +AB$, $E = +AC$, $F = +BC$ and $G = +ABC$ for the principle fraction or using the design generators $D = -AB$, $E = -AC$, $F = -BC$ and $G = -ABC$ for the alternate fraction.

The last option involves fewer tests than the first. As to be expected the last option will result in the loss of more information on how to control the process compared to the first option. The **resolution** of the experimental design summarises how much information is lost. The lower the resolution the more information that is lost. Table 4.4 shows that the last option above is only resolution III, whilst the first option is resolution VII. The concept and meaning of resolution will be discussed in more detail in the next chapter. There is therefore a clear trade off between the number of tests to carry out and the amount of information lost and Table 4.4 shows those experiments that maximises the amount of information that can be obtained from the shown number of tests.

4.4 A 2_{IV}^{7-3} DESIGN FOR THE AUSFORMING PROCESS

G. Taguchi[1] has had an influential impact on the use of experiments in engineering design. He and his supporters make use of so called **orthogonal designs**. There is a vast selection of such designs but the two most popular are the so-called $L_8(2^7)$ and $L_{16}(2^{15})$. The first allows up to seven variables to be studied at two levels and involves eight tests. The second allows up to 15 variables to be studied at two levels and requires 16 tests. Both these designs are nothing more than highly fractionated factorials. These orthogonal designs will be described in more detail in Section 4.5. below but for the moment it is important to become familiar with some highly fractionated designs. To this end reconsider the ausforming process discussed in Sections 2.1 and 3.3. In Section 3.3, five factors were studied but the description of the ausforming process given in Section 2.1 suggested that a further two factors may also be important in influencing the strength of the manufactured steel. These being the austenitising time, (factor G), and the isothermal incubation time, (factor F). These seven factors basically encapsulate the complete ausforming process.

To model the ausforming process the high and low levels for each factor were set at the values shown in Table 4.5. These high and low levels were picked because of the belief that over such ranges all the measured relationships would be linear. The difference between this study and that of Section 3.3 is that now the full set of ausforming process factors is being considered.

A full unreplicated 2^7 factorial would require $2^7 = 128$ tests and for most experimental programmes this is likely to be to prohibitive in terms of time and money. Suppose resource constraints limit the number of tests that can be carried out to 16. Then Table 4.4 reveals that with 16 tests and seven factors the highest resolution design available is the one-eighth fractional factorial, i.e. the 2^{7-3}_{IV} design. Hence $p = 3$ and the design is resolution IV. To find out the test level combinations required for this experiment the base design needs to be identified. This will be a full 2^{7-3} or 2^4 factorial design and this is shown in Table 4.6. This is constructed by writing out the $2^4 = 16$ test conditions in standard order form.

The design generators for the principle fraction are, from Table 4.4, $E = ABC$, $F = BCD$ and $G = ACD$. Factors E to G are therefore reintroduced into the base design using these design generators. Thus the elements in column E of Table 4.7 are obtained by multiplying the elements in columns A, B and C of Table 4.6 together. The elements in column F of Table 4.7 are obtained by multiplying the elements in columns B, C and D of Table 4.6 together. Finally, the elements in column G of Table 4.7 are obtained by multiplying the elements in columns A, C and D of Table 4.6 together. Notice that the first test of the 2^{7-3} design involves setting all seven factors at their low levels, whilst the last test will involve setting all seven factors at their high levels. Again it is important to stress that the actual sixteen tests should not be carried out in the order shown – they should in fact be randomised. The results obtained from this experiment are those shown in the last column of Table 4.7. The tensile strength measurements are in MPa.

4.5 TAGUCHI'S ORTHOGONAL ARRAYS

Taguchi does not design his experiments using the steps outlined in Sections 4.2 and 4.3 for a fractional factorial experiment. This is despite the fact that most of his designs are nothing more than fractional factorials. Instead he and his supporters use one or more of a fixed set of designs that were originally 'constructed' by Taguchi. For a full set of such designs the reader is referred to Phadke.[15] It is as a result of using these designs that many industrial experiments carried out in the past have been extremely inefficient.

One of his most widely used and popular designs is the $L_8(2^7)$ orthogonal array that was briefly mentioned in Section 4.4. This design is capable of analysing up to seven variables at two different levels, (hence 2^7), and always involves eight different test

Table 4.5 Test conditions for the complete ausforming process.

Levels	Austenitising Temperature	Deformation Temperature	Amount of Deformation	Quench Rate	Rate of Deformation	Isothermal Incubation Time	Austenitising Time
	Factor A	Factor B	Factor C	Factor D	Factor E	Factor F	Factor G
High	1040°C	550°C	80%	7 K s^{-1}	5 mm min^{-1}	10 min	120 min
Low	930°C	400°C	50%	1 K s^{-1}	0.3 mm min^{-1}	3 min	30 min

Table 4.6 The base design for the 2_{IV}^{7-3} fractional factorial.

Test	Coded Test Conditions			
	A	B	C	D
(1)	−1	−1	−1	−1
a	1	−1	−1	−1
b	−1	1	−1	−1
ab	1	1	−1	−1
c	−1	−1	1	−1
ac	1	−1	1	−1
bc	−1	1	1	−1
abc	1	1	1	−1
d	−1	−1	−1	1
ad	1	−1	−1	1
bd	−1	1	−1	1
abd	1	1	−1	1
cd	−1	−1	1	1
acd	1	−1	1	1
bcd	−1	1	1	1
abcd	1	1	1	1

conditions, (hence L_8). The $L_8(2^7)$ orthogonal array is shown in Table 4.8 below. (Ignore the coded factors at the bottom of this Table for the moment. These are not part of the published $L_8(2^7)$ design).

This $L_8(2^7)$ design is nothing more than a fractional factorial design. Unfortunately, the resolution of this design depends upon how many factors are being studied with this design. For example, if seven factors are being studied this $L_8(2^7)$ design is actually a

Table 4.7 The 2_{IV}^{7-3} design for the ausforming process.

Coded Test Conditions							Tests	Response
A	B	C	D	E = ABC	F = BCD	G = ACD		MPa
−1	−1	−1	−1	−1	−1	−1	(1)	2331
1	−1	−1	−1	1	−1	1	aeg	2275
−1	1	−1	−1	1	1	−1	bef	2180
1	1	−1	−1	−1	1	1	abfg	1731
−1	−1	1	−1	1	1	1	cefg	2910
1	−1	1	−1	−1	1	−1	acf	2461
−1	1	1	−1	−1	−1	1	bcg	2714
1	1	1	−1	1	−1	−1	abce	2611
−1	−1	−1	1	−1	1	1	dfg	2380
1	−1	−1	1	1	1	−1	adef	2100
−1	1	−1	1	1	−1	1	bdeg	2384
1	1	−1	1	−1	−1	−1	abd	1687
−1	−1	1	1	1	−1	−1	cde	2893
1	−1	1	1	−1	−1	1	acdg	2428
−1	1	1	1	−1	1	−1	bcdf	2541
1	1	1	1	1	1	1	abcdefg	2447

very unusual 2_{III}^{7-4} fractional factorial. In this sense there is nothing new about any of Taguchi's orthogonal arrays as they can always be expressed and interpreted as a fractional factorial. To see the equivalence of these two designs simply construct a 2_{III}^{7-4} fractional factorial using the steps outlined in Section 4.3. This initially requires setting up a base design given by a full 2^3 factorial (see Table 4.9) and then adding the remaining four factors using a mixture of the design generators shown in Table 4.4, i.e., $D = -AB$, $E = -AC$, $F = -BC$ and $G = ABC$ (see Table 4.10). Notice that the $L_8(2^7)$ design is therefore a mixture of a principle and alternate fractional factorial design in that some of the design generators have minus signs and the others positive signs. It is this that makes Taguchi's designs unusual and often wasteful and inefficient. The necessary calculations are illustrated in Tables 4.9 and 4.10 below.

Now compare Table 4.10 with Table 4.8. Although the columns are in a different order, the two designs are the same. That is, if factor A is assigned to column 4 of the $L_8(2^7)$ design, factor B to column 2 of the $L_8(2^7)$ design, factor C to column 1 of the $L_8(2^7)$ design, factor D to column 6 of the $L_8(2^7)$ design, factor E to column 5 of the $L_8(2^7)$ design, factor F to column 3 of the $L_8(2^7)$ design and factor G to column 7 of the $L_8(2^7)$ design then the designs shown in Tables 4.8 and 4.10 are identical. See the factor codings at the bottom of Table 4.8 for a confirmation of this.

The 2^{k-p} Experimental Design: Fractional Factorials

Table 4.8 Taguchi's $L_8(2^7)$ orthogonal array.

Test	Factors							Response
	1	2	3	4	5	6	7	
1	−1	−1	−1	−1	−1	−1	−1	Y_1
2	−1	−1	−1	1	1	1	1	Y_2
3	−1	1	1	−1	−1	1	1	Y_3
4	−1	1	1	1	1	−1	−1	Y_4
5	1	−1	1	−1	1	−1	1	Y_5
6	1	−1	1	1	−1	1	−1	Y_6
7	1	1	−1	−1	1	1	−1	Y_7
8	1	1	−1	1	−1	−1	1	Y_8
	C	B	F	A	E	D	G	

Table 4.9 Base design for a 2_{III}^{7-4} fractional factorial.

Tests	Coded Test Conditions		
	A	B	C
(1)	−1	−1	−1
a	1	−1	−1
b	−1	1	−1
ab	1	1	−1
c	−1	−1	1
ac	1	−1	1
bc	−1	1	1
abc	1	1	1

Table 4.10 The 2_{III}^{7-4} fractional factorial using the unusual design generators.

Coded Test Conditions							Tests
A	B	C	D = −AB	E = −AC	F = −BC	G = ABC	
−1	−1	−1	−1	−1	−1	−1	(1)
1	−1	−1	1	1	−1	1	adeg
−1	1	−1	1	−1	1	1	bdfg
1	1	−1	−1	1	1	−1	abef
−1	−1	1	−1	1	1	1	cefg
1	−1	1	1	−1	1	−1	acdf
−1	1	1	1	1	−1	−1	bcde
1	1	1	−1	−1	−1	1	abcg

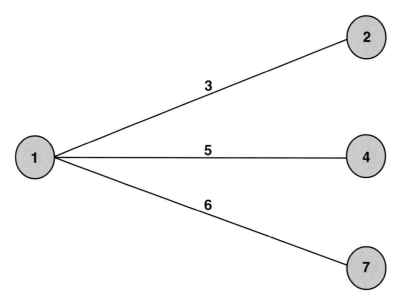

Fig. 4.1 Linear graph for the $L_8(2^7)$ design.

Having seen the foundation of the $L_8(2^7)$ design consider the spring free height experiment first discussed in Section 2.7. The engineers involved with this experiment decided to use the $L_8(2^7)$ orthogonal array to analyse the design factors of the process. Recall that this process has four design factors but the $L_8(2^7)$ design being used to represent it contains seven columns. At first sight this suggests that three columns will remain empty. This is not quite true. The three surplus columns are used to measure the way some of the factors influence the response being measured. More will be said about this in the next chapter. To find out which columns contain the levels for each factor and which of them are redundant as far as defining the tests conditions are concerned, Taguchi supporters make use of **linear graphs**. There is a linear graph for each type of Taguchi design. The linear graph for the $L_8(2^7)$ design is shown in Figure 4.1.

The numbers in the circles give the columns that factors must be assigned to, whilst the numbers on the lines refer to the columns that are redundant in the sense described above. To study four factors, simply assign these four factors to columns 1, 2, 4 and 7 of the $L_8(2^7)$ design shown in Table 4.8. The minus and plus ones of these columns alone then define the eight tests to carry out. But what if five factors needed to be studied? In this situation four of the factors would be assigned to columns 1, 2, 4 and 7 and the fifth factor to the highest ranked unfilled column, i.e. column 6. Now the plus and minus ones in columns 1, 2, 4, 6 and 7 define the eight tests to carry out.

Table 4.11 Results for the spring free height experiment.

Test	Design Factors							Noise Factor		
	1	2	3	4	5	6	7	E^- (low)	E^+ (high)	
	C	B		A			D	Responses		Test Conditions for Design Factors
1	−1	−1	−1	−1	−1	−1	−1	7.79	7.29	(1)
2	−1	−1	−1	1	1	1	1	8.07	7.73	ad
3	−1	1	1	−1	−1	1	1	7.52	7.52	bd
4	−1	1	1	1	1	−1	−1	7.63	7.65	ab
5	1	−1	1	−1	1	−1	1	7.94	7.4	cd
6	1	−1	1	1	−1	1	−1	7.95	7.62	ac
7	1	1	−1	−1	1	1	−1	7.54	7.2	bc
8	1	1	−1	1	−1	−1	1	7.69	7.63	abcd

Hence in the spring free height experiment, factors A to D must be assigned to columns 1, 2, 4 and 7 of the $L_8(2^7)$ design shown in Table 4.8. It does not matter which factor goes into which of these columns. The engineers working on this programme of research actually carried out the experiment with factor A assigned to column 4, factor B to column 2, factor C to column 1 and the last design factor, D, to column 7. This assignment of factors to columns is shown in Table 4.11 above, which shows the experiment together with the results obtained from running all the tests. Notice that the response column is presented in a way not seen before. There is one set of responses obtained when the noise factor (factor E) is set low and another when the noise factor is set high. Whilst this noise factor is not controllable at the point of manufacture it was controllable in the laboratory where this experiment was actually carried out using a small-scale model of the manufacturing process.

Note how the design factor level settings for each test shown in the last column of Table 4.11 are obtained. That is, by reading the signs on the elements in columns 1, 2, 4 and 7 only. Thus the first test involves setting design factors A to D low as all the numbers are negative in columns 1, 2, 4 and 7. At these settings a response was recorded when the noise factor was low and again when it was set high. The second test involves setting only the design factors A and D high, as there are positive elements in columns 4 and 7.

A major shortcoming of the Taguchi approach to designing experiments is that they are usually inefficient. To understand this, the above structure needs to be converted into the more familiar fractional factorial framework. Converting the two response columns in Table 4.11 into a single column will do this. To achieve this, take columns 1, 2, 4 and 7

Table 4.12 An equivalent representation of the experiment in Table 4.11.

	Factors					Response	Test Conditions
	A	B	C	E	D		
1	−1	−1	−1	−1	−1	7.79	(1)
2	1	−1	−1	−1	1	8.07	ad
3	−1	1	−1	−1	1	7.52	bd
4	1	1	−1	−1	−1	7.63	ab
5	−1	−1	1	−1	1	7.94	cd
6	1	−1	1	−1	−1	7.95	ac
7	−1	1	1	−1	−1	7.54	bc
8	1	1	1	−1	1	7.69	abcd
9	−1	1	−1	1	−1	7.29	e
10	1	−1	−1	1	1	7.73	ade
11	−1	1	−1	1	1	7.52	bde
12	1	1	−1	1	−1	7.65	abe
13	−1	−1	1	1	1	7.4	cde
14	1	−1	1	1	−1	7.62	ace
15	−1	1	1	1	−1	7.2	bce
16	1	1	1	1	1	7.63	abcde

of Table 4.11, replicate them and place these additional eight rows below the original eight rows of Table 4.11. The result of Table 4.12 containing 16 rows. Now add a further column, column E to this construction. Let the first eight rows of column E contain −1s and the last eight rows +1s. That is, the first eight rows correspond to factor E being low, the next eight to factor E being high. After placing factors A, B and C in alphabetical order and swapping the ordering for columns D and E, the design in Table 4.12 is obtained. This is equivalent to the one in Table 4.11, except the twin response column has been transformed to the more familiar structure of a factorial design where the noise factor is just another column in the fractional factorial design.

Notice that the design involves five factors but there are only sixteen tests. Thus the design shown in Table 4.12 is in fact a one half fractional factorial in five variables, i.e., 2^{5-1}. This involves $2^{5-1} = 16$ tests. But what resolution is it? This can best be seen by ignoring factor D for the minute and just consider factors A, B, C and E. These four factors together with the sixteen tests define a full factorial in these four variables. This can be clearly seen by dropping the d from the test condition column of Table 4.12 and noting that the resulting test conditions are in standard order form for factors A, B, C and E. The elements in column D are then found using the generator $D = ABC$ to give the 2^{5-1} fractional factorial. For example, take the first row of column D in Table 4.12. The

−1 is obtained by multiplying the elements in the first row of columns A, B and C, $-1 \times -1 \times -1 = -1$.

Table 4.4 shows that for these number of factors and tests, a one-half fractional factorial can be set up which is resolution V. The design generator that will achieve this is $D = ABCE$ instead of the $D = ABC$ implicit in the above Taguchi type design that used a $L_8(2^7)$ orthogonal array for the design factors. Because the design generator $D = ABCE$ extracts the maximum amount of information concerning process optimisation from just sixteen tests, the Taguchi experiment must be less than resolution V and so wasteful and inefficient. This very simple example demonstrates quite forcibly that Taguchi experimental designs that use his orthogonal arrays often lead to poor resolution or excessive testing. It is therefore much better to include noise factors as part of a fractional factorial design, and set up a design using the two-step procedure outlined in Section 4.3.

PART III
OPTIMISATION OF LINEAR PROCESSES

5. Controlling the Mean: Location Effects in Linear Designs

This chapter studies the mean values for a product's quality characteristics and how the process variables of a manufacturing process can be manipulated so that such mean values can be set equal to the target requirements of customers. Location effects estimated from a factorial or fractional factorial experiment can be used to put the mean of a products quality characteristic onto a target specification, provided that the process is linear in nature. This chapter starts of by defining a main location effect and then progresses to define the meaning of first and higher order interaction location effects. This is then followed by a discussion on how to use these location effects to manipulate the mean quality characteristics of a company's product so that they are on target. Next, the Yates and least squares procedures for estimating location effects in larger designs are discussed and this is followed by a section on quantifying the loss of information that results from using fractional factorial designs. All these concepts are then illustrated using the results from the ausforming experiments given in the previous two chapters.

5.1 DEFINITION OF LOCATION EFFECTS

5.1.1 CONTROL OF A MEAN QUALITY CHARACTERISTIC USING MAIN LOCATION EFFECTS

In this chapter those factors or process variables that can be varied so as to move the mean quality characteristics of a finished product closer to the target requirements set by customers are identified. For quality control engineers to be able to manipulate the mean quality characteristics of the product being manufactured they must know the exact way in which these mean characteristics vary with the process variables. For a linear process this can be measured by comparing the average quality characteristic obtained when a factor is set at its high level to the average quality characteristic obtained when that factor is set at its low level. This defines the **main location effect** for a factor.

More specifically, the main location effect for factor A is defined as the change in the average response that results when factor A in increased from its low to its high level. If $\bar{Y}_{A+}$ represents the average response when factor A is set high, and $\bar{Y}_{A-}$ the average response when factor A is set low then the main location effect for factor A, named A_{Loc}, is measured as

$$A_{Loc} = \overline{Y}_{A+} - \overline{Y}_{A-} \tag{5.1}$$

Similarly, the main location effect for factor B is measured as,

$$B_{Loc} = \overline{Y}_{B+} - \overline{Y}_{B-} \tag{5.2}$$

Such main location effects can obviously be calculated for any process variable from the results obtained from a factorial of fractional factorial design. These simple statistics provide some limited information on how to control the mean quality characteristic of a manufactured product. For example, the value for A_{Loc} will tell the quality control manager the change in the mean response that is to be expected if factor A is changed from its low level to its high level. These high and low levels for factor A are of course those of the factorial experiment used to gather the results from which these main location effects are estimated.

5.1.2 Control of a Mean Quality using First Order Interaction Location Effects

Unfortunately, controlling the mean is not quite as straight forward as suggested above because for many processes the change in average response following a change in the level of a factor depends also on the level set for one or more of the other factors in the process. When this is the case an interaction location effect is said to exist. If $\overline{Y}_{A+(B+)}$ is the mean response when both factors A and B are set at their high levels and $\overline{Y}_{A-(B+)}$ the mean response when factor A is set at its low level and factor B at its high level, then $\overline{Y}_{A+(B+)} - \overline{Y}_{A-(B+)}$ measures the change in average response following a change in the level of factor A under the condition that factor B is kept at its high level. Similarly, if $\overline{Y}_{A+(B-)}$ is the mean response when factor A is set at its high level and factor B at its low level and $\overline{Y}_{A-(B-)}$ the mean response when both factors A and B are at their low levels, then $\overline{Y}_{A+(B-)} - \overline{Y}_{A-(B-)}$ measures the change in average response following a change in the level of factor A under the condition that factor B is kept at its low level.

If $\overline{Y}_{A+(B+)} - \overline{Y}_{A-(B+)}$ and $\overline{Y}_{A+(B-)} - \overline{Y}_{A-(B-)}$ have the same value, then no interaction between these two factors is present and a change in the level for factor A generates the same change in the average response irrespective of the level set for factor B. In this situation and in a two factor process, A_{Loc} provides all the information that can be used to control the mean quality characteristic through the manipulation of factor A. However, if those two terms are not of the same value, an interaction between factors A and B exists and the size of this interaction is given by the difference between these two terms. In fact, the AB interaction location effect is traditionally defined as half this difference. The reason for this will become clear below. Thus

$$AB_{Loc} = \frac{1}{2}\left\{\left[\overline{Y}_{A+(B+)} - \overline{Y}_{A+(B-)}\right] - \left[\overline{Y}_{A-(B+)} - \overline{Y}_{A-(B-)}\right]\right\} \quad (5.3)$$

AB_{Loc} is an example of an interaction between two factors and is termed a **first order** interaction. First order interactions are always between two factors. In this situation a change in the level for factor A generates a different change in the average response depending upon the level set for factor B. In the presence of such an interaction, and for a two factor process it is A_{Loc} together with AB_{Loc} that provides all the information required to control the mean quality characteristic through factor A. To see this more clearly notice what happens when the A_{Loc} and AB_{Loc} effects are added up

$$A_{Loc} + AB_{Loc} = \left[\overline{Y}_{A+} - \overline{Y}_{A-}\right] + \frac{1}{2}\left\{\left[\overline{Y}_{A+(B+)} - \overline{Y}_{A-(B+)}\right] - \left[\overline{Y}_{A+(B-)} - \overline{Y}_{A-(B-)}\right]\right\}$$

Expanding the first square bracket gives, when there are two factors only,

$$A_{Loc} + AB_{Loc} = \frac{1}{2}\left[\overline{Y}_{A+(B+)} + \overline{Y}_{A+(B-)} - \overline{Y}_{A-(B+)} - \overline{Y}_{A-(B-)}\right] + \frac{1}{2}\left\{\left[\overline{Y}_{A+(B+)} - \overline{Y}_{A-(B+)}\right] - \left[\overline{Y}_{A+(B-)} - \overline{Y}_{A-(B-)}\right]\right\}$$

Cancelling out like terms then yields

$$A_{Loc} + AB_{Loc} = \left[0.5\overline{Y}_{A+(B+)} - 0.5\overline{Y}_{A-(B+)}\right] + \left\{\left[0.5\overline{Y}_{A+(B+)} - 0.5\overline{Y}_{A-(B+)}\right]\right\} = \overline{Y}_{A+(B+)} - \overline{Y}_{A-(B+)}$$

Most terms cancel out because of the inclusion of ½ in the definition of an interaction location effect. So $A_{Loc} + AB_{Loc}$ measures the change in average response as factor A is changed from its low to its high level when factor B is held fixed at its high level. By analogy, $A_{Loc} - AB_{Loc}$ will measure the change in average response as factor A is changed from its low to its high level when factor B is held fixed at its low level. In a two factor process there are thus two statistics yielding information on how to manipulate the mean response by changing factor A. One for when factor B is set high and one for when factor B is set low.

5.1.3 Higher Order Interaction Location Effects

When three factors interact a **second order** interaction location effect is said to exist and when four factors interact a **third order** interaction location effect exists. This can be generalised to any number of factors. Thus ABC_{Loc} is the interaction between factors A, B and C in the determination of the mean response and $ABCD_{Loc}$ is the interaction between factors A, B, C and D in the determination of the mean response. However, these higher order interactions are not likely to be important in determining the mean response because what they actually measure is the ability, for example, of a third factor to influence the size of a first order interaction location effect and not the mean response itself. Thus, the value for ABC_{Loc} measures how the value for AB_{Loc} varies with the level set for factor C. Typically, therefore such interactions are usually found to be quite small in value and can often be ignored when trying to control the mean quality characteristic of a manufactured product.

Table 5.1 Control matrix for factor A in a two factor manufacturing process.

When:	Change in Average Response as Factor A is Changed from its Low to High Level
Factor B High	$A_{Loc} + AB_{Loc}$
Factor B Low	$A_{Loc} - AB_{Loc}$

Table 5.2 Control matrix for factor B in a two factor manufacturing process.

When:	Change in Average Response as Factor B is Changed from its Low to High Level
Factor A High	$B_{Loc} + AB_{Loc}$
Factor A Low	$B_{Loc} - AB_{Loc}$

5.2 METHODS OF CONTROLLING THE MEAN QUALITY CHARACTERISTICS

The above definitions on location effects now allow for a discussion of two techniques that can be used to model and manipulate the mean quality characteristic of a product being manufactured. The first makes use of a simple control matrix whose size depends upon the number of process variables being considered. The second makes use of a first order response surface model for the mean quality characteristic. Both techniques work on the assumption that all second and higher order interaction location effects can be ignored, i.e. are equal zero in value.

5.2.1 A Control Matrix for the Mean of a Quality Characteristic

The discussions above have lead to the conclusion that a change in the level of one factor can have different effects on the mean response depending upon the levels set for all the other factors of the manufacturing process. This complexity was studied using an interaction location effect. Thus if a manufacturing process has only two factors, Table 5.1 gives the **control matrix** that summarises all the information needed to control the mean quality characteristic of a manufactured product using factor A.

Similarly, Table 5.2 shows the control matrix summarising all the information needed to control the mean of the process using factor B.

When factor A is set high the change to be expected in the mean response, following a change in the level of factor B from its low to its high level, is given by the value of

Controlling the Mean: Location Effects in Linear Designs

Table 5.3 Control matrix for factor A in a three factor manufacturing process.

	Change in Average Response as Factor A is Changed from its Low to High Level	
When:	Factor C High	Factor C Low
Factor B High	$A_{Loc} + AB_{Loc} + AC_{Loc}$	$A_{Loc} + AB_{Loc} - AC_{Loc}$
Factor B Low	$A_{Loc} - AB_{Loc} + AC_{Loc}$	$A_{Loc} - AB_{Loc} - AC_{Loc}$

$B_{Loc} + AB_{Loc}$. But when factor A is set low, the change to be expected in the mean response, following a change in the level of factor B from its low to its high level, is given by the value of $B_{Loc} - AB_{Loc}$.

Now if a manufacturing process has three factors, Table 5.3 shows the control matrix summarising all the information needed to control the mean quality characteristic of the manufactured product using factor A.

There are now four possibilities to consider when using factor A to control the mean response. When factors B and C are both set high, the value of $A_{Loc} + AB_{Loc} + AC_{Loc}$ shows how the mean response changes as factor A is changed from its low to its high level. When factors B and C are both set low then the value of $A_{Loc} - AB_{Loc} - AC_{Loc}$ shows how the mean response changes as factor A is changed from its low to its high level. When factor B is set low and factor C high then the value of $A_{Loc} - AB_{Loc} + AC_{Loc}$ shows how the mean response changes as factor A is changed from its low to its high level. Finally, when factor B is set high and factor C low then the value of $A_{Loc} + AB_{Loc} - AC_{Loc}$ shows how the mean response changes as factor A is changed from its low to its high level. Clearly factor A will be most effective at altering the mean response at some particular setting for the levels of all the other factors and the matrix above will help the quality control engineer find what these settings are. Similar control matrices can be constructed that summarise all the information needed to control the mean response of the process using factor B and factor C – see Tables 5.4 and 5.5.

If a manufacturing process has four factors the matrix shown in Table 5.6 summarises all the information needed to control the mean quality characteristic of the manufactured product using factor A.

There are now eight possibilities to consider when using factor A to control the mean response. Take some examples. When factors B, C and D are all set high then the value of $A_{Loc} + AB_{Loc} + AC_{Loc} + AD_{Loc}$ shows how the mean response changes as factor A is changed from its low to its high level. When factors B, C and D are all set low then the value of $A_{Loc} - AB_{Loc} - AC_{Loc} - AD_{Loc}$ shows how the mean response changes as factor A is changed from its low to its high level. As a final example, when factor B is set high and factors C and D low then the value of $A_{Loc} + AB_{Loc} - AC_{Loc} - AD_{Loc}$ shows how the mean response changes as factor A is changed from its low to its high level. Again it is the case that factor A will be most effective at altering the mean response at some particular

Table 5.4 Control matrix for factor B in a three factor manufacturing process.

	Change in Average Response as Factor B is Changed from its Low to High Level	
When:	Factor C High	Factor C Low
Factor A High	$B_{Loc} + AB_{Loc} + BC_{Loc}$	$B_{Loc} + AB_{Loc} - BC_{Loc}$
Factor A Low	$B_{Loc} - AB_{Loc} + BC_{Loc}$	$B_{Loc} - AB_{Loc} - BC_{Loc}$

Table 5.5 Control matrix for factor C in a three factor manufacturing process.

	Change in Average Response as Factor C is Changed from its Low to High Level	
When:	Factor B High	Factor B Low
Factor A High	$C_{Loc} + BC_{Loc} + BC_{Loc}$	$C_{Loc} - BC_{Loc} + BC_{Loc}$
Factor A Low	$C_{Loc} + BC_{Loc} - BC_{Loc}$	$C_{Loc} - BC_{Loc} - BC_{Loc}$

Table 5.6 Control matrix for factor A in a four factor manufacturing process.

When:	Change in Average Response as Factor A is Changed from its Low to High Level			
	Factor C High		Factor C Low	
	Factor D High	Factor D Low	Factor D High	Factor D Low
Factor B High	$A_{Loc} + AB_{Loc}$ $+ AC_{Loc} + AD_{Loc}$	$A_{Loc} + AB_{Loc}$ $+ AC_{Loc} - AD_{Loc}$	$A_{Loc} + AB_{Loc}$ $- AC_{Loc} + AD_{Loc}$	$A_{Loc} + AB_{Loc}$ $- AC_{Loc} - AD_{Loc}$
Factor B Low	$A_{Loc} - AB_{Loc}$ $+ AC_{Loc} + AD_{Loc}$	$A_{Loc} - AB_{Loc}$ $+ AC_{Loc} - AD_{Loc}$	$A_{Loc} - AB_{Loc}$ $- AC_{Loc} + AD_{Loc}$	$A_{Loc} - AB_{Loc}$ $- AC_{Loc} - AD_{Loc}$

setting for the levels of all the other factors and the matrix in Table 5.6 will help the quality control engineer find what these settings are. Similar control matrices can be constructed that summarises all the information needed to alter the mean response of the process by manipulating factors B, C and D.

5.2.2 A First Order Response Surface Model for the Mean

A **first order response surface** model for the mean response takes on the following general form

$$Y_j = f(\text{Main location effects, first order interaction location effects}). \tag{5.4}$$

In equation (5.4) the subscript j refers to a particular test condition, so that Y_j is the recorded response at test condition j. $f()$ is a linear functional form. For a manufacturing process with just two factors the first order interaction AB_{Loc} is the only interaction so that the first order response surface model for the mean response is given by

$$Y_j = \beta_0 + \beta_A A_j + \beta_B B_j + \beta_{AB} A_j B_j \tag{5.5}$$

In equation (5.5) the subscript j again refers to a particular test condition, so that Y_j is the recorded response at test condition j. A_j is the value of the coded level (see Section 3.1.2 for a recap on coded test conditions) for factor A at test condition j and B_j is the value of the coded level for factor B at test condition j. $A_j B_j$ is the product of the coded levels for factors A and B at test condition j.

The β values in equation (5.5) are related to the main and interaction location effects in the following way. $\beta_A = A_{Loc}/2$, $\beta_B = B_{Loc}/2$ and $\beta_{AB} = AB_{Loc}/2$. Half the location effects are used, because the β values in equation (5.5) measure the change in mean response following a unit change in the coded level of each factor, whilst the location effects themselves measure the change in mean response following a two unit (-1 to $+1$) change in the coded level of each factor. Finally, β_0 is the mean response over all test conditions.

For a manufacturing process with three factors the second order interaction is the highest order interaction. However, a first order response surface model assumes that all interaction effects above order one are zero on the average. This is modelled through an **prediction error** term, ε_j, whose value on the average equals zero. Thus in a three factor process ε_j equals the value for the term $\beta_{ABC} A_j B_j C_j$, which is left out of the first order response surface model. However, ABC_{Loc} ($= \beta_{ABC}/2$) is assumed to be approximately zero in value and thus ε_j is on average equal to zero. Thus

$$Y_j = \beta_0 + \beta_A A_j + \beta_B B_j + \beta_C C_j + \beta_{AB} A_j B_j + \beta_{AC} A_j C_j + \beta_{BC} B_j C_j + \varepsilon_j \tag{5.6}$$

In equation (5.6) C_j represent the value of the coded level for factor C at test condition j, $A_j C_j$ is the product of the coded levels for factors A and C at test condition j and $B_j C_j$ is the product of the coded levels for factors B and C at test condition j. As in equation (5.5) $\beta_C = C_{Loc}/2$ and $\beta_{AC} = AC_{Loc}/2$ and $\beta_{BC} = BC_{Loc}/2$.

Because ε_j averages out to zero over many test conditions, the equation

$$Y_j = \beta_0 + \beta_A A_j + \beta_B B_j + \beta_C C_j + \beta_{AB} A_j B_j + \beta_{AC} A_j C_j + \beta_{BC} B_j C_j$$

will predict the response correctly on the average. That is, this last equation may incorrectly predict the response at various test conditions, but the average error in these predictions (averaged over many test conditions) will be zero.

Clearly, a first order response surface model can be written out for a process with any number of factors. For example, take a process that has four factors

$$Y_j = \beta_0 + \beta_A A_j + \beta_B B_j + \beta_C C_j + \beta_D D_j + \beta_{AB} A_j B_j + \beta_{AC} A_j C_j + \beta_{AD} A_j D_j +$$
$$\beta_{BC} B_j C_j + \beta_{BD} B_j D_j + \beta_{CD} C_j D_j + \varepsilon_j \tag{5.7}$$

Equation (5.7) contains all the main location effects plus all the first order interaction location effects. ε_j therefore picks up all the second and third order interaction effects whose values are all close to zero, i.e., $\varepsilon_j = \beta_{ABC}A_jB_jC_j + \beta_{ABD}A_jB_jD_j + \beta_{BCD}B_jC_jD_j + \beta_{ACD}A_jC_jD_j + \beta_{ABCD}A_jB_jC_jD_j \cong 0$.

These first order response surface models can be used to control the mean quality characteristic of a manufactured product. For example, take a two factor manufacturing process that is currently being operated with factor B at its high level, i.e. at $+1$ in coded units. Suppose also that the company operating this process wishes the product to have a mean quality characteristic given by Y^*. Then this can be achieved by operating the process with the following level for factor A

$$Y^* = \beta_0 + \beta_A A_j + \beta_B(+1) + \beta_{AB}A_j(+1)$$

Rearranging for A_j gives,

$$A_j = \frac{Y^* - \beta_0 - \beta_B}{\beta_A + \beta_{AB}} \tag{5.8}$$

This value for A_j together with $B_j = +1$ will give a mean quality characteristic of Y^* for the manufactured product. To achieve such control, quality control engineers need to know values for all the main and first order interaction location effects, (i.e. all the β values). These values come from running a factorial or fractional factorial experimental design. An important point to note about this approach to control is that the levels set for each factor must never be outside of the levels used in that experiment. It is safe to **interpolate** in a linear manufacturing process but it is not so safe to **extrapolate**.

5.3 THE YATES AND LEAST SQUARES PROCEDURES FOR ESTIMATING LOCATION EFFECTS IN FULL FACTORIALS

When the number of factors being studied is quite large the **Yates** and **least squares** procedure offer a very quick way to execute equations (5.1) to (5.3). These techniques can quickly calculate the main and interaction location effects for all the factors in an experimental design.

5.3.1 THE YATES TECHNIQUE

This technique involves the creation of a simple table by progressing through the following five steps. First write out the standard order form in column one of the table. Thus each row of the created table must represent a particular test condition. Next create a series of columns with headings identical to, and in the same order as, the row headings – ignoring test (1). For example, ignoring test (1) the next test in standard order form is a. So the

Table 5.7 The Yates procedure for the 2^2 design.

Test	Coded Test Conditions			Mean Response	Factor Contrasts		
	A	B	AB		A	B	AB
(1)	−1	−1	1	$\overline{Y}_{(1)}$	$-\overline{Y}_{(1)}$	$-\overline{Y}_{(1)}$	$+\overline{Y}_{(1)}$
a	1	−1	−1	$\overline{Y}_a$	$+\overline{Y}_a$	$-\overline{Y}_a$	$-\overline{Y}_a$
b	−1	1	−1	$\overline{Y}_b$	$-\overline{Y}_b$	$+\overline{Y}_b$	$-\overline{Y}_b$
ab	1	1	1	$\overline{Y}_{ab}$	$+\overline{Y}_{ab}$	$+\overline{Y}_{ab}$	$+\overline{Y}_{ab}$
				Contrasts	Sum	Sum	Sum
				Location Effect	$\dfrac{\text{Contrast}}{2^{k-1}}$	$\dfrac{\text{Contrast}}{2^{k-1}}$	$\dfrac{\text{Contrast}}{2^{k-1}}$

next column of the table is headed A. The next test in standard order form is b and so the next column of the table is headed B and so on. Thirdly, fill in these headed columns with the correctly signed elements (−1 or +1). The columns headed with single letters will contain the coded test condition for that factor. Thus for columns A, B, C, D etc. insert a −1 if that factor is to be set low and +1 if it is to be set high. For all the other columns, such as those headed AB, AC, ABC, ABCD etc., put in the number that results from multiplying together the numbers in the columns that form that columns heading. For example in the column headed AB insert the product of the numbers in columns A and B. Fourthly, create another column in the table and fill it with the mean response corresponding to each test. Finally, multiply the response column by all the other columns to create the factor contrast columns. The numbers in these new columns are then summed and averaged to give all main and higher order interaction location effects.

As an example consider a 2^2 experiment carried out on a manufacturing process. From this experiment, quality control engineers wish to estimate all of the location effects. Going through the five steps above will result in the creation of Table 5.7.

The first column of Table 5.7 contains the shorthand test descriptors for the 2^2 design in standard order form. Then three columns are created with headings identical to, and in the same order as, the row headings when test (1) is ignored. The first row is headed a, so the first column is headed A, the final row is headed ab and so the third column is headed AB. The columns headed A and B are used to write out the coded test conditions for the 2^2 design. Thus for test (1) both factors A and B are at their low level so that a minus one is entered in the A and B columns for this row. Likewise, for test ab both factors are high so a plus one is entered in each column for this row. The next column, the *AB* column, is found by multiplying together the plus and minus one's in the *A* and *B* columns.

The next column to create contains the mean response obtained under each of these test conditions. Thus $\bar{Y}_{(1)}$, the mean response obtained from test (1), is shown in the first row of this column and $\bar{Y}_{ab}$, the mean response obtained for test *ab*, is shown in the last row of this column. When there is no replication, the actual response should be placed in this column. The mean response column is then multiplied by each of the other columns to yield the numbers shown in the last three columns of Table 5.7. Finally, the sum of the numbers in each of these last three columns is shown at the bottom of the table. Such a sum is called the **factor contrast**.

The Yates technique works by putting the right sign in front of each mean response. For example, take the contrast associated with factor *A*. From Table 5.7 this is

$$A_{Contrast} = -\bar{Y}_{(1)} + \bar{Y}_a - \bar{Y}_b + \bar{Y}_{ab} = [\bar{Y}_a + \bar{Y}_{ab}] - [\bar{Y}_b + \bar{Y}_{(1)}] \tag{5.9}$$

If each bracketed term in equation (5.9) is divided through by two, then we have the main effect for factor *A*. That is, a comparison of the average response when factor *A* is set at its high levels to when factor *A* is set at its low level. Finally, in Table 5.7 the main and first order interaction location effects are shown. For a full factorial these are always found by dividing each contrast by 2^{k-1}

$$\text{Location effect} = \frac{\text{Contrast}}{2^{k-1}} \tag{5.10}$$

For a fractional factorial these are always found by dividing each contrast by $2^{k-p}/2$

$$\text{Location effect} = \frac{\text{Contrast}}{\left(\frac{2^{k-p}}{2}\right)}. \tag{5.11}$$

5.3.2 THE LEAST SQUARES TECHNIQUE

To illustrate this technique take equation (5.6). In the least squares procedure, each location effect is given a value that results in the prediction error from a first order response surface model being as small as possible. Each β term, (which represents half the location effects), in equation (5.6) is therefore given a value that results in the prediction error, ε_j, being as small as possible. That is, so equation (5.6) best fits the experimental data. In particular least squares chooses values for all the β parameters so that

$$\sum_{j=1}^{2^{k-p}} \varepsilon_j^2,$$

is as small as possible. k is the number of factors in the experiment and p equals zero in the full design, one in a half fractional design and so on. Thus 2^{k-p} represents the total number of test conditions. The values for β which minimise this total squared prediction error are given by

$$B = \frac{1}{2^{k-p}} I[X'\bar{Y}], \qquad (5.12)$$

for a full factorial design and by

$$B = \frac{1}{\left(\frac{2^{k-p}}{2}\right)} I[X'\bar{Y}], \qquad (5.13)$$

for a fractional factorial design.

In equations (5.12) and (5.13), B is a $2^{k-p} \times 1$ ($p = 0$ for a full factorial) vector containing the β coefficients (all of the location effects halved), $\bar{Y}$ is a $2^{k-p} \times 1$ vector containing the 2^{k-p} average responses in standard order form (there is an average response for each test condition). I is an identity matrix and k is the number of factors used in the experiment. X is a $2^{k-p} \times 2^{k-p}$ matrix made up of the -1 and $+1$ columns of the Yates table with an initial column of one's at the beginning. X' is the transposition of X.

For a 2^2 design these matrices look like

$$\bar{Y} = \begin{bmatrix} \bar{Y}_{(1)} \\ \bar{Y}_a \\ \bar{Y}_b \\ \bar{Y}_{ab} \end{bmatrix}; \quad X = \begin{bmatrix} & A & B & AB \\ 1 & -1 & -1 & 1 \\ 1 & 1 & -1 & -1 \\ 1 & -1 & 1 & -1 \\ 1 & 1 & 1 & 1 \end{bmatrix}; \quad B = \begin{bmatrix} \beta_0 \\ \beta_A \\ \beta_B \\ \beta_{AB} \end{bmatrix}$$

The first column of X is always a column of one's and is there to allow a value for β_0 to be made. The second column of X corresponds to the column headed A in the Yates procedure of Table 5.7. The third column of X corresponds to column headed B and the last column of X corresponds to the column headed AB in the Yates procedure of Table 5.7. The β values in matrix B will always come out in standard order form if $\bar{Y}$ and X are constructed in this way. It can be shown that the resulting values for β are the same as those obtained using the Yates procedure when the design is linear. (That is for all 2^k and 2^{k-p} designs).

This least squares procedure is very easy to implement in Excel. For the 2^2 design there are four β values to estimated so just highlight a block of four columns by one row and type into this block the formula = LINEST(R1, R2, FALSE, FALSE) where R1 is the range of cells down a single column containing the values for the mean responses (matrix $\bar{Y}$) and R2 is the range of cells down the columns containing values for the matrix X. The function is executed by hitting the ctrl shift return buttons simultaneously. If the initial column of X (the column of ones) is left out of the range R2, a value for β_0 can still be obtained by typing in the following version of the LINEST function, LINEST(R1, R2, TRUE, FALSE). The important point to note about this function is that the estimated parameters are presented in reverse order to that shown in the response surface models (e.g. in equation (5.5)). That is, in reverse order to the parameters shown in the B matrix above.

5.4 LOCATION EFFECTS ESTIMATED FROM FRACTIONAL DESIGNS

In Section 4.3 the concept of resolution was introduced. Resolution measures the amount of information (needed for process control) that is lost when using a fractional factorial design instead of the full factorial design. The lower the resolution the more information that is lost. This section now examines exactly what type of information is lost in these fractional factorial designs.

To illustrate the type of information that is lost consider again the simplest full factorial design. This is the 2^2 design with responses $Y_{(1)}$, Y_a, Y_b and Y_{ab}. Now suppose that only half of these tests are actually carried out. This would give a one half fraction of the 2^2 design or simply a 2^{2-1} design. Consequently, it would require only half the number of tests, $2^{2-1} = 2$. Suppose it is decided to undertake the (1), ab half of the full 2^2 design. Then, the only estimate of the main location effect for factor A that can be made from the data is

$$A_{Loc}^{Ali} = (Y_{ab} - Y_{(1)}) \tag{5.14}$$

The only possible estimate of the main location effect for factor B that can be made from the data is also

$$B_{Loc}^{Ali} = (Y_{ab} - Y_{(1)}) \tag{5.15}$$

Clearly the effects of changes in factors A and B on the response variable can not be distinguished from one another and are therefore said to be **aliased**. When two effects are aliased in this way, the resulting estimated effect from the fractional design is actually the sum of the two individual effects that would be made from the results obtained from the full design. The only way that one effect can be identified is to assume the other is zero, i.e. not important. The main location effect estimates above are therefore not estimates of the true A_{Loc} and B_{Loc} effects that would be obtained from the full 2^2 design. Instead they are estimates of the **aliased effects**. Hence the distinguishing superscripts in equations (5.14) and (5.15). All location effects estimated from a fractional factorial design are in fact aliased effects and will be given this distinguishing superscript.

The AB_{Loc} interaction effect as defined in equation (5.3) can be rewritten as

$$AB_{Loc} = \frac{[Y_{ab} + Y_{(1)}] - [Y_b + Y_a]}{2} \tag{5.16}$$

So in carrying out just the two tests, ab and (1), only one half of the interaction effect can be determined and so no estimate of this interaction location effect can be made from the fractional design. Notice that the first two terms in equation (5.16) defines the tests to be carried out in the half fractional factorial experiment discussed above. When part of an interaction effect defines the tests making up the experiment, that interaction is said to be **confounded**. Fractional factorials are therefore designed by taking interactions that are most likely to be zero and confounding them, i.e. taking one section of the interaction as the experiment. To illustrate further, take the following hypothetical responses from a 2^2 design

$Y_{(1)} = 1$, $Y_a = 4$, $Y_b = 2$ and $Y_{ab} = 5$,

with $N = 1$. Then from equations (5.1), (5.2) and (5.16).

$$A_{Loc} = \frac{[4+5]-[1+2]}{2} = 3,\ B_{Loc} = \frac{[2+5]-[1+4]}{2} = 1,\ AB_{Loc} = \frac{[5+1]-[4+2]}{2} = 0$$

But suppose only one–half of this experiment is carried out using tests *ab* and (1). Now the aliased location effects of factors *A* and *B* are estimated simultaneously as $(5-1)/1 = 4$. These are the aliased A_{Loc} and the aliased B_{Loc} effects. The AB_{Loc} effect can't be estimated but is zero anyway. Notice that this aliased effect of four is actually the sum of the individual A_{Loc} and B_{Loc} effects from the full design. The A_{Loc} effect can only be quantified if the B_{Loc} effect is assumed zero and visa versa. This aliasing pattern is written as

$$A_{Loc}^{Ali} = A_{Loc} + B_{Loc} \tag{5.17}$$

or $A = B$ for short and reads the A_{Loc} effect is aliased with B_{Loc} effect. Thus the value obtained for the main location effect of factor *A*, based on the results obtained from a fractional design, is not the same as the value obtained for the main location effect of factor *A* based on the results obtained from the full design. It will actually equal the sum of values obtained for both main location effects based on the results from the full fractional design. The A_{Loc} effect can only be quantified from the fractional design if the B_{Loc} effect is zero. It is this information (needed for process control) that is lost when carrying out fractional designs.

If the second half of the AB_{Loc} interaction effect was taken to define the tests of the experiment, i.e. tests a and b, then the $A_{Loc}^{Ali} = (4-2)/1 = 2$ and the $B_{Loc}^{Ali} = (2-4)/1 = -2$. Notice now that the aliased A_{Loc} effect of 2 is found by subtracting the B_{Loc} effect in the full design from the A_{Loc} effect – $(3-1 = 2)$. Likewise, the aliased B_{Loc} effect of –2 is found by subtracting the A_{Loc} effect in the full design from the B_{Loc} effect –$(1-3 = -2)$. Hence this aliasing pattern is written as

$$A_{Loc}^{Ali} = A_{Loc} - B_{Loc} \tag{5.18}$$

$$B_{Loc}^{Ali} = B_{Loc} - A_{Loc} \tag{5.19}$$

or $A = -B$ for short. Taking the first part of the AB_{Loc} interaction as the fractional factorial defines the **principal fraction**, whilst taking the second part defines the **alternate fraction**. In either fraction, the estimated aliased effect can only be taken as a correct estimate of the true effect from the full design if one of the effects is actually zero.

The trick is to confound those interactions that minimise the loss of important information. For example, one that results in no main effects being aliased with any other main effect, only with higher order interactions which can safely be assumed zero.

Under such reasonable assumptions the main location effects are identified and measurable from a fractional design.

5.4.1 ALIASING ALGEBRA

The actual aliasing structure of the 2^{k-p} design can be found using **alias algebra**. In alias algebra all squared terms are eliminated from the algebraic expression. The alias for each location effect is found by multiplying all the confounded interaction effects by the relevant factor and eliminating all squared terms. In turn the confounded interaction effects are found from the design generators used to construct the fractional factorial (see Table 4.4). The number of location effects that are confounded is derived by first taking the confounded interactions defined by the generators and then adding to this list the interactions obtained by multiplying together all possible combinations of these confounded interactions.

As an example of how this works consider the 2^{5-1} fractional factorial. The design generator shown in Table 4.4 is $E = +ABCD$ for the principal design. Thus the only confounded interaction effect is the $ABCDE_{Loc}$ effect. This is found by taking E over to the otherside of the equation. Abbreviating, by letting A represent the A_{Loc} effect, B the B_{Loc} effect etc. of the 2^{5-1} design, the alias structure for factor A is therefore

$$A = A \cdot ABCDE = A^2BCDE = BCDE$$

That is, the A_{Loc} effect is aliased with the 3rd order interaction effect, $BCDE_{Loc}$. So in the principle design the estimate made for the A_{Loc} effect is and will in fact equal the A_{Loc} effect plus the $BCDE_{Loc}$ effect obtainable from the full design

$$A_{Loc}^{Ali} = A_{Loc} + BCDE_{Loc}$$

In the alternate design the estimate made for the A_{Loc} effect is A_{Loc}^{Ali} and will equal the A_{Loc} effect minus the $BCDE_{Loc}$ effect obtainable from the full design

$$A_{Loc}^{Ali} = A_{Loc} - BCDE_{Loc}$$

Only under the assumption that second and higher order interactions are zero can the estimated A_{Loc} effect from the fractional design be interpreted as the A_{Loc} effect itself.

The complete aliasing structure for the 2^{5-1} design can be found in the same way:

$A = A^2BCDE = BCDE$ $AB = A^2B^2CDE = CDE$ $BD = AB^2CD^2E = ACE$
$B = AB^2CDE = ACDE$ $AC = A^2BC^2DE = BDE$ $BE = AB^2CDE^2 = ACD$
$C = ABC^2DE = ABDE$ $AD = A^2BCD^2E = BCE$ $CD = ABC^2D^2E = ABE$
$D = ABCD^2E = ABCE$ $AE = A^2BCDE^2 = BCD$ $CE = ABC^2DE^2 = ABD$
$E = ABCDE^2 = ABCD$ $BC = AB^2C^2DE = ADE$ $DE = ABCD^2E^2 = ABC$

It should be clear from the above example, that in any 2^{k-p} design there are exactly $(2^{k-p}-1)$ unique aliased effects to be worked out. When working out the aliasing structure always proceed through the effects in alphabetical order. First, go through all the main location effects alphabetically, then the first order interactions alphabetically and so on until all $(2^{k-p}-1)$ aliased effects have been identified. Note that in the 2^{5-1} design no more calculations beyond the DE_{Loc} aliased effect have been worked out. This is because there are no more unique aliased effects. Sixteen tests are only capable of identifying 15 effects. Alphabetically, the next effect beyond DE_{Loc} is the second order interaction ABC_{Loc}. It's alias structure is $ABC = A^2B^2C^2DE = DE$. Yet this has already been worked out - the last aliased effect shown above.

The meaning of resolution can now be more fully defined. In the resolution III design, the p generators shown in Table 4.4 are chosen so that no main location effect is aliased with any other main location effect, but main location effects are aliased with first order interactions and these first order interactions are aliased with each other. This design is written as 2^{k-p}_{III}. This is a very low resolution design requiring an engineer to make the strong assumption that all first order interactions are negligible before the main location effects can be quantified from the fractional design.

In the resolution IV design the design generators in Table 4.4 were chosen so that no main location effect is aliased with any other main location effect or with any first order interaction location effects. However, first order interaction location effects are aliased with each other. Main location effects can now be identified without assuming zero values for first order interaction location effects. This is a modest resolution design because first order interactions still cannot be identified and quantified without arbitrarily assuming some of the first order interaction effects are small. An example of such a design will be given in the next section. This design is written as 2^{k-p}_{IV}.

In the resolution V design the generators are chosen so that no main or first order interaction location effect is aliased with any other main or first order interaction location effect, but first order interaction location effects are aliased with second and higher order interaction location effects. Consequently, all main and first order interaction location effects can be identified and quantified by making the plausible assumption that second and higher order interaction effects are zero in value. This design is written as 2^{k-p}_{V}.

This three fold classification of resolution is easily extended. For example, in a resolution VI design main location effects would be aliased with fourth order interactions and first order interactions would be aliased with third and higher order interactions.

It follows from this concept of resolution that fractional factorial designs involve a trade off between the number of tests and the amount of information about location effects that can be obtained from the design. The lower the resolution, the smaller the number of tests that are required, but very strong assumptions need to be made to obtain information on the main location effects. If possible it is advisable to aim for at least a resolution IV design or have some prior theoretical knowledge that will untangle the aliased effects in a low resolution design. The aliasing structure for the 2^{5-1} design derived above shows that this design is in fact resolution V as confirmed in Table 4.4.

As a final example consider the 2^{6-3}_{III} principle design shown in Table 4.4. Because it is a one eighth fractional factorial there are three design generators for the principal design, $D = AB$, $E = AC$ and $F = BC$. That is, the base design is a $2^{6-3} = 2^3$ factorial containing factors A to C and the remaining three factors are obtained by setting the levels for factor D equal to the product of columns A and B, the levels for factor E equal to the product of columns A and C and the levels for factor F equal to the product of columns B and C. This will yield a resolution III design with seven confounded interactions

From Generators	Combinations of Two	Combinations of Three
ABD ACE BCF	$ABD \cdot ACE = A^2BCDE = BCDE$ $ABD \cdot BCF = AB^2CDF = ACDF$ $ACE \cdot BCF = ABC^2EF = ABEF$	$ABD \cdot ACE \cdot BCF = A^2B^2C^2DEF = DEF$

The alias structure for the A_{Loc} effect is then found by multiplying each confounded interaction by A,

$$A = A \cdot ABD = A \cdot ACE = A \cdot BCF = A \cdot BCDE = A \cdot ACDF = A \cdot ABEF = A \cdot DEF$$

$$A = A^2BD = A^2CE = ABCF = ABCDE = A^2CDF = A^2BEF = ADEF$$

Or

$$A = BD = CE = ABCF = ABCDE = CDF = BEF = ADEF$$

That is, the A_{Loc} effect is aliased with two first order interactions, (the BD and CE effects), two second order interactions, (the CDF and BEF interactions), two third order interaction effects, (ADEF and ABCF), and one fourth order interaction effect (ABCDE). So in this principle design the estimated aliased A_{Loc} effect is in fact an estimate of the A_{Loc} effect plus all these higher order interaction effects from the full design.

$$A^{Ali}_{Loc} = A_{Loc} + BD_{Loc} + CE_{Loc} + CDF_{Loc} + BEF_{Loc} + ABCF_{Loc} + ADEF_{Loc} + ABCDE_{Loc}$$

Only under the assumption that first and higher order interactions are zero can the estimated aliased A_{Loc} effect from this fractional design be interpreted as the A_{Loc} effect alone. This is a strong assumption to make as first order interactions are often present in engineering problems.

The total number of aliased effects to be found in any fractional design is given by $(2^{k-p} - 1)$. So for 2^{6-3}_{III} design there are $2^{6-3} - 1 = 7$ unique aliased effects to be found. Progressing through all the effects alphabetically gives

```
A = BD   = CE   = ABCF = ABCDE = CDF   =   BEF   = ADEF
B = AD   = ABCE = CF   = CDE   = ABCDF =   AEF   = BDEF
C = ABCD = AE   = BF   = BDE   = ADF   =   ABCEF = CDEF
D = AB   = ACDE = BCDF = BCE   = ACF   =   ABDEF = EF
E = ABDE = AC   = BCEF = BCD   = ACDEF =   ABF   = DF
F = ABDF = ACEF = BC   = BCDEF = ACD   =   ABE   = DE
```

Note that in the above structure the only first order interaction that is not contained in any of the aliasing structures is the CD_{Loc} effect. Thus

$$CD = ABC = ADE = BDF = BE = AF = ABCDEF = CEF$$

All the other effects are contained within the above aliasing structure.

Table 5.8a Interaction Table for the $L_8(2^7)$ design.

			Interaction Columns				
(1)	3	2	5	4	7	6
	(2)	1	6	7	4	5
		(3)	7	6	5	4
			(4)	1	2	3
				(5)	3	2
					(6)	1
						(7)

5.4.2 Taguchi Designs and Aliasing

Finally, there is the question of how to identify the aliasing structure in a Taguchi orthogonal experiment. The $L_8(2^7)$ used to study the spring free height problem in Section 4.5 is an example of a popular Taguchi design. The tests carried out were in this experiment shown in Table 4.11 and they were identified using the linear graph of Figure 4.1. It was found in Section 4.5 that for this experiment three of the columns in the $L_8(2^7)$ were redundant columns, with the other four columns defining the tests to be carried out. In fact these columns are used to measure the values for three interaction location effects. The numbers on the lines of the linear graph in Figure 4.1 give the interaction effects measured by these columns. For example, redundant column 3 (in Table 4.11) of the design measures the interaction between the factors assigned to columns one and two. Thus column 3 measures the interaction between factors B and C. Column five of the design measures the interaction between the factors assigned to columns one and four. Thus column five measures the interaction between factors A and C. Then column six of the design measures the interaction between the factors assigned to columns one and seven. Thus column six measures the interaction between factors C and D.

A major problem with this Taguchi approach to experimental design is that the full aliasing structure is not immediately apparent. Part of the aliasing structure can be obtained from an interaction table. There is a separate **interaction table** for each Taguchi design. The interaction table for the $L_8(2^7)$ design is shown in Table 5.8a. However, such tables only show the interactions present between the design factors. The interaction table works as follows. The first number above the diagonal line is 3. This lies between the numbers (1) and (2) below the diagonal. Thus column three is said to contain the interaction

74   OPTIMISATION OF MANUFACTURING PROCESSES

between the factors in columns one and two, i.e. the BC_{Loc} interaction location effect in the spring free height experiment.

This much is already known from the linear graph. But note also the number 3 above the diagonal line in Table 5.8a also occurs at the intersection of numbers (4) and (7) and again at the intersection of numbers (5) and (6) below the diagonal line. Thus column three of the L_8 design also measures the interaction between the factors in columns four and seven, i.e. the AD_{Loc} effect in the spring free height experiment, and the interaction between factors in columns five and six, i.e. the ACD_{Loc} effect in the spring free height experiment. This is thus one aliased effect, $BC = AD = ACD$. Clearly this is a resolution IV design because first order interactions are aliased with each other.

This interaction table does not tell us what the aliasing structure is for the main location effects (i.e. the A_{Loc} effect, B_{Loc} effect etc.) nor what the aliasing structure is between the design factors and the noise factors. The only way to find this out is to incorporate the noise factor fully into a fractional factorial design in the way shown in Section 4.5. From this format the design generators can be discovered and from the generators the alias structure can be obtained. Given that this is so, the question remains as to why bother at all with these Taguchi designs and their associated linear graphs and interaction tables? This is a major shortcoming of the Taguchi approach to experimental design.

5.4.3 YATES TECHNIQUE FOR FRACTIONAL FACTORIALS

The Yates procedure can be used to estimate the location effects from a fractional design and this procedure is very similar to that for the full design. Again it involves the creation of a simple table where the first column of this table contains all the tests for the fractional design in standard order form.

Then a number of additional columns are created. The first group of columns will contain the coded test values for each of the k factors of the design. The next group of columns will contain the remaining aliased effects that can be calculated from the fractional design as determined by the alias algebra discussed in Section 5.4.1. They are headed as such, so if AB_{Loc}^{Ali} is an aliased effect than can be estimated from the fractional design a column headed AB is created and the elements in this column will be the product of the elements in columns A and B. Then comes a column containing the mean response associated with each test condition and finally the factor contrast columns are calculated by multiplying this response column by all the other columns. The numbers in these factor contrast columns are then summed and averaged to give all the main and higher order aliased interaction effects.

For example, take the principal 2^{4-1} design. Table 5.8b applies the above Yates procedure to this design. Table 4.4. gives a design generator of $D = ABC$ for the principal fraction of this design so that the eight tests making up this design are, in standard order form, (1) *ad, bd, ab, cd, ac, bc* and *abcd* (To identify these write a base design for factors A to C, introduce D equal to ABC and read the above tests off). Thus the first column of

Table 5.8b The Yates procedure for the 2^{4-1} design.

	Coded Test Conditions							Mean Response
	A	B	C	D	AB	AC	AD	
(1)	−1	−1	−1	−1	1	1	1	$\overline{Y}_{(1)}$
ad	1	−1	−1	1	−1	−1	1	$\overline{Y}_{ad}$
bd	−1	1	−1	1	−1	1	−1	$\overline{Y}_{bd}$
ab	1	1	−1	−1	1	−1	−1	$\overline{Y}_{ab}$
cd	−1	−1	1	1	1	−1	−1	$\overline{Y}_{cd}$
ac	1	−1	1	−1	−1	1	−1	$\overline{Y}_{ac}$
bc	−1	1	1	−1	−1	−1	1	$\overline{Y}_{bc}$
abcd	1	1	1	1	1	1	1	$\overline{Y}_{abcd}$

	Factor Contrasts						
	A	B	C	D	AB	AC	AD
(1)	$-\overline{Y}_{(1)}$	$-\overline{Y}_{(1)}$	$-\overline{Y}_{(1)}$	$-\overline{Y}_{(1)}$	$\overline{Y}_{(1)}$	$\overline{Y}_{(1)}$	$\overline{Y}_{(1)}$
ad	$\overline{Y}_{ad}$	$-\overline{Y}_{ad}$	$-\overline{Y}_{ad}$	$\overline{Y}_{ad}$	$-\overline{Y}_{ad}$	$-\overline{Y}_{ad}$	$\overline{Y}_{ad}$
bd	$-\overline{Y}_{bd}$	$\overline{Y}_{bd}$	$-\overline{Y}_{bd}$	$\overline{Y}_{bd}$	$-\overline{Y}_{bd}$	$\overline{Y}_{bd}$	$-\overline{Y}_{bd}$
ab	$\overline{Y}_{ab}$	$\overline{Y}_{ab}$	$-\overline{Y}_{ab}$	$-\overline{Y}_{ab}$	$\overline{Y}_{ab}$	$-\overline{Y}_{ab}$	$-\overline{Y}_{ab}$
cd	$-\overline{Y}_{cd}$	$-\overline{Y}_{cd}$	$\overline{Y}_{cd}$	$\overline{Y}_{cd}$	$\overline{Y}_{cd}$	$-\overline{Y}_{cd}$	$-\overline{Y}_{cd}$
ac	$\overline{Y}_{ac}$	$-\overline{Y}_{ac}$	$\overline{Y}_{ac}$	$-\overline{Y}_{ac}$	$-\overline{Y}_{ac}$	$\overline{Y}_{ac}$	$-\overline{Y}_{ac}$
bc	$-\overline{Y}_{bc}$	$\overline{Y}_{bc}$	$\overline{Y}_{bc}$	$-\overline{Y}_{bc}$	$-\overline{Y}_{bc}$	$-\overline{Y}_{bc}$	$\overline{Y}_{bc}$
abcd	$\overline{Y}_{abcd}$	$\overline{Y}_{abcd}$	$\overline{Y}_{abcd}$	$\overline{Y}_{abcd}$	$\overline{Y}_{abcd}$	$\overline{Y}_{abcd}$	$\overline{Y}_{abcd}$
Contrasts	Sum	Sum	Sum	Sum	Sum	Sum	Sum
Aliased Effects	Sum/$(2^{k-p}/2)$	Sum/$(2^{k-p}/2)$	Sum/$(2^{k-p}/2)$	Sum/$(2^{k-p}/2)$	Sum/$(2^{k-p}/2)$	Sum/$(2^{k-p}/2)$	Sum/$(2^{k-p}/2)$

Table 5.8b contained these test descriptors and because this design is used to study four factors the first four columns of Table 5.8b are headed A to D. These columns are filled with the coded test conditions for each factor. Thus the first row of Table 5.8b describes test (1) (all factors held low) and so a − 1 is placed in all four columns. The last first row of Table 5.8b described test *abcd* (all factors held high) and so a + 1 is placed in columns A to D.

With eight tests there are seven aliased effects that can be estimated from this design. The first four are the main location effects for factors A to D and the remaining three are derived using alias algebra. From the design generator, the confounded interaction is $ABCD$. So the next group (after the D effect) of aliased effects are, in alphabetical order,

$AB = AB \cdot ABCD = CD$

$AC = AC \cdot ABCD = BD$

$AD = AD \cdot ABCD = BC$

There are no more unique aliased effects. (The next aliased effect in alphabetical order is BC, but that has been shown above to be aliased with AD). Thus the next three columns of Table 5.8b are given these headings and the numbers going into them are obtained by multiplying the elements in the columns that form their headings. Thus the column headed AB in Table 5.8b is filled with number obtained by multiplying together the numbers in columns A and B – row by row. Thus for the first row, the number going into column AB is $-1 \times -1 = 1$.

The next column in Table 5.8b contains the mean response measured at each of the eight test conditions and the remaining factor contrast columns are obtained by multiplying all the previous columns by this mean response column. Finally, these factor contrast columns are summed to give the factor contrasts and these contrasts divided through by $2^{k-p}/2 = 2^{4-1}/2 = 4$ to obtain the aliased effects.

5.5 LOCATION EFFECTS IN THE AUSFORMING PROCESS

5.5.1 THE FIRST TWO FACTORS ONLY

A study of the first two factors of the ausforming process can now be made using a 2^2 factorial design with the remaining five factors all held fixed at their low levels. The test conditions and the results obtained from this experiment were given in Table 3.2. The main location effect for the austenitising temperature (factor A) can now be worked out using the data in Table 3.2 and equation (5.1).

$$A_{Loc} = \bar{Y}_{A+} - \bar{Y}_{A-} = \left[\frac{Y_a + Y_{ab}}{2}\right] - \left[\frac{Y_{(1)} + Y_b}{2}\right] = \left[\frac{2100 + 1668}{2}\right] - \left[\frac{2331 + 2146}{2}\right] = -354.5\,\text{MPa}$$

This suggests that raising the austenitising temperature from 930 to 1040°C will decrease average tensile strength by 354.5 MPa. This is pictured geometrically in Figure 5.1 as the difference between the average response on the right hand side of the square and the average response on the left hand side of the square.

The main location effect for factor B can be calculated as

$$B_{Loc} = \bar{Y}_{B+} - \bar{Y}_{B-} = \left[\frac{Y_b + Y_{ab}}{2}\right] - \left[\frac{Y_{(1)} + Y_a}{2}\right] = \left[\frac{2146 + 1668}{2}\right] - \left[\frac{2331 + 2100}{2}\right] = -308.5\,\text{MPa}$$

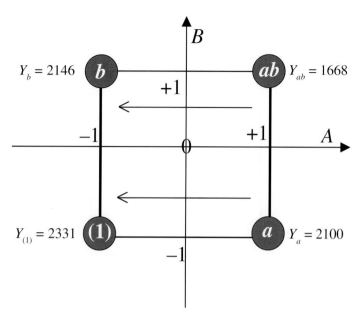

Fig. 5.1 The effect of austenitising temperature (the A_{Loc} effect).

This suggests that raising the deformation temperature from 400 to 550°C will decrease average tensile strength by 308.5 MPa. Again this can be pictured geometrically in Figure 5.2 as the difference between the average response on the top of the square and the average response on the bottom of the square.

One of the main advantages of a 2^2 factorial design over the traditional approach to experimentation should now be apparent. In the traditional experiment shown in Table 3.1 the effect of austenitising temperature was found by comparing just two strength measurements, $2100 - 2331 = -231$ MPa. But as the previous two calculations have shown the effect of the austenitising and deformation temperatures in a 2^2 design can be found by comparing two strength values which are themselves an average of two strength measurements. The estimated effect of -354.5 MPa is therefore likely to be more accurate than the -231 MPa from the traditional experiment as any experimental error is averaged out to a greater extent.

The size of the interaction between the austenitising and deformation temperature can now be estimated as

$$AB_{Loc} = \frac{1}{2}\{[\bar{Y}_{A+(B+)} - \bar{Y}_{A-(B+)}] - [\bar{Y}_{A+(B-)} - \bar{Y}_{A-(B-)}]\} = \frac{[Y_{ab} - Y_b] - [Y_a - Y_{(1)}]}{2}$$

$$= \frac{-478 - (-231)}{2} = \frac{-247}{2} = -123.5 \text{ MPa}$$

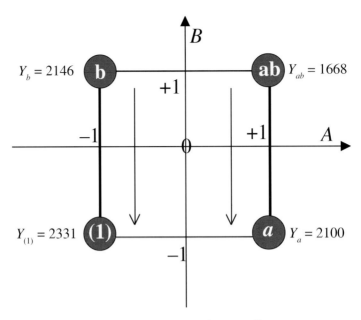

Fig. 5.2 The effect of deformation temperature (the B_{Loc} effect).

Note also that in the traditional approach to experimentation no estimate of this interaction effect can be obtained. This is a further benefit obtained from taking every possible combination of the test levels.

This interaction location effect for the ausforming process can be visualised through the use of a two-way diagram. In Figure 5.3 the response (strength) is plotted on the vertical axis and the level of the austenitising temperature, (factor A), on the horizontal axis. One line then joins the two values for strength when factor B is high ($B+$) and the other line joins the two strength readings when factor B is low ($B-$). It is clearly seen from the resulting graph that when factor A is increased from its low to high level the change in strength is only -231 MPa when factor B is at its low level. However, the change in strength for the same change in the level of factor A is -478 MPa when factor B is at its high level. The two lines are not parallel and this is an indication that a strong interaction effect is at play.

This two-way diagram can also be expressed in the form of a control matrix for each factor. These are constructed on the basis that all the other factors [C to G] are held fixed at their low levels, (Table 5.9).

Thus factor A will have the biggest impact on tensile strength if it is changed when factor B is set high. Here when factor A changed from its low level of 930°C to its high level of 1040°C the change in mean tensile strength will be -478 MPa. Notice also that the two calculations shown in Table 5.9 give the gradients of the two lines in Figure 5.3.

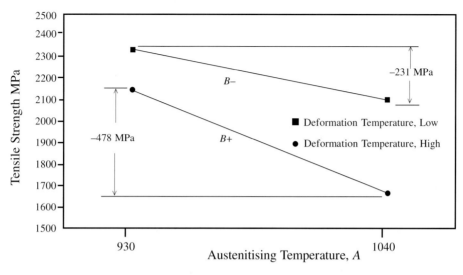

Fig. 5.3 Effect of austenitising temperature for a varying deformation temperature.

Table 5.9 Control matrix for factor A in a two factor manufacturing process.

When:	Change in Average Response as Factor A is Changed from its Low to High Level
Factor B High	$A_{Loc} + AB_{Loc} = -354.5 + (-123.5) = -478$ MPa
Factor B Low	$A_{Loc} - AB_{Loc} = -354.5 - (-123.5) = -231$ MPa

Factor B will have the biggest impact on tensile strength if it is changed when factor A is set high. Here when factor B changed from its low level of 400°C to its high level of 550°C the change in mean tensile strength will be –432 MPa – see Table 5.10.

The first order response surface model for the mean under the condition that all other factors [C to G] are held low is given by

$$Y_j = \beta_0 + \beta_A A_j + \beta_B B_j + \beta_{AB} A_j B_j = 2061.25 - \frac{354.5}{2} A_j - \frac{308.5}{2} B_j - \frac{123.5}{2} A_j B_j$$

β_0 is the mean response over all test conditions. From Table 3.2 this equals (2331 + 2100 + 2146 + 1668)/4 = 2061.25.

As an illustration on how to control the ausforming process, suppose that currently the process is operating with factors C to G at their low levels, factor A at its low level

Table 5.10 Control matrix for factor B in a two factor manufacturing process.

When:	Change in Average Response as Factor B is Changed from its Low to High Level
Factor A High	$B_{Loc} + AB_{Loc} = -308.5 + (-123.5) = -432$ MPa
Factor A Low	$B_{Loc} - AB_{Loc} = -308.5 - (-123.5) = -185$ MPa

and B at its high level. At these conditions suppose further that the process is producing a steel with a strength that on average is not equal to a target requirement of 2100 MPa. To meet this target, factor A needs to be set at the level (see equation 5.8 above).

$$A_j = \frac{Y^* - \beta_0 - \beta_B}{\beta_A + \beta_{AB}} = \frac{2100 - 2061.25 - (-308.5/2)}{(-354.5/2) + (-123.5/2)} = \frac{193}{-239} = -0.808$$

From equation (3.2) this corresponds to a temperature of 940.6°C. Factor A therefore has to be increased slightly from its low level of 930°C to meet the target strength of 2100 MPa.

5.5.2 THE FIRST THREE FACTORS ONLY

A study of the first three factors of the ausforming process can be made using a 2^3 factorial design with the remaining four factors all held fixed at their low levels. The test conditions and the results obtained from this experiment were given in Table 3.3. The Yates procedure is best suited to the derivation of all the location effects in this and larger designs. The results from carrying out this procedure are shown in Table 5.11.

This table was constructed by going through the steps discussed in Section 5.3 above.

i. Write out the standard order form for the eight tests in the first column of the table.
ii. Find headings for the next seven columns of the table. The ordering of the headings given to these columns of Table 5.11 are derived by realising that the ordering of the tests rows and column headings are the same. The first column is given the heading A because a occurs first in test column – ignoring test (1). The second column is headed B because b occurs next in the test column. Finally, the last column is headed ABC because abc occurs in the last row of the test column.
iii. Fill in the headed test condition columns with the correct minus and plus ones. The numbers in the A, B and C columns correspond to the levels set for each factor and for each of the eight tests. So consider the following examples. The first row of Table 5.11 corresponding to test (1) where all factors are set low. So for this row -1 is entered into columns A, B and C. For test bc factors B and C are high, so $+1$ is placed in columns B and C and -1 in column A for this row. The last row of Table 5.11

Table 5.11 The Yates procedure for the 2^3 design of the ausforming process.

Test	Coded Test Conditions							Mean Response	Factor Contrasts						
	A	B	AB	C	AC	BC	ABC		A	B	AB	C	AC	BC	ABC
(1)	−1	−1	1	−1	1	1	−1	2331	−2331	−2331	2331	−2331	2331	2331	−2331
a	1	−1	−1	−1	−1	1	1	2100	2100	−2100	−2100	−2100	−2100	2100	2100
b	−1	1	−1	−1	1	−1	1	2146	−2146	2146	−2146	−2146	2146	−2146	2146
ab	1	1	1	−1	−1	−1	−1	1668	1668	1668	1668	−1668	−1668	−1668	−1668
c	−1	−1	1	1	−1	−1	1	2717	−2717	−2717	2717	2717	−2717	−2717	2717
ac	1	−1	−1	1	1	−1	−1	2594	2594	−2594	−2594	2594	2594	−2594	−2594
bc	−1	1	−1	1	−1	1	−1	2717	−2717	2717	−2717	2717	−2717	2717	−2717
abc	1	1	1	1	1	1	1	2486	2486	2486	2486	2486	2486	2486	2486
							Contrasts		−1063	−725	−355	2269	355	509	139

Table 5.12 Control matrix for factor A in a three factor manufacturing process.

Deformation Temperature, B	Amount of Deformation, C	
	C– (Low)	C+ (High)
B– (Low)	$A_{Loc} - AB_{Loc} - AC_{Loc}$ –265.75 – (–88.75) –88.75 = –265.75	$A_{Loc} - AB_{Loc} + AC_{Loc}$ –265.75– (–88.75) + 88.75 = –88.25
B+ (High)	$A_{Loc} + AB_{Loc} - AC_{Loc}$ –265.75 + (–88.75) –88.75 = –443.25	$A_{Loc} + AB_{Loc} + AC_{Loc}$ –265.75 + (–88.75) + 88.75 = –265.75

corresponds to test abc where all factors are set high, so for this row +1 is entered into columns A, B and C.

The numbers in the AB column are found as the product of the numbers in the A and B columns, the numbers in the BC column are found as the product of the numbers in the B and C columns and the numbers in the ABC are found as a product of the numbers in the A, B and C columns.

iv. Fill in the response column of the table. The response column contains the **mean** responses obtained at each test condition. As there is only one result at each test condition this turns out to be the actual response.

v. Derive the factor contrast columns of the table. Multiplying each coded test condition column by the response column gives the last seven columns in Table 5.11. The sums of each column are shown at the bottom and are in fact equal to the factor contrasts, (including the first and second order interactions). Using equation (5.10) these factor contrasts can be converted into location effects as follows

$$A_{Loc} = \frac{-1063}{2^{3-1}} = -265.75, B_{Loc} = \frac{-725}{2^{3-1}} = -181.25, C_{Loc} = \frac{2269}{2^{3-1}} = 567.25$$

$$AB_{Loc} = \frac{-355}{2^{3-1}} = -88.75, AC_{Loc} = \frac{+355}{2^{3-1}} = +88.75, BC_{Loc} = \frac{509}{2^{3-1}} = 127.25$$

$$ABC_{Loc} = \frac{139}{2^{3-1}} = 34.75$$

Still assuming there is no important second order interaction, (that is if the experiment were repeated again and again the average value for ABC_{Loc} would equal zero) a control matrix for factor A can be derived in the way described in Section 5.2. This is shown in Table 5.12.

Table 5.13 Control matrix for factor B in a three factor manufacturing process.

Austenitising Temperature, A	Amount of Deformation, C	
	C – (Low)	C + (High)
A – (Low)	$B_{Loc} - AB_{Loc} - BC_{Loc}$ –181.25 – (–88.75) –127.25 = –219.75	$B_{Loc} - AB_{Loc} + BC_{Loc}$ –181.25 – (–88.75) + 127.25 = 34.75
A + (High)	$B_{Loc} + AB_{Loc} - BC_{Loc}$ –181.25 – 88.75 –127.25 = –397.25	$B_{Loc} + AB_{Loc} + BC_{Loc}$ –181.25 – 88.75 + 127.25 = –142.75

The calculations shown in Table 5.12 are straightforward. When factor B is high and factor C is low a decrease in austenitising temperature will add 443.25 MPa to the strength of the finished steel. It would appear that under such conditions austenitising temperature has its greatest impact on strength. Note also that when factors B and C are both set low or both set high the interaction effects offset each other leaving just the main location effect for factor A.

A similar analysis can be carried out for the deformation temperature. Assuming the second order interaction is not important, Table 5.13 shows the four separate effects of changes in deformation temperature on the strength of the finished steel.

Notice that deformation temperature has its greatest impact on the steels strength when austenitising temperature is high and the amount of deformation is low. Note also that increases in deformation temperature actually increase strength when factor A is low and C high, i.e. the direction of causation changes under these test conditions.

A similar analysis can be carried out for the amount of deformation. Assuming the second order interaction effect is not important, Table 5.14 shows the four separate effects of changes in the amount of deformation on the average strength of the finished steel. Notice that the amount of deformation has its greatest impact on the steels strength when austenitising and deformation temperature are set high.

Each of these interaction location effects can be given a graphical representation. The interaction between factors A and B is shown in Figure 5.4 for example. Notice how the end points of the two lines in Figure 5.4 are derived by averaging over the two test results associated with each of the four test conditions. For example, there are two data points associated with both factors A and B being at their low levels, $Y_{(1)}$ and Y_c. The first point defining the upper line in Figure 5.4 is therefore (2331 + 2717)/2 = 2524 MPa. Similarly, there are two results that were obtained when factor A was at its high level and factor B at its low level, Y_a and Y_{ac}. Hence the second point defining the upper line in Figure 5.4 is (2100 + 2594)/2 = 2347 MPa. Joining up these two points gives the B low [B–] line in Figure 5.4. There are again two data points associated with both factors A and B

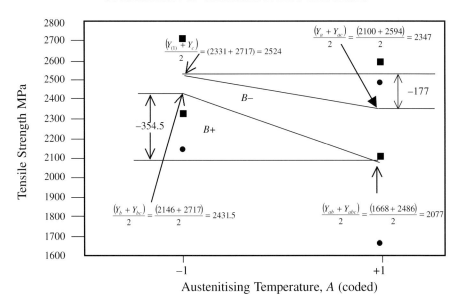

Fig. 5.4 The AB_{Loc} interaction effect.

Table 5.14 Control matrix for factor C in a three factor manufacturing process.

Austenitising Temperature, A	Deformation Temperature, B	
	$B-$ (Low)	$B+$ (High)
$A-$ (Low)	$C_{Loc} - AC_{Loc} - BC_{Loc}$	$C_{Loc} - AC_{Loc} + BC_{Loc}$
	$567.27 - 88.75 - 127.25 = 351.27$	$567.27 - 88.75 + 127.25 = 605.77$
$A+$ (High)	$C_{Loc} + AC_{Loc} - BC_{Loc}$	$C_{Loc} + AC_{Loc} + BC_{Loc}$
	$567.27 + 88.75 - 127.25 = 528.77$	$567.27 + 88.75 + 127.25 = 783.27$

being at their high levels, Y_{ab} and Y_{abc}. Thus the first point defining the lower line in Figure 5.4 is $(1668 + 2486)/2 = 2077$ MPa. Finally, there are two data points obtained from a combination of factor A low and factor B high, Y_b and Y_{bc}. Thus the second point defining the lower line in Figure 5.4 is $(2146 + 2717)/2 = 2431.5$ MPa. Joining up these two points gives the B high [$B+$] line in Figure 5.4.

All these location effects can be placed into the following first order response surface model that is applicable when factors D to G are fixed at their low levels

$$Y_j = \beta_0 + \beta_A A_j + \beta_B B_j + \beta_C C_j + \beta_{AB} A_j B_j + \beta_{AC} A_j C_j + \beta_{BC} B_j C_j + \varepsilon_j$$

The prediction error, ε_j, captures all the higher order interaction effects left out of the model which are assumed to equal zero on the average. So a prediction for average tensile strength can be found by using the following equation,

$$Y_j = 2344.875 - 132.875 A_j - 90.625 B_j + 283.625 C_j - 44.375 A_j B_j + 44.375 A_j C_j + 63.625 B_j C_j$$

As usual 2344.875 is the mean response over all eight tests. The remaining numbers are half the respective location effects. Coded levels for factors A to C can be placed into this equation and the predicted mean response noted. In this way, the levels required of each factor to ensure the mean strength meets the target strength can be found.

For example, to work out the average strength when austenitising temperature is low, deformation temperature high and the amount of deformation low, simply substitute -1 for A_j, $+1$ for B_j and -1 for C_j into the above equation. Note that this corresponds to test condition b (which is the third test when arranged in standard order form). Thus $j = 3$ and $A_j B_j = -1 \times +1 = -1$, $A_j C_j = -1 \times -1 = +1$ and $B_j C_j = +1 \times -1 = -1$. Substituting all these values into the last equation yields,

$$Y_3 = 2344.875 + 132.875 - 90.625 - 283.625 + 44.375 + 44.375 - 63.625 = 2128.625$$

Note that this prediction differs from the actual strength at test condition b of 2146 MPa (see Table 5.11). This difference of 2146–2128.625 = 17.375 exactly equals half the ABC_{Loc} effect left out from the first order response surface model (β_{ABC} = 17.375).

The β values of the response surface model shown above (the ABC_{Loc} effect) can be obtained directly using the least squares procedure. The image in Figure 5.5 shows how this can be done in Excel using the LINEST function described in Section 5.3.2.

Notice first the structure of the formula shown in the formula box towards the top of the image. The first range in brackets is the column containing the mean responses of the 2^3 design in standard order form and the second range are the headed coded test condition columns from the table containing the Yates procedure for the 2^3 design – with a column of one's in front (Table 5.11). The formula is typed in the formula box when the shaded cells in the image above have been highlighted. Then hitting the shift ctrl return buttons simultaneously yields the β estimates shown in the image. Note they are in reverse order to the headed columns. Also note that the β estimates are not in the same order as the first order response surface model above.

5.5.3 THE FIRST FIVE FACTORS ONLY

A study of the first five factors of the ausforming process can be made using a 2^5 factorial design with the remaining two factors all held fixed at their low levels. The test conditions and the results obtained from this experiment were given in Table 3.5. The Yates and least squares procedures are used next to derive all of the location effects in this design. The results from carrying out these procedures are shown in Tables 5.15 and 5.16.

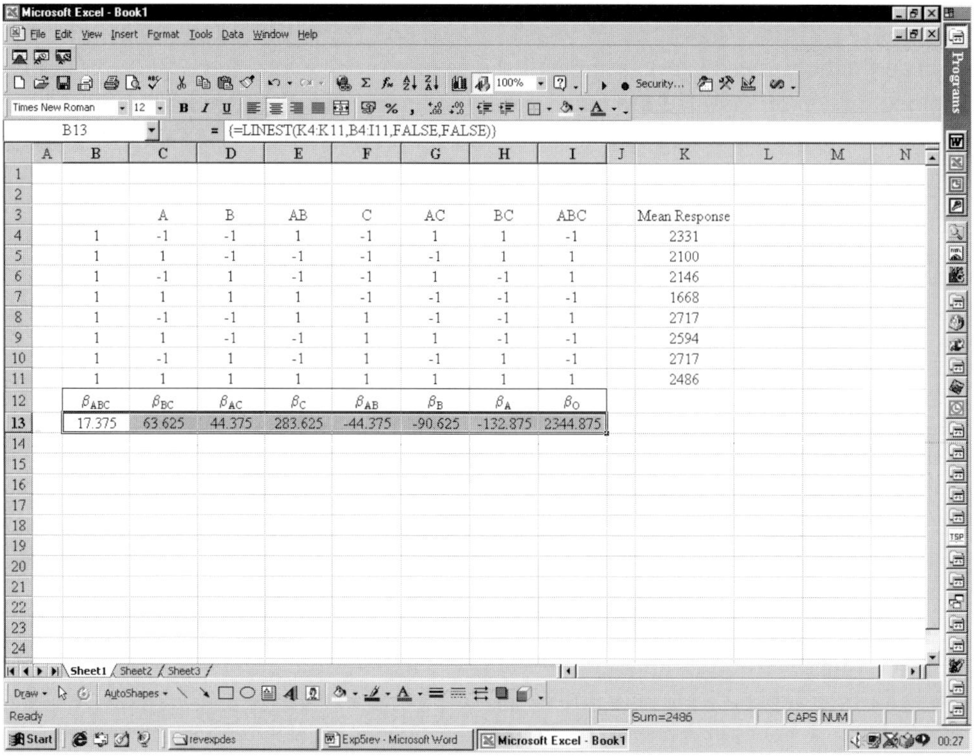

B13 = {=LINEST(K4:K11,B4:I11,FALSE,FALSE)}

	B	C	D	E	F	G	H	I	K
3		A	B	AB	C	AC	BC	ABC	Mean Response
4	1	-1	-1	1	-1	1	1	-1	2331
5	1	1	-1	-1	-1	-1	1	1	2100
6	1	-1	1	-1	-1	1	-1	1	2146
7	1	1	1	1	-1	-1	-1	-1	1668
8	1	-1	-1	1	1	-1	-1	1	2717
9	1	1	-1	-1	1	1	-1	-1	2594
10	1	-1	1	-1	1	-1	1	-1	2717
11	1	1	1	1	1	1	1	1	2486
12	β_{ABC}	β_{BC}	β_{AC}	β_C	β_{AB}	β_B	β_A	β_O	
13	17.375	63.625	44.375	283.625	-44.375	-90.625	-132.875	2344.875	

Fig. 5.5 Excel screen shot showing LINEST function for a 2^3 design.

As usual the first column of Table 5.15 contains the tests to be carried out in standard order form. The ordering down to abc should by now be fully familiar. After test *abc*, (which is the last test for the 2^3 design), comes test *d* which is identified from the multiplication of *d* by (1). Then comes test ad found by multiplying *d* by the second test *a*. This procedure can be repeated to obtain the final test, *abcde*. The next 31 columns in Table 5.15 are in the same order as the tests shown in the first column and the numbers in columns *A* to *E* are found by inserting –1 into a column when its letter is absent from the test description shown in the first column of the table and +1 when it is present. For example, take test abe. For this test, (row), column *A* contains +1 (because a is present in the description abe), column *B* +1, column *C* –1 (because c is missing from abe), column *D* –1 and column *E* +1. The other columns are found as particular product combinations of these five columns. Thus the numbers in column *ABC* are found by multiplying the numbers in columns *A*, *B* and *C*. So for test abe the number in column *ABC* is $1 \times 1 \times -1 = -1$.

Table 5.15 A 2^5 Design and the Yates procedure for the ausforming process.

Tests	\multicolumn{15}{c}{Coded Test Condition}

Tests	A	B	AB	C	AC	BC	ABC	D	AD	BD	ABD	CD	ACD	BCD	ABCD	E
(–1)	–1	–1	1	–1	1	1	–1	–1	1	1	–1	1	–1	–1	1	–1
a	1	–1	–1	–1	–1	1	1	–1	–1	1	1	1	1	–1	–1	–1
b	–1	1	–1	–1	1	–1	1	–1	1	–1	1	1	–1	1	–1	–1
ab	1	1	1	–1	–1	–1	–1	–1	–1	–1	–1	1	1	1	1	–1
c	–1	–1	1	1	–1	–1	1	–1	1	1	–1	–1	1	1	–1	–1
ac	1	–1	–1	1	1	–1	–1	–1	–1	1	1	–1	–1	1	1	–1
bc	–1	1	–1	1	–1	1	–1	–1	1	–1	1	–1	1	–1	1	–1
abc	1	1	1	1	1	1	1	–1	–1	–1	–1	–1	–1	–1	–1	–1
d	–1	–1	1	–1	1	1	–1	1	–1	–1	1	–1	1	1	–1	–1
ad	1	–1	–1	–1	–1	1	1	1	1	–1	–1	–1	–1	1	1	–1
bd	–1	1	–1	–1	1	–1	1	1	–1	1	–1	–1	1	–1	1	–1
abd	1	1	1	–1	–1	–1	–1	1	1	1	1	–1	–1	–1	–1	–1
cd	–1	–1	1	1	–1	–1	1	1	–1	–1	1	1	–1	–1	1	–1
acd	1	–1	–1	1	1	–1	–1	1	1	–1	–1	1	1	–1	–1	–1
bcd	–1	1	–1	1	–1	1	–1	1	–1	1	–1	1	–1	1	–1	–1
abcd	1	1	1	1	1	1	1	1	1	1	1	1	1	1	1	–1
e	–1	–1	1	–1	1	1	–1	–1	1	1	–1	1	–1	–1	1	1
ae	1	–1	–1	–1	–1	1	1	–1	–1	1	1	1	1	–1	–1	1
be	–1	1	–1	–1	1	–1	1	–1	1	–1	1	1	–1	1	–1	1
abe	1	1	1	–1	–1	–1	–1	–1	–1	–1	–1	1	1	1	1	1
ce	–1	–1	1	1	–1	–1	1	–1	1	1	–1	–1	1	1	–1	1
ace	1	–1	–1	1	1	–1	–1	–1	–1	1	1	–1	–1	1	1	1
bce	–1	1	–1	1	–1	1	–1	–1	1	–1	1	–1	1	–1	1	1
abce	1	1	1	1	1	1	1	–1	–1	–1	–1	–1	–1	–1	–1	1
de	–1	–1	1	–1	1	1	–1	1	–1	–1	1	–1	1	1	–1	1
ade	1	–1	–1	–1	–1	1	1	1	1	–1	–1	–1	–1	1	1	1
bde	–1	1	–1	–1	1	–1	1	1	–1	1	–1	–1	1	–1	1	1
abde	1	1	1	–1	–1	–1	–1	1	1	1	1	–1	–1	–1	–1	1
cde	–1	–1	1	1	–1	–1	1	1	–1	–1	1	1	–1	–1	1	1
acde	1	–1	–1	1	1	–1	–1	1	1	–1	–1	1	1	–1	–1	1
bcde	–1	1	–1	1	–1	1	–1	1	–1	1	–1	1	–1	1	–1	1
abcde	1	1	1	1	1	1	1	1	1	1	1	1	1	1	1	1

Table 5.15 Continued.

Tests	Coded Test Condition															Mean Response Strength, (MPa)
	AE	BE	ABE	CE	ACE	BCE	ABCE	DE	ADE	BDE	ABDE	CDE	ACDE	BCDE	ABCDE	
(−1)	1	1	−1	1	−1	−1	1	1	−1	−1	1	−1	1	1	−1	2331
a	−1	1	1	1	1	−1	−1	1	1	−1	−1	−1	−1	1	1	2100
b	1	−1	1	1	−1	1	−1	1	−1	1	−1	−1	1	−1	1	2146
ab	−1	−1	−1	1	1	1	1	1	1	1	1	−1	−1	−1	−1	1668
c	1	1	−1	−1	1	1	−1	1	−1	−1	1	1	−1	−1	1	2717
ac	−1	1	1	−1	−1	1	1	1	1	−1	−1	1	1	−1	−1	2594
bc	1	−1	1	−1	1	−1	1	1	−1	1	−1	1	−1	1	−1	2717
abc	−1	−1	−1	−1	−1	−1	−1	1	1	1	1	1	1	1	1	2486
d	1	1	−1	1	−1	−1	1	−1	1	1	−1	1	−1	−1	1	2038
ad	−1	1	1	1	1	−1	−1	−1	−1	1	1	1	1	−1	−1	1498
bd	1	−1	1	1	−1	1	−1	−1	1	−1	1	1	−1	1	−1	1930
abd	−1	−1	−1	1	1	1	1	−1	−1	−1	−1	1	1	1	1	1529
cd	1	1	−1	−1	1	1	−1	−1	1	1	−1	−1	1	1	−1	2779
acd	−1	1	1	−1	−1	1	1	−1	−1	1	1	−1	−1	1	1	1899
bcd	1	−1	1	−1	1	−1	1	−1	1	−1	1	−1	1	−1	1	2594
abcd	−1	−1	−1	−1	−1	−1	−1	−1	−1	−1	−1	−1	−1	−1	−1	1853
e	−1	−1	1	−1	1	1	−1	−1	1	1	−1	1	−1	−1	1	2517
ae	1	−1	−1	−1	−1	1	1	−1	−1	1	1	1	1	−1	−1	2084
be	−1	1	−1	−1	1	−1	1	−1	1	−1	1	1	−1	1	−1	2285
abe	1	1	1	−1	−1	−1	−1	−1	−1	−1	−1	1	1	1	1	1992
ce	−1	−1	1	1	−1	−1	1	−1	1	1	−1	−1	1	1	−1	2609
ace	1	−1	−1	1	1	−1	−1	−1	−1	1	1	−1	−1	1	1	1837
bce	−1	1	−1	1	−1	1	−1	−1	1	−1	1	−1	1	−1	1	2285
abce	1	1	1	1	1	1	1	−1	−1	−1	−1	−1	−1	−1	−1	1652
de	−1	−1	1	−1	1	1	−1	1	−1	−1	1	−1	1	1	−1	2887
ade	1	−1	−1	−1	−1	1	1	1	1	−1	−1	−1	−1	1	1	2455
bde	−1	1	−1	−1	1	−1	1	1	−1	1	−1	−1	1	−1	1	2656
abde	1	1	1	−1	−1	−1	−1	1	1	1	1	−1	−1	−1	−1	2362
cde	−1	−1	1	1	−1	−1	1	1	−1	−1	1	1	−1	−1	1	2980
acde	1	−1	−1	1	1	−1	−1	1	1	−1	−1	1	1	−1	−1	2208
bcde	−1	1	−1	1	−1	1	−1	1	−1	1	−1	1	−1	1	−1	2656
abcde	1	1	1	1	1	1	1	1	1	1	1	1	1	1	1	2023

Table 5.15 Continued.

Location Contrasts and Effects

	A	B	AB	C	AC	BC	ABC	D	AD	BD	ABD	CD	ACD	BCD	ABCD
(−1)	−2331	−2331	2331	−2331	2331	2331	−2331	−2331	2331	2331	−2331	2331	−2331	−2331	2331
a	2100	−2100	−2100	−2100	−2100	2100	2100	−2100	−2100	2100	2100	2100	2100	−2100	−2100
b	−2146	2146	−2146	−2146	2146	−2146	2146	−2146	2146	−2146	2146	2146	−2146	2146	−2146
ab	1668	1668	1668	−1668	−1668	−1668	−1668	−1668	−1668	−1668	−1668	1668	1668	1668	1668
c	−2717	−2717	2717	2717	−2717	−2717	2717	−2717	2717	2717	−2717	−2717	2717	2717	−2717
ac	2594	−2594	−2594	2594	2594	−2594	−2594	−2594	−2594	2594	2594	−2594	−2594	2594	2594
bc	−2717	2717	−2717	2717	−2717	2717	−2717	−2717	2717	−2717	2717	−2717	2717	−2717	2717
abc	2486	2486	2486	2486	2486	2486	2486	−2486	−2486	−2486	−2486	−2486	−2486	−2486	−2486
d	−2038	−2038	2038	−2038	2038	2038	−2038	2038	−2038	−2038	2038	−2038	2038	2038	−2038
ad	1498	−1498	−1498	−1498	−1498	1498	1498	1498	1498	−1498	−1498	−1498	−1498	1498	1498
bd	−1930	1930	−1930	−1930	1930	−1930	1930	1930	−1930	1930	−1930	−1930	1930	−1930	1930
abd	1529	1529	1529	−1529	−1529	−1529	−1529	1529	1529	1529	1529	−1529	−1529	−1529	−1529
cd	−2779	−2779	2779	2779	−2779	−2779	2779	2779	−2779	−2779	2779	2779	−2779	−2779	2779
acd	1899	−1899	−1899	1899	1899	−1899	−1899	1899	1899	−1899	−1899	1899	1899	−1899	−1899
bcd	−2594	2594	−2594	2594	−2594	2594	−2594	2594	−2594	2594	−2594	2594	−2594	2594	−2594
abcd	1853	1853	1853	1853	1853	1853	1853	1853	1853	1853	1853	1853	1853	1853	1853
e	−2517	−2517	2517	−2517	2517	2517	−2517	−2517	2517	2517	−2517	2517	−2517	−2517	2517
ae	2084	−2084	−2084	−2084	−2084	2084	2084	−2084	−2084	2084	2084	2084	2084	−2084	−2084
be	−2285	2285	−2285	−2285	2285	−2285	2285	−2285	2285	−2285	2285	2285	−2285	2285	−2285
abe	1992	1992	1992	−1992	−1992	−1992	−1992	−1992	−1992	−1992	−1992	1992	1992	1992	1992
ce	−2609	−2609	2609	2609	−2609	−2609	2609	−2609	2609	2609	−2609	−2609	2609	2609	−2609
ace	1837	−1837	−1837	1837	1837	−1837	−1837	−1837	−1837	1837	1837	−1837	−1837	1837	1837
bce	−2285	2285	−2285	2285	−2285	2285	−2285	−2285	2285	−2285	2285	−2285	2285	−2285	2285
abce	1652	1652	1652	1652	1652	1652	1652	−1652	−1652	−1652	−1652	−1652	−1652	−1652	−1652
de	−2887	−2887	2887	−2887	2887	2887	−2887	2887	−2887	−2887	2887	−2887	2887	2887	−2887
ade	2455	−2455	−2455	−2455	−2455	2455	2455	2455	2455	−2455	−2455	−2455	−2455	2455	2455
bde	−2656	2656	−2656	−2656	2656	−2656	2656	2656	−2656	2656	−2656	−2656	2656	−2656	2656
abde	2362	2362	2362	−2362	−2362	−2362	−2362	2362	2362	2362	2362	−2362	−2362	−2362	−2362
cde	−2980	−2980	2980	2980	−2980	−2980	2980	2980	−2980	−2980	2980	2980	−2980	−2980	2980
acde	2208	−2208	−2208	2208	2208	−2208	−2208	2208	2208	−2208	−2208	2208	2208	−2208	−2208
bcde	−2656	2656	−2656	2656	−2656	2656	−2656	2656	−2656	2656	−2656	2656	−2656	2656	−2656
abcde	2023	2023	2023	2023	2023	2023	2023	2023	2023	2023	2023	2023	2023	2023	2023
Loc Contrast	−7887	−2699	479	3411	−1683	−15	139	327	−1499	417	631	−137	−1035	−663	−137
Loc Effect	−492.9375	−168.6875	29.9375	213.1875	−105.1875	−0.9375	8.6875	20.4375	−93.6875	26.0625	39.4375	−8.5625	−64.6875	−41.4375	−8.5625
β	−246.46875	−84.34375	14.96875	106.59375	−52.59375	−0.46875	4.34375	10.21875	−46.84375	13.03125	19.71875	−4.28125	−32.34375	−20.71875	−4.28125

Table 5.15 Continued

Location Contrasts and Effects

	E	AE	BE	ABE	CE	ACE	BCE	ABCE	DE	ADE	BDE	ABDE	CDE	ACDE	BCDE	ABCDE
(-1)	-2331	2331	2331	-2331	2331	-2331	-2331	2331	2331	-2331	-2331	2331	-2331	2331	2331	-2331
a	-2100	-2100	2100	2100	2100	2100	-2100	-2100	2100	2100	-2100	-2100	-2100	-2100	2100	2100
b	-2146	2146	-2146	2146	2146	-2146	2146	-2146	2146	-2146	2146	-2146	2146	-2146	-2146	2146
ab	-1668	-1668	-1668	-1668	1668	1668	1668	1668	1668	1668	1668	1668	-1668	-1668	-1668	-1668
c	-2717	2717	2717	-2717	-2717	2717	2717	-2717	2717	-2717	-2717	2717	2717	-2717	-2717	2717
ac	-2594	-2594	2594	2594	-2594	-2594	2594	2594	2594	2594	-2594	-2594	2594	2594	-2594	-2594
bc	-2717	2717	-2717	2717	-2717	2717	-2717	2717	2717	-2717	2717	-2717	2717	-2717	2717	-2717
abc	-2486	-2486	-2486	-2486	-2486	-2486	-2486	-2486	2486	2486	2486	2486	2486	2486	2486	2486
d	-2038	2038	2038	-2038	2038	-2038	-2038	2038	-2038	2038	2038	-2038	2038	-2038	-2038	2038
ad	-1498	-1498	1498	1498	1498	1498	-1498	-1498	-1498	-1498	1498	1498	1498	1498	-1498	-1498
bd	-1930	1930	-1930	1930	1930	-1930	1930	-1930	-1930	1930	-1930	1930	1930	-1930	1930	-1930
abd	-1529	-1529	-1529	-1529	1529	1529	1529	1529	-1529	-1529	-1529	-1529	1529	1529	1529	1529
cd	-2779	2779	2779	-2779	-2779	2779	2779	-2779	-2779	2779	2779	-2779	-2779	2779	2779	-2779
acd	-1899	-1899	1899	1899	-1899	-1899	1899	1899	-1899	-1899	1899	1899	-1899	-1899	1899	1899
bcd	-2594	2594	-2594	2594	-2594	2594	-2594	2594	-2594	2594	-2594	2594	-2594	2594	-2594	2594
abcd	-1853	-1853	-1853	-1853	-1853	-1853	-1853	-1853	-1853	-1853	-1853	-1853	-1853	-1853	-1853	-1853
e	2517	-2517	-2517	2517	-2517	2517	2517	-2517	-2517	2517	2517	-2517	2517	-2517	-2517	2517
ae	2084	2084	-2084	-2084	-2084	-2084	2084	2084	-2084	-2084	2084	2084	2084	2084	-2084	-2084
be	2285	-2285	2285	-2285	-2285	2285	-2285	2285	-2285	2285	-2285	2285	2285	-2285	2285	-2285
abe	1992	1992	1992	1992	-1992	-1992	-1992	-1992	-1992	-1992	-1992	-1992	1992	1992	1992	1992
ce	2609	-2609	-2609	2609	2609	-2609	-2609	2609	-2609	2609	2609	-2609	-2609	2609	2609	-2609
ace	1837	1837	-1837	-1837	1837	1837	-1837	-1837	-1837	-1837	1837	1837	-1837	-1837	1837	1837
bce	2285	-2285	2285	-2285	2285	-2285	2285	-2285	-2285	2285	-2285	2285	-2285	2285	-2285	2285
abce	1652	1652	1652	1652	1652	1652	1652	1652	-1652	-1652	-1652	-1652	-1652	-1652	-1652	-1652
de	2887	-2887	-2887	2887	-2887	2887	2887	-2887	2887	-2887	-2887	2887	-2887	2887	2887	-2887
ade	2455	2455	-2455	-2455	-2455	-2455	2455	2455	2455	2455	-2455	-2455	-2455	-2455	2455	2455
bde	2656	-2656	2656	-2656	-2656	2656	-2656	2656	2656	-2656	2656	-2656	-2656	2656	-2656	2656
abde	2362	2362	2362	2362	-2362	-2362	-2362	-2362	2362	2362	2362	2362	-2362	-2362	-2362	-2362
cde	2980	-2980	-2980	2980	2980	-2980	-2980	2980	2980	-2980	-2980	2980	2980	-2980	-2980	2980
acde	2208	2208	-2208	-2208	2208	2208	-2208	-2208	2208	2208	-2208	-2208	2208	2208	-2208	-2208
bcde	2656	-2656	2656	-2656	2656	-2656	2656	-2656	2656	-2656	2656	-2656	2656	-2656	2656	-2656
abcde	2023	2023	2023	2023	2023	2023	2023	2023	2023	2023	2023	2023	2023	2023	2023	2023
Loc Contrast	2609	-637	-633	633	-5387	-1033	-725	-139	5605	1499	-417	-635	141	1035	663	141
Loc Effect	163.0625	-39.8125	-39.5625	39.5625	-336.6875	-64.5625	-45.3125	-8.6875	350.3125	93.6875	-26.0625	-39.6875	8.8125	64.6875	41.4375	8.8125
β	81.53125	-19.9063	-19.7813	19.7813	-168.3438	-32.28125	-22.6563	-4.34375	175.1563	46.84375	-13.0313	-19.8438	4.40625	32.3438	20.7188	4.40625

Table 5.16 Least squares procedure for the 2^5 design of the ausforming process.

$\bar{Y}=$	Tests		A	B	AB	C	AC	BC	ABC	D	AD	BD	ABD	CD	ACD	BCD	ABCD	E	AE
2331	= (1)		-1	-1	1	-1	1	1	-1	-1	1	1	-1	1	-1	-1	1	-1	1
2100	= a		1	-1	-1	-1	-1	1	1	-1	-1	1	1	1	1	-1	-1	-1	-1
2146	= b		-1	1	-1	-1	1	-1	1	-1	1	-1	1	1	-1	1	-1	-1	1
1668	= ab		1	1	1	-1	-1	-1	-1	-1	-1	-1	-1	1	1	1	1	-1	-1
2717	= c		-1	-1	1	1	-1	-1	1	-1	1	1	-1	-1	1	1	-1	-1	1
2594	= ac		1	-1	-1	1	1	-1	-1	-1	-1	1	1	-1	-1	1	1	-1	-1
2717	= bc		-1	1	-1	1	-1	1	-1	-1	1	-1	1	-1	1	-1	1	-1	1
2486	= abc		1	1	1	1	1	1	1	-1	-1	-1	-1	-1	-1	-1	-1	-1	-1
2038	= d		-1	-1	1	-1	1	1	-1	1	-1	-1	1	-1	1	1	-1	-1	1
1498	= ad		1	-1	-1	-1	-1	1	1	1	1	-1	-1	-1	-1	1	1	-1	-1
1930	= bd		-1	1	-1	-1	1	-1	1	1	-1	1	-1	-1	1	-1	1	-1	1
1529	= abd		1	1	1	-1	-1	-1	-1	1	1	1	1	-1	-1	-1	-1	-1	-1
2779	= cd		-1	-1	1	1	-1	-1	1	1	-1	-1	1	1	-1	-1	1	-1	1
1899	= acd		1	-1	-1	1	1	-1	-1	1	1	-1	-1	1	1	-1	-1	-1	-1
2594	= bcd		-1	1	-1	1	-1	1	-1	1	-1	1	-1	1	-1	1	-1	-1	1
1853	= abcd		1	1	1	1	1	1	1	1	1	1	1	1	1	1	1	-1	-1
2517	= e		-1	-1	1	-1	1	1	-1	-1	1	1	-1	1	-1	-1	1	1	-1
2084	= ae		1	-1	-1	-1	-1	1	1	-1	-1	1	1	1	1	-1	-1	1	1
2285	= be		-1	1	-1	-1	1	-1	1	-1	1	-1	1	1	-1	1	-1	1	-1
1992	= abe		1	1	1	-1	-1	-1	-1	-1	-1	-1	-1	1	1	1	1	1	1
2609	= ce		-1	-1	1	1	-1	-1	1	-1	1	1	-1	-1	1	1	-1	1	-1
1837	= ace		1	-1	-1	1	1	-1	-1	-1	-1	1	1	-1	-1	1	1	1	1
2285	= bce		-1	1	-1	1	-1	1	-1	-1	1	-1	1	-1	1	-1	1	1	-1
1652	= abce		1	1	1	1	1	1	1	-1	-1	-1	-1	-1	-1	-1	-1	1	1
2887	= de		-1	-1	1	-1	1	1	-1	1	-1	-1	1	-1	1	1	-1	1	-1
2455	= ade		1	-1	-1	-1	-1	1	1	1	1	-1	-1	-1	-1	1	1	1	1
2656	= bde		-1	1	-1	-1	1	-1	1	1	-1	1	-1	-1	1	-1	1	1	-1
2362	= abde		1	1	1	-1	-1	-1	-1	1	1	1	1	-1	-1	-1	-1	1	1
2980	= cde		-1	-1	1	1	-1	-1	1	1	-1	-1	1	1	-1	-1	1	1	-1
2208	= acde		1	-1	-1	1	1	-1	-1	1	1	-1	-1	1	1	-1	-1	1	1
2656	= bcde		-1	1	-1	1	-1	1	-1	1	-1	1	-1	1	-1	1	-1	1	-1
2023	= abcde		1	1	1	1	1	1	1	1	1	1	1	1	1	1	1	1	1

$X =$

Table 5.16 Continued.

$$X'\bar{Y} = \begin{bmatrix} 72367 \\ -7887 \\ -2699 \\ 479 \\ 3411 \\ -1683 \\ -15 \\ 139 \\ 327 \\ -1499 \\ 417 \\ 631 \\ -137 \\ -1035 \\ -663 \\ -137 \\ 2609 \\ -637 \\ -633 \\ 633 \\ -5387 \\ -1033 \\ -725 \\ -139 \\ 5605 \\ 1499 \\ -417 \\ -635 \\ 141 \\ 1035 \\ 663 \\ 141 \end{bmatrix}$$

$$IX'\bar{Y} = \begin{bmatrix} 72367 \\ -7887 \\ -2699 \\ 479 \\ 3411 \\ -1683 \\ -15 \\ 139 \\ 327 \\ -1499 \\ 417 \\ 631 \\ -137 \\ -1035 \\ -663 \\ -137 \\ 2609 \\ -637 \\ -633 \\ 633 \\ -5387 \\ -1033 \\ -725 \\ -139 \\ 5605 \\ 1499 \\ -417 \\ -635 \\ 141 \\ 1035 \\ 663 \\ 141 \end{bmatrix}$$

$$B = \begin{bmatrix} 2261.47 \\ -246.47 \\ -84.344 \\ 14.9688 \\ 106.594 \\ -52.594 \\ -0.4688 \\ 4.34375 \\ 10.2188 \\ -46.844 \\ 13.0313 \\ 19.7188 \\ -4.2813 \\ -32.344 \\ -20.719 \\ -4.2813 \\ 81.5313 \\ -19.906 \\ -19.781 \\ 19.7813 \\ -168.34 \\ -32.281 \\ -22.656 \\ -4.3438 \\ 175.156 \\ 46.8438 \\ -13.031 \\ -19.844 \\ 4.40625 \\ 32.3438 \\ 20.7188 \\ 4.40625 \end{bmatrix} = \begin{bmatrix} \beta_0 \\ \beta_A \\ \beta_B \\ \beta_{AB} \\ B_C \\ B_{AC} \\ B_{BC} \\ B_{ABC} \\ B_D \\ B_{AD} \\ B_{BD} \\ B_{ABD} \\ B_{CD} \\ B_{ACD} \\ B_{BCD} \\ B_{ABCD} \\ B_E \\ B_{AE} \\ B_{BE} \\ B_{ABE} \\ B_{CE} \\ B_{ACE} \\ B_{BCE} \\ B_{ABCE} \\ B_{DE} \\ B_{ADE} \\ B_{BDE} \\ B_{ABDE} \\ B_{CDE} \\ B_{ACDE} \\ B_{BCDE} \\ B_{ABCDE} \end{bmatrix}$$

Table 5.16 Continued.

$$X' =$$

Table 5.16 Continued.

The remainder of Table 5.15 is found from multiplying the column containing the mean response by each of the columns before the response column. At the bottom of this table are the sums of each column, which are the factor contrasts. Each location effect is then found from these contrasts using equation (5.10). The β values in Table 5.15 are then just half these location effects.

These β values can also be obtained directly using the least squares technique summarised in equation (5.12). Table 5.16 shows how to construct each of the matrices of this equation for a 2^5 design, together with the calculations required to execute equation (5.12). Notice that the vector $\bar{Y}$ contains all the 32 test results in standard order form. The X matrix is identical to the headed columns containing the –1 and +1 elements shown in Table 5.15, with a column of 1s added in at the beginning. Notice that the $X'\bar{Y}$ vector gives, in the same order, the contrasts shown towards the bottom of Table 5.15. Finally, the B matrix contains, in the same order, half the location effects shown towards the bottom of Table 5.15.

Tables 5.15 and 5.16 have 31 estimated location effects so that a lot of information is available on how to control the mean strength of the steel. In the next chapter a technique is discussed that identifies which of these effects can be ignored so making the act of control much simpler. Thus a further discussion on how to use this information to put the mean quality characteristic on target will be left until Chapter 8.

However, a provisional interpretation of the results can be achieved through the construction of two-way diagrams. Figure 5.6 shows the first order interaction between factors A and C. The two lines diverge and so an interaction between these two factors seems to be present. The end points defining each of the two linear lines are now an average of eight strength recordings. The exact calculations required to draw these lines are shown in Figure 5.6. For example, when both factors A and C are low there are eight strength readings available under this test condition, $Y_{(1)}$, Y_b, Y_d, Y_{bd}, Y_e, Y_{be}, Y_{de} and Y_{bde}. The average of these strength readings is 2348.8 MPa and this defines the first point of the lower line in Figure 5.6.

Ignoring all location effects except the A_{Loc} and AC_{Loc} effects for the moment it follows that when factor C is at its high level a decrease in austenitising temperature will increase the steels strength by the $A_{Loc} + AC_{Loc} = -492.9375 - 105.1875 = 598.125$ MPa. (The minus sign in front of 598.125 MPa has been dropped as the above illustration looks at a decrease in austenitising temperature). On the other hand, when factor C is low the same decrease in austenitising temperature will raise the steels strength by $A_{Loc} - AC_{Loc} = 387.75$ MPa. (The minus sign has again been dropped). Thus austenitising temperature has a slightly bigger impact on the material strength when the amount of deformation is large.

In the absence of any important second, third and fourth order interactions, the A_{Loc} effect can be interpreted as follows. When all factors are high the A_{Loc} effect becomes $A_{Loc} + AB_{Loc} + AC_{Loc} + AD_{Loc} + AE_{Loc} = -492.93 + 29.9375 \pm 105.188 \pm 93.6875 \pm 39.8125 = -701.6885$ MPa. If on the other hand all factors are low A_{Loc} effect becomes $A_{Loc} - AB_{Loc} - AC_{Loc} - AD_{Loc} - AE_{Loc} = -492.938 - 29.9375 - 105.188 - 93.6875 - 39.8125 = -284.1875$ MPa.

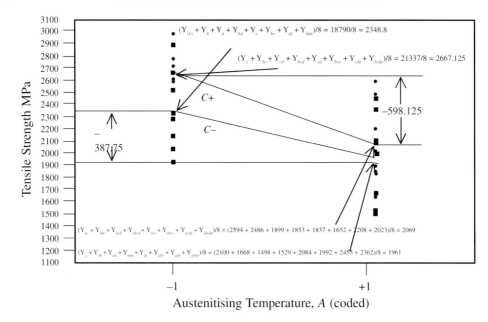

Fig. 5.6 Effect of austenitising temperature for varying deformation amounts.

In Section 4.2.2 the tests required for the principal half of a 2^{5-1} fractional factorial experiment were derived. These test were shown in Table 4.2 with the first test being test *e* and the last being test *abcde*. Suppose that instead of carrying out the full 2^5 experiment for the ausforming process shown above, only half of these test were carried out. In particular, that half given by the 2^{5-1} design. Taking those tensile strength measurements from Table 5.15 corresponding to half the test required for the 2^{5-1} design gives the results shown in Table 5.17a. Notice the tensile strength measurements shown in the last colum of this table are not new results – they are 16 of the 32 measurements shown in Table 5.15.

From fhe design generator $E = ABCD$, the confounded interaction in this experiment is $ABCDE$ – see Section 5.4 for a recap. Thus the alias structure for factor A in this 2^{5-1} design is:

$$A = A \cdot ABCDE = A^2 BCDE = BCDE.$$

That is, the A_{Loc} effect is aliased with the 3rd order interaction effect, $BCDE_{Loc}$. So in the principle fractional design the estimate made for the A_{Loc} effect is A_{Loc}^{Ali} and will infact equal the A_{Loc} effect plus the $BCDE_{Loc}$ effect obtainable from the full design

$$A_{Loc}^{Ali} \text{ effect} = A_{Loc} \text{ effect} + BCDE_{Loc} \text{ effect},$$

Only under the assumption that second and higher order interactions are zero can the estimated A_{Loc} effect from the fractional design be interpreted as the A_{Loc} effect itself.

Table 5.17a A 2^{5-1} design for the first 5 factors of the ausforming process.

Tests	A	B	C	D	E = ABCD	Tensile Strength, MPa
e	−1	−1	−1	−1	1	2517
a	1	−1	−1	−1	−1	2100
b	−1	1	−1	−1	−1	2146
abe	1	1	−1	−1	1	1992
c	−1	−1	1	−1	−1	2717
ace	1	−1	1	−1	1	1837
bce	−1	1	1	−1	1	2285
abc	1	1	1	−1	−1	2486
d	−1	−1	−1	1	−1	2038
ade	1	−1	−1	1	1	2455
bde	−1	1	−1	1	1	2656
abd	1	1	−1	1	−1	1529
cde	−1	−1	1	1	1	2980
acd	1	−1	1	1	−1	1899
bcd	−1	1	1	1	−1	2594
abcde	1	1	1	1	1	2023

The complete aliasing structure for the 2^{5-1} was derived in Section 5.4 and the results are duplicated below:

$A = A^2BCDE = BCDE.$ $AB = A^2B^2CDE = CDE.$ $BD = AB^2CD^2E = ACE.$
$B = AB^2CDE = ACDE.$ $AC = A^2BC^2DE = BDE.$ $BE = AB^2CDE^2 = ACD.$
$C = ABC^2DE = ABDE.$ $AD = A^2BCD^2E = BCE.$ $CD = ABC^2D^2E = ABE.$
$D = ABCD^2E = ABCE.$ $AE = A^2BCDE^2 = BCD.$ $CE = ABC^2DE^2 = ABD.$
$E = ABCDE^2 = ABCD.$ $BC = AB^2C^2DE = ADE.$ $DE = ABCD^2E^2 = ABC.$

To shed further light on the meaning of these aliased effects it is helpful. to compare the estimated effects from the full 2^5 and the fractional 2^{5-1} experiments.

Table 5.17b shows the Yates procedure in operation for a 2^{5-1} design of the ausforming process. In Table 5.17b the response column, (containing tensile strengths), is obtained by reading off the responses shown in Table 5.17a. Look next at the columns of plus and minus ones in Table 5.17b. For each aliased effect there is a column containing plus and minus ones. The ordering of these columns is the same as the ordering in the derived aliasing structure above. First comes the main alaised effects in alphabetical order, then the first order aliased interaction effects in alphabetical order, then the second order interaction effects and so until all 15 aliased effects are included in the Yates Table.

Table 5.17b Yates procedure for the 2^{5-1} design of the ausforming process.

Test	Coded Test Conditions					Response
	A	B	C	D	E	
e	−1	−1	−1	−1	1	2517
a	1	−1	−1	−1	−1	2100
b	−1	1	−1	−1	−1	2146
abe	1	1	−1	−1	1	1992
c	−1	−1	1	−1	−1	2717
ace	1	−1	1	−1	1	1837
bce	−1	1	1	−1	1	2285
abc	1	1	1	−1	−1	2486
d	−1	−1	−1	1	−1	2038
ade	1	−1	−1	1	1	2455
bde	−1	1	−1	1	1	2656
abd	1	1	−1	1	−1	1529
cde	−1	−1	1	1	1	2980
acd	1	−1	1	1	−1	1899
bcd	−1	1	1	1	−1	2594
abcde	1	1	1	1	1	2023

Coded Test Conditions				
AB	AC	AD	AE	BC
1	1	1	−1	1
−1	−1	−1	−1	1
−1	1	1	1	−1
1	−1	−1	1	−1
1	−1	1	1	−1
−1	1	−1	1	−1
−1	−1	1	−1	1
1	1	−1	−1	1
1	1	−1	1	1
−1	−1	1	1	1
−1	1	−1	−1	−1
1	−1	1	−1	−1
1	−1	−1	−1	−1
−1	1	1	−1	−1
−1	−1	−1	1	1
1	1	1	1	1

Table 5.17b Continued.

	Coded Test Conditions				
	BD	BE	CD	CE	DE
	1	−1	1	−1	−1
	1	1	1	1	1
	−1	−1	1	1	1
	−1	1	1	−1	−1
	1	1	−1	−1	1
	1	−1	−1	1	−1
	−1	1	−1	1	−1
	−1	−1	−1	−1	1
	−1	1	−1	1	−1
	−1	−1	−1	−1	1
	1	1	−1	−1	1
	1	−1	−1	1	−1
	−1	−1	1	1	1
	−1	1	1	−1	−1
	1	−1	1	−1	−1
	1	1	1	1	1

	Location Contrasts and Effects				
	A_{Loc}^{Ali} effect	B_{Loc}^{Ali} effect	C_{Loc}^{Ali} effect	D_{Loc}^{Ali} effect	E_{Loc}^{Ali} effect
Test	A = BCDE	B = ACDE	C = ABDE	D = ABCE	E = ABCD
e	−2517	−2517	−2517	−2517	−2517
a	2100	−2100	−2100	−2100	−2100
b	−2146	2146	−2146	−2146	−2146
abe	1992	1992	−1992	−1992	1992
c	−2717	−2717	2717	−2717	−2717
ace	1837	−1837	1837	−1837	1837
bce	−2285	2285	2285	−2285	2285
abc	2486	2486	2486	−2486	−2486
d	−2038	−2038	−2038	2038	−2038
ade	2455	−2455	−2455	2455	2455
bde	−2656	2656	−2656	2656	2656
abd	1529	1529	−1529	1529	−1529
cde	−2980	−2980	2980	2980	2980
acd	1899	−1899	1899	1899	−1899
bcd	−2594	2594	2594	2594	−2594
abcde	2023	2023	2023	2023	2023
Loc Contrast	−3612	−832	1388	94	1236
Loc Effect	−451.5	−104	173.5	11.75	154.5

Table 5.17b Continued.

	Location Contrasts and Effects				
	AB_{Loc}^{Ali} effect	AC_{Loc}^{Ali} effect	AD_{Loc}^{Ali} effect	AE_{Loc}^{Ali} effect	BC_{Loc}^{Ali} effect
Test	AB = CDE	AC = BDE	AD = BCE	AE = BCD	BC = ADE
e	2517	2517	2517	−2517	2517
a	−2100	−2100	−2100	−2100	2100
b	−2146	2146	2146	2146	2146
abe	1992	−1992	−1992	1992	−1992
c	2717	−2717	2717	2717	−2717
ace	−1837	1837	−1837	1837	−1837
bce	−2285	−2285	2285	−2285	2285
abc	2486	2486	−2486	−2486	2486
d	2038	2038	−2038	2038	2038
ade	−2455	−2455	2455	2455	2455
bde	−2656	2656	−2656	−2656	−2656
abd	1529	−1529	1529	−1529	−1529
cde	2980	−2980	−2980	−2980	−2980
acd	−1899	1899	1899	−1899	−1899
bcd	−2594	−2594	−2594	2594	2594
abcde	2023	2023	2023	2023	2023
Loc contrast	**310**	**−1050**	**−1112**	**−650**	**742**
Loc effect	**38.75**	**−131.25**	**−139**	**−81.25**	**92.75**
	Location Contrasts and Effects				
	BD_{Loc}^{Ali} effect	BE_{Loc}^{Ali} effect	CD_{Loc}^{Ali} effect	CE_{Loc}^{Ali} effect	DE_{Loc}^{Ali} effect
Test	BD = ACE	BE = ACD	CD = ABE	CE = ABD	DE = ABC
e	2517	−2517	2517	−2517	−2517
a	2100	2100	2100	2100	2100
b	−2146	−2146	2146	2146	2146
abe	−1992	1992	1992	−1992	−1992
c	2717	2717	−2717	−2717	2717
ace	1837	−1837	−1837	1837	1837
bce	−2285	2285	−2285	2285	−2285
abc	−2486	−2486	−2486	−2486	2486
d	−2038	2038	−2038	2038	2038
ade	−2455	−2455	−2455	−2455	2455
bde	2656	2656	−2656	−2656	2656
abd	1529	−1529	−1529	1529	−1529
cde	−2980	−2980	2980	2980	2980
acd	−1899	1899	1899	−1899	−1899
bcd	2594	−2594	2594	−2594	−2594
abcde	2023	2023	2023	2023	2023
Loc Contrast	**−308**	**−834**	**248**	**−2378**	**2872**
Loc Effect	**−38.5**	**−104.25**	**31**	**−297.25**	**359**

The values in columns A to E are derived from the test descriptors in the test column. Thus the first row of these columns is associated with test *e*. Hence − 1 is inserted into columns A to D and + into column E. The last row of these columns is associated with test *abcde* so +1 is entered into columns A to E. The next column, AB, is derived by multiplying the elements in columns A and B, whilst the last column is derived by multiplying the elements in columns D and E.

Now turn to the columns of Table 5.17b that contain the factor contrast columns. To work out the aliased main location contrasts simply multiply each coded test condition column by the response colunm and then sum. Each aliased main location effect is then found by dividing each contrast by eight. To work out the aliased AB_{Loc}, contrast, first multiply together the A and B columns of coded test conditions and then multiply the resulting column by the response column. Finally, sum the obtained column and divide through by eight again to obtain the aliased AB_{Loc} effect. The remaining effects are found in a similar way.

From the aliasing structure above it is to be expected that the aliased A_{Loc} effect from the 2^{5-1} design should equal the A_{Loc} effect plus the $BCDE_{Loc}$ effect from the full 2^5 design in Table 5.15. This is indeed the case. From Table 5.15, A_{Loc} effect from + $BCDE_{Loc}$ effect = –492.9375 MPa + 41.4375 MPa = –451.5 MPa. This value is indeed the estimated aliased A_{Loc} effect from the 2^{5-1} design in Table 5.17b.

A_{Loc}^{Ali} effect = − 451.5 MPa

Comparing Tables 5.15 with 5.17b does indeed confirm the full aliasing structure shown above.

5.5.4 ALL SEVEN FACTORS OF THE AUSFORMING PROCESS

In the above section five factors were studied, but the description of the ausforming process given in Section 2.1 suggested that a further two factors may also be important in influencing the steels strength. These factors being the austenitising time (factor *G*) and the isothermal incubation time (factor *F*). The seven factors basically encapsulate the complete ausforming process. To model the ausforming process the high and low levels for each factor were set at the values shown in Table 4.5. These high and low levels were picked because of the belief that over such ranges all the measured relationships would be linear – see Section 4.4 for a recap.

A full unreplicated 2^7 factorial would require $2^7 = 128$ tests and for most experimental programs this is likely to be to prohibitive in terms of time and money. For illustration purposes, suppose resource constraints limit the number of tests that can be carried out to sixteen. Then Table 4.4 reveals that with sixteen tests and seven variables the highest resolution design available is the one eighth fractional factorial, i.e. the 2^{7-3}_{IV} design. Hence *p* = 3 and the design is resolution IV. The test level combinations required for this experiment were derived in Section 4.4 and the results obtained at each test level combination were given in Table 4.7.

The aliasing structure for this design can be obtained from the following confounded interactions:

From Generators	Combinations of Two	Combinations of Three
ABCE	ABCE · BCDF = ADEF	ABCE · BCDF · ACDG = CEFG
BCDF	ABCE · ACDG = BDEG	
ACDG	BCDF · ACDG = ABFG	

The aliasing structure of this design involves $2^{7-3} - 1 = 15$ aliased effects and using alias algebra these are

$A = A^2BCE = ABCDF = A^2CDG = A^2DEF = ABDEG = A^2BFG = ACEFG.$

or,

A	= BCE	= ABCDF	= CDG	= DEF	= ABDEG	= BFG	= ACEFG
B	= ACE	= CDF	= ABCDG	= ABDEF	= DEG	= AFG	= BCEGF
C	= ABE	= BDF	= ADG	= ACDEF	= BCDEG	= ABCG	= EFG
D	= ABCDE	= BCF	= ACG	= AEF	= BEG	= ABDFG	= CDEFG
E	= ABC	= BCDEF	= ACDEG	= ADF	= BDG	= ABEFG	= CFG
F	= ABCEF	= BCD	= ACDFG	= ADE	= BDEFG	= ABG	= CEG
G	= ABCEG	= BCDFG	= ACD	= ADEFG	= BDE	= ABF	= CEF
AB	= CE	= ACDF	= BCDG	= BDEF	= ADEG	= FG	= ABCEFG
AC	= BE	= ABDF	= DG	= CDEF	= ABCDEG	= BCFG	= AEFG
AD	= BCDE	= ABCF	= CG	= EF	= ABEG	= BDFG	= ACDEFG
AE	= BC	= ABCDEF	= CDEG	= DF	= ABDG	= BEFG	= ACFG
AF	= BCEF	= ABCD	= CDFG	= DE	= ABDEFG	= BG	= ACEG
AG	= BCEG	= ABCDFG	= CD	= DEFG	= ABDE	= BF	= ACEF

The next unique aliased effect is the BD_{Loc} effect, i.e., BC is already present in the above alias structure. So

$BD = ACDE = CF = ABCG = ABEF = EG = ADFG = ACDEFG.$

Finally, the only remaining unique aliased effect is the ABD_{Loc} effect

$ABD = CDE = ACF = BEF = BCG = AEG = DFG = ABCDEFG.$

Notice that none of the main location effects are aliased with any of the first order interactions but the first order interactions are aliased with each other. The design is indeed resolution IV.

Table 5.18 shows the Yates procedure for this 2^{7-3} design. The technique works in the usual way. Look first at the first 15 columns of plus and minus ones and notice that the ordering of these columns is the same as the ordering in the derived aliasing structure above. That is, main effects in alphabetical order followed by first order interaction in alphabetical order, followed by second order and so on until all the aliased effects are covered.

As usual, the values in columns A to G are derived from the test descriptors in the test column. Thus the first row of these columns is associated with test (1). Hence –1 is

Table 5.18 Yates procedure for the 2_{IV}^{7-3} design of the ausforming process.

Tests	A	B	C	D	E = ABC	F = BCD	G = ACD	Response
(1)	−1	−1	−1	−1	−1	−1	−1	2331
aeg	1	−1	−1	−1	1	−1	1	2275
bef	−1	1	−1	−1	1	1	−1	2180
abfg	1	1	−1	−1	−1	1	1	1731
cefg	−1	−1	1	−1	1	1	1	2910
acf	1	−1	1	−1	−1	1	−1	2461
bcg	−1	1	1	−1	−1	−1	1	2714
abce	1	1	1	−1	1	−1	−1	2611
dfg	−1	−1	−1	1	−1	1	1	2380
adef	1	−1	−1	1	1	1	−1	2100
bdeg	−1	1	−1	1	1	−1	1	2384
abd	1	1	−1	1	−1	−1	−1	1687
cde	−1	−1	1	1	1	−1	−1	2893
acdg	1	−1	1	1	−1	−1	1	2428
bcdf	−1	1	1	1	−1	1	−1	2541
abcdefg	1	1	1	1	1	1	1	2447
							Mean =	2379.56

Tests	AB	AC	AD	AE	AF	AG
(1)	1	1	1	1	1	1
aeg	−1	−1	−1	1	−1	1
bef	−1	1	1	−1	−1	1
abfg	1	−1	−1	−1	1	1
cefg	1	−1	1	−1	−1	−1
acf	−1	1	−1	−1	1	−1
bcg	−1	−1	1	1	1	−1
abce	1	1	−1	1	−1	−1
dfg	1	1	−1	1	−1	−1
adef	−1	−1	1	1	1	−1
bdeg	−1	1	−1	−1	1	−1
abd	1	−1	1	−1	−1	−1
cde	1	−1	−1	−1	1	1
acdg	−1	1	1	−1	−1	1
bcdf	−1	−1	−1	1	−1	1
abcdefg	1	1	1	1	1	1

Tests	BD	ABD
(1)	1	−1
aeg	1	1
bef	−1	1
abfg	−1	−1
cefg	1	−1
acf	1	1
bcg	−1	1
abce	−1	−1
dfg	−1	1
adef	−1	−1
bdeg	1	−1
abd	1	1
cde	−1	1
acdg	−1	−1
bcdf	1	−1
abcdefg	1	1

Table 5.18 Continued.

	Location Contrasts and Effects						
Tests	A_{Loc}^{Ali} $A = BCE =$ $DEF =$ $CDG = BFG$	B_{Loc}^{Ali} $B = ACE =$ $CDF =$ $DEG = AFG$	C_{Loc}^{Ali} $C = ABE =$ $BDF =$ $ADG = EFG$	D_{Loc}^{Ali} $D = BCF =$ $AEF =$ $ACG = BEG$	E_{Loc}^{Ali} $E = ABC =$ $ADF =$ $BDG = CFG$	F_{Loc}^{Ali} $F = BCD =$ $ADE =$ $ABG = CEG$	G_{Loc}^{Ali} $G = ACD =$ $BDE =$ $ABF = CEF$
(1)	−2331	−2331	−2331	−2331	−2331	−2331	−2331
aeg	2275	−2275	−2275	−2275	2275	−2275	2275
bef	−2180	2180	−2180	−2180	2180	2180	−2180
abfg	1731	1731	−1731	−1731	−1731	1731	1731
cefg	−2910	−2910	2910	−2910	2910	2910	2910
acf	2461	−2461	2461	−2461	−2461	2451	−2461
bcg	−2714	2714	2714	−2714	−2714	−2714	2714
abce	2611	2611	2611	−2611	2611	−2611	−2611
dfg	−2380	−2380	−2380	2380	−2380	2380	2380
adef	2100	−2100	−2100	2100	2100	2100	−2100
bdeg	−2384	2384	−2384	2384	2384	−2384	2384
abd	1687	1687	−1687	1687	−1687	−1687	−1687
cde	−2893	−2893	2893	2893	2893	−2893	−2893
acdg	2428	−2428	2428	2428	−2428	−2428	2428
bcdf	−2541	2541	2541	2541	−2541	2541	−2541
abcdefg	2447	2447	2447	2447	2447	2447	2447
Loc Contrast	−2593	−1483	3937	−353	1527	−573	465
Loc Effect	−324.125	−185.375	492.125	−44.125	190.875	−71.625	58.125

Table 5.18 Continued.

	Location Contrasts and Effects						
	AB_{Loc}^{Ali}	AC_{Loc}^{Ali}	AD_{Loc}^{Ali}	AE_{Loc}^{Ali}	AF_{Loc}^{Ali}	AG_{Loc}^{Ali}	
Tests	$AB=CE=FG$	$AC=BE=DG$	$AD=CG=EF$	$AE=BC=DF$	$AF=DE=BG$	$AG=CD=BF$	
(1)	2331	2331	2331	2331	2331	2331	
aeg	−2275	−2275	−2275	2275	−2275	2275	
bef	−2180	2180	2180	−2180	−2180	2180	
abfg	1731	−1731	−1731	−1731	1731	1731	
cefg	2910	−2910	2910	−2910	−2910	−2910	
acf	−2461	2461	−2461	−2461	2461	−2461	
bcg	−2714	−2714	2714	2714	2714	−2714	
abce	2611	2611	−2611	2611	−2611	−2611	
dfg	2380	2380	−2380	2380	−2380	−2380	
adef	−2100	−2100	2100	2100	2100	−2100	
bdeg	−2384	2384	−2384	−2384	2384	−2384	
abd	1687	−1687	1687	−1687	−1687	−1687	
cde	2893	−2893	−2893	−2893	2893	2893	
acdg	−2428	2428	2428	−2428	−2428	2428	
bcdf	−2541	−2541	−2541	2541	−2541	2541	
abcdefg	2447	2447	2447	2447	2447	2447	
Loc Contrast	−93	371	−479	725	49	−421	
Loc Effect	−11.625	46.375	−59.875	90.625	6.125	−52.625	

Table 5.18 Continued.

	Location Contrasts and Effects	
	BD_{Loc}^{Ali}	ABD_{Loc}^{Ali}
Tests	$BD = CF = EG$	$ABD = CDE = ACF =$ $BEF = BCG = AEG = DFG$
(1)	2331	−2331
aeg	2275	2275
bef	−2180	2180
abfg	−1731	−1731
cefg	2910	−2910
acf	2461	2461
bcg	−2714	2714
abce	−2611	−2611
dfg	−2380	2380
adef	−2100	−2100
bdeg	2384	−2384
abd	1687	1687
cde	−2893	2893
acdg	−2428	−2428
bcdf	2541	−2541
abcdefg	2447	2447
Loc Contrast	−1	1
Loc Effect	−0.125	0.125

inserted into columns A to G. The last row of these columns is associated with test abcdefg so +1 is entered into columns A to G. The next column, AB, is derived by multiplying the elements in columns A and B, whilst the last column is derived by multiplying the elements in columns A, B and D.

Now turn to the columns of Table 5.18 that contain the factor contrasts. Multiplying the column headed A by the response column and summing the resulting numbers gives the contrast for factor A. When this is then divided through by $2^{7-3}/2 = 8$, the A_{Loc}^{Ali} effect is obtained. This can be repeated for the other columns to yield the remaining main location effects. Remember these estimates can be interpreted as measuring the main location effects under the assumption that all second and higher order interaction effects are zero.

To get the AB^{Ali}_{Loc} effect, (which is actually the sum of the *AB*, *CE* and *FG* effects under the assumption that all higher order interactions are zero), multiply the coded values in columns headed *A* and *B*. Then multiply the resulting *AB* column by the response column and sum the resulting numbers to give the factor contrast. The AB_{Loc} effect is then found by dividing this contrast through by eight. This procedure can be repeated for the other interaction location effects.

Table 5.18 has 15 estimated aliased location effects so that a lot of information is available on how to control the mean strength of the steel. A discussion on how to use this information to put the mean quality characteristic on target will be left until the techniques necessary to simplify this information are discussed in the next chapter. These results will therefore be looked at again in Chapter 8.

6. Testing the Importance of Location Effects in the 2^k Design

In Section 5.5.3, a full 2^5 factorial design for the ausforming process was carried out and some 31 location effects estimated. Many of these location effects were second and higher order interactions which are often so small in magnitude as to be of very little consequence. That is, most effects of this order can often be ignored allowing the quality control engineer to construct simplified and more parsimonious models for the process being studied. What is now needed is some scientific test procedure that can be used to identify whether an estimated effect is for all intents and purposes zero. Such a test has to be statistical in nature and forms the main content of this chapter. Section 6.1 illustrates the idea of an effect having a distribution of values. Then in Section 6.2 it is shown that each and every location effect estimated from an orthogonal design will have the same standard deviation and a formula for estimating this standard deviation is derived. In Section 6.3 a simple Student t test is constructed and applied to the ausforming process. Then in Section 6.4 this t test is presented within the least squares framework of analysis. A simple graphical test is derived in Section 6.5 and applied to both the high strength steel case study of Section 2.2 and the 2^5 experiment of the ausforming process shown in Section 5.5.3.

6.1 A DISTRIBUTION OF EFFECT ESTIMATES

In the previous chapter location effects were estimated from a variety of statistically designed experiments. It is important to realise that each of the estimated location effects, as a consequence of the way they are constructed, actually follow an approximate **normal distribution**. In the unreplicated 2^3 design of the ausforming process given in Section 5.5.2, eight tests were conducted under eight different test conditions and values for seven location effects were estimated. The other four factors were held fixed at their low levels during this experiment. If this experiment were to be repeated, the likelihood is that the same strength values would not be recorded under the same eight test conditions - the explanation for this being process variability in the form of common cause variation. (The noise factors are fixed in this experiment). To confirm this the 2^3 design for the ausforming process of Section 5.5.2 is repeated and the results from this second replicate are shown, (together with the first replicate), in Table 6.1. Notice that the strength measurement is not always the same at each test condition.

Table 6.1 A replicated 2^3 design for the experiment of Section 3.2.3.

Test	Coded Test Conditions			Replicate 1	Replicate 2
	A	B	C	Strength, MPa	Strength, MPa
(1)	−1	−1	−1	2331	2331
a	1	−1	−1	2100	2270
b	−1	1	−1	2146	2270
ab	1	1	−1	1668	1559
c	−1	−1	1	2717	2810
ac	1	−1	1	2594	2455
bc	−1	1	1	2717	2671
abc	1	1	1	2486	2440

Using only the second replicate of data and equation (5.1) the main location effect for factor A is estimated as

$$A_{2Loc} = \overline{Y}_{A+} - \overline{Y}_{A-} = \frac{2270+1559+2455+2440}{4} - \frac{2331+2270+2810+2671}{4} =$$

$$-339.5 \text{ MPa}$$

The subscript 2 in the expression above has been added to make clear the fact that this location effect has been estimated using the $i = 2$ replicate only. This differs slightly from the estimate made in Section 5.5 of the A_{Loc} effect using the first replicate, i.e. $A_{1Loc} = -265.75$ MPa. Now if a very large number of replicates are made and the A_{iLoc} effect computed for each replicate, a very large number of A_{iLoc} effect values would be obtained. The range of such values can be plotted out along a horizontal axis such as that in the top half of Figure 6.1. Some estimated A_{iLoc} effect values would occur more often than others during this replication process so that a distribution of the A_{iLoc} effect estimates would emerge such as the one shown in the top half of Figure 6.1. That is, each estimated A_{iLoc} effect makes up a proportion of all the estimated A_{iLoc} effects and this proportion is shown on the vertical axis. It is termed the probability of observing each A_{iLoc} effect estimate - $P(A_{iLoc})$.

As each A_{iLoc} effect is made up of a series of averages, the **central limit theorem** guarantees that this distribution will be approximately normal. For such a distribution the mean of all the estimated A_{iLoc} effects, μ_A, is the most likely estimate to be made of the A_{Loc} effect from any one replicate. Put differently, μ_A is the value for the A_{Loc} effect obtained by averaging all the replicates of the estimated A_{Loc} effects obtained using equation (5.1) when N is very large. Because N is large this is the most reliable estimate that can be made for the A_{Loc} effect. In practise it is not economically feasible to carry out a large

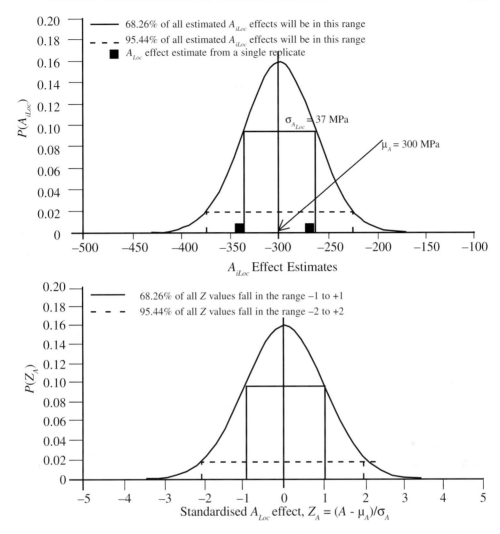

Fig. 6.1 Hypothetical normal and standard normal distribution for the estimated A_{Loc} effect.

number of replicates. What is required is an approach that will enable the engineer, with only a small number of replicates, to say whether the A_{Loc} effect that could be obtained from a large number of replicates is zero, i.e. not important.

To know for certain the true mean of these A_{iLoc} effects, μ_A, would require an infinite number of replications from which the average of an infinite number of A_{iLoc} effects could be computed. In order to illustrate some properties of the normal distribution, assume for the moment, as the Figure 6.1 suggests, that the true average value for the A_{Loc} effect is $\mu_A = -300$ MPa. The spread of the A_{iLoc} effect estimates around this average is measured using the standard deviation of the A_{iLoc} effects, σ_{ALoc}. A precise definition of this is given in Section 6.2. Again this value can only be found from an infinite number of replications. So for illustrative purposes assume this standard deviation is known to be 37 MPa. Because the A_{iLoc} effects are normally distributed, 68.26% of all possible estimated A_{iLoc} effects must be in the range mean plus and minus one of these standard deviations. That is, in the range −337 to −263 MPa. Furthermore, 95.44% of all possible measures of the A_{iLoc} effect must be in the range mean plus and minus two standard deviations, i.e. −374 to −226 MPa.

When working out such ranges it is useful to **standardise** the estimated A_{iLoc} effect using the formula

$$Z_{iA} = \frac{(A_{iLoc} - \mu_A)}{\sigma_{ALoc}} \tag{6.1}$$

Thus when the estimated A_{iLoc} effect equals the mean value plus one standard deviation the resulting Z_{iA} value equals

$$\frac{(-263 - (-300))}{37} = +1.$$

Similarly, when the estimated A_{iLoc} effect equals the mean minus one standard deviation, the Z_{iA} value equals

$$\frac{(-337 - (-300))}{37} = -1.$$

It therefore follows that 68.26% of all Z_{iA} values must fall in the range −1 to +1. Again 95.44% of all Z_{iA} values fall in the range −2 to +2. Such Z values follow a standard normal distribution and areas under this distribution for any value of Z_{iA} have been tabulated – see Table 6.2. For example, 95% of all Z_{iA} values fall in the range −1.96 to +1.96.

These concepts can be used to test hypothesised values for the value of the A_{Loc} effect that would be obtained from a very large number of replicates. That is, the value that would be obtained from an application of equation (5.1) when N approaches infinity. Indeed, these concepts can be used to test an hypothesised value for any location effect that would be obtained from an infinitely large number of replicates without having to carry out such a large number of replicates.

Statisticians tend to work on the rule that if an hypothesised effect value has less than a 5% chance of actually occurring, i.e. being estimated, then that hypothesised value can be assumed to never occur. That is, the values for the A_{iLoc} effect obtained from a small number of replicates that lie within the range of mean plus and minus two standard deviations have a sufficiently high probability of actually being obtained from an experiment as to be considered as plausible values for the A_{Loc} effect that would be obtained from an experiment with N close to infinity. The remainder are therefore implausible values for the A_{Loc} effect. It follows from this that an A_{iLoc} effect estimate is only rejected as plausible if its value is outside the mean plus and minus two standard deviations range. That is, if the standardised Z_{iA} value is outside of the range −2 to +2 or more

Table 6.2 Areas under the standard normal distribution.

Z	0	0.01	0.02	0.03	0.04	0.05	0.06	0.07	0.08	0.09
0	0.5	0.496	0.492	0.488	0.484	0.4801	0.4761	0.4721	0.4681	0.4641
0.1	0.4602	0.4562	0.4522	0.4483	0.4443	0.4404	0.4364	0.4325	0.4286	0.4247
0.2	0.4207	0.4168	0.4129	0.409	0.4052	0.4013	0.3974	0.3936	0.3897	0.3856
0.3	0.3821	0.3783	0.3745	0.3707	0.3669	0.3632	0.3594	0.3557	0.352	0.3483
0.4	0.3446	0.3409	0.3372	0.3336	0.33	0.3264	0.3228	0.3192	0.3156	0.3121
0.5	0.3085	0.305	0.3015	0.2981	0.2946	0.2912	0.2877	0.2843	0.281	0.2776
0.6	0.2743	0.2709	0.2676	0.2643	0.2611	0.2578	0.2546	0.2514	0.2483	0.2451
0.7	0.242	0.2389	0.2358	0.2327	0.2296	0.2266	0.2236	0.2206	0.2177	0.2148
0.8	0.2119	0.209	0.2061	0.2033	0.2005	0.1977	0.1949	0.1922	0.1894	0.1867
0.9	0.1841	0.1814	0.1788	0.1762	0.1736	0.1711	0.1685	0.166	0.1635	0.1611
1	0.1587	0.1562	0.1539	0.1515	0.1492	0.1469	0.1446	0.1423	0.1401	0.1379
1.1	0.1357	0.1335	0.1314	0.1292	0.1271	0.1251	0.123	0.121	0.119	0.117
1.2	0.1151	0.1131	0.1112	0.1093	0.1075	0.1056	0.1038	0.102	0.1003	0.0985
1.3	0.0968	0.0951	0.0934	0.0918	0.0901	0.0885	0.0869	0.0853	0.0838	0.0823
1.4	0.0808	0.0793	0.0778	0.0764	0.0749	0.0735	0.0721	0.0708	0.0694	0.0681
1.5	0.0668	0.0655	0.0643	0.063	0.0618	0.0606	0.0594	0.0582	0.0571	0.0559
1.6	0.0548	0.0537	0.0526	0.0516	0.0505	0.0495	0.0485	0.0475	0.0465	0.0455
1.7	0.0446	0.0436	0.0427	0.0418	0.0409	0.0401	0.0392	0.0384	0.0375	0.0367
1.8	0.0359	0.0351	0.0344	0.0336	0.0329	0.0322	0.0314	0.0307	0.0301	0.0294
1.9	0.0287	0.0281	0.0274	0.0268	0.0262	0.0256	0.025	0.0244	0.0239	0.0233
2	0.0228	0.0222	0.0217	0.0212	0.0207	0.0202	0.0197	0.0192	0.0188	0.0183
2.1	0.0179	0.0174	0.017	0.0166	0.0162	0.0158	0.0154	0.015	0.0146	0.0143
2.2	0.0139	0.0136	0.0132	0.0129	0.0125	0.0122	0.0119	0.0116	0.0113	0.011
2.3	0.0107	0.0104	0.0102	0.0099	0.0096	0.0094	0.0091	0.0089	0.0087	0.0084
2.4	0.0082	0.008	0.0078	0.0075	0.0073	0.0071	0.0069	0.0068	0.0066	0.0064
2.5	0.0062	0.006	0.0059	0.0057	0.0055	0.0054	0.0052	0.0051	0.0049	0.0048
2.6	0.0047	0.0045	0.0044	0.0043	0.0041	0.004	0.0039	0.0038	0.0037	0.0036
2.7	0.0035	0.0034	0.0033	0.0032	0.0031	0.003	0.0029	0.0028	0.0027	0.0026
2.8	0.0026	0.0025	0.0024	0.0023	0.0023	0.0022	0.0022	0.0021	0.002	0.0019
2.9	0.0019	0.0018	0.0018	0.0017	0.0016	0.0016	0.0015	0.0015	0.0014	0.0014
3	0.0013	0.0013	0.0013	0.0012	0.0012	0.0011	0.0011	0.0011	0.001	0.001

Source: Fisher and Yates.[16]

precisely –1.96 to +1.96. For example, suppose it is hypothesised that the true value for the average A_{Loc} effect is –210 MPa (this is the value that would be obtained from a very large number of replicates) and from the single replicate the estimated A_{Loc} effect is –339.5 MPa. Then substituting –210 for μ_A and –339.5 MPa for A_{iLoc} effect in equation (6.1) gives

$$Z_{iA} = \frac{-339.5 - (-210)}{37} = -3.5$$

This is clearly outside the –1.96 cut off point and so we reject this as a correct hypothesised value for the average A_{Loc} effect. Put differently, it can be stated with only a 5% chance of being wrong, that if an infinitely large number of replicates are made for the ausforming experiment, the average value obtained for the A_{Loc} effect will not be –210 MPa. This is not surprising as –210 lies outside the 95% range for plausible A_{Loc} effect values shown in Figure 5.1.

When trying to identify which location effects can be ignored the hypothesis of interest becomes $\mu_A = 0.0$. That is, whether the average location effect estimated from a very large number of replicates is actually zero and therefore unimportant.

6.2 THE STANDARD DEVIATION OF A LOCATION EFFECT ESTIMATE

The test conclusions drawn above were based on a belief that the standard deviation for location effect A was 37 MPa. This value must be known in order to apply this test. This issue now needs to be addressed. For any 2^k or 2^{k-p} design (or any other orthogonal design) each of the estimated location effects will have the same standard deviation irrespective of whether process variability depends on the level set for each process variable. This is most clearly seen be referring back to the 2^2 design where the A_{Loc} effect was given by

$$A_{Loc} = \frac{1}{2}\left[Y_a + Y_{ab} - Y_{(1)} - Y_b\right]$$

The **standard deviation of this location effect** can be found from the statistical theorem

Standard deviation $(\alpha Y_1 \pm \alpha Y_2) =$

$$\sqrt{\alpha^2 \left[[\text{standard deviation}(Y_1)]^2 + [\text{standard deviation}(Y_2)]^2\right]},$$

where Y_1 and Y_2 are two different responses and α is some constant. Applying this statistical rule to the A_{Loc} effect (with $\alpha = \frac{1}{2}$ and $Y_1 = Y_a$, $Y_2 = Y_{ab}$ etc.) gives

$$\sigma_{A_{Loc}} = \sqrt{\frac{1}{2^2}\left[\sigma_{Y_a}^2 + \sigma_{Y_{ab}}^2 + \sigma_{Y_{(1)}}^2 + \sigma_{Y_b}^2\right]} \tag{6.2a}$$

In equation (6.2a), σ_{Y_a} is the standard deviation in the responses recorded at test condition a, σ_{Y_b} is the standard deviation in the responses recorded at test condition b, $\sigma_{Y_{ab}}$ is the standard deviation in the responses recorded at test condition ab and $\sigma_{Y_{(1)}}$ is the standard deviation in the responses recorded at test condition (1). These standard deviations can be estimated from a small number of replicates and this will be fully discussed in Section 6.3 below. For the moment note that the B_{Loc} effect is given by

$$B_{Loc} = \frac{1}{2}\left[Y_b + Y_{ab} - Y_{(1)} - Y_a\right]$$

Applying the same rule for the calculation of its standard deviation will give the same answer as that for the A_{Loc} effect, namely

$$\sigma_{B_{Loc}} = \sqrt{\frac{1}{2^2}\left[\sigma_{Y_a}^2 + \sigma_{Y_{ab}}^2 + \sigma_{Y_{(1)}}^2 + \sigma_{Y_b}^2\right]} \tag{6.2b}$$

This occurs because each location effect in a 2^{k-p} design is a linear combination of the same responses. The signs differ in each combination, but signs are not important in the determination of a standard deviation. In general any 2^{k-p} design will have $l = 1, M{-}1$ location effects associated with it, where $M = 2^{k-p}$. These location effects will be aliased effects for $p > 0$. Hence each location effect has the same standard deviation

$$\sigma_{1Loc} = \sigma_{2Loc} = \ldots\ldots \sigma_{M\text{-}1Loc} = \sigma_{Loc} \tag{6.3a}$$

With this in mind equation (6.1) can be generalised to give the standardised value for any location effect estimate made from the ith replicate as follows

$$Z_{il} = \frac{(Loc_{il} - \mu_l)}{\sigma_{Loc}} \tag{6.3b}$$

where μ_l is the mean of the lth location effect and σ_{Loc} the standard deviation associated with all the location effects.

This result holds irrespective of whether process variability depends on the level set for each process variable. In equations (6.2a) and (6.2b) it does not matter whether the variability in the response is the same at each test condition,

i.e. whether $\sigma_{Y_a}^2 = \sigma_{Y_{ab}}^2 = \sigma_{Y_{(1)}}^2 = \sigma_{Y_b}^2$ or not.

These same four terms are present in both equations so that the standard deviation of the A_{Loc} and B_{Loc} effect will be the same even if these variances differ. That is, even if the variability in response differs at each test condition.

The advantage of having process variability that is independent of the test conditions comes from the simplifications that can be made to equations (6.2a) and (6.2b). If $\sigma_{Y_a}^2 = \sigma_{Y_{ab}}^2 = \sigma_{Y_{(1)}}^2 = \sigma_{Y_b}^2 = \sigma_Y^2$, then the formula for the standard deviation of any location effects in a 2^2 design simplifies to

$$\sigma_{Loc} = \sqrt{\frac{1}{2^2}\left[4\sigma_Y^2\right]} \tag{6.4a}$$

In this equation σ_Y is the standard deviation in the response at any one of the test conditions. This generalises to any 2^k design with N replicates

$$\sigma_{Loc} = \sqrt{\left(\frac{2}{2^k N}\right)^2 \left(N 2^k \sigma_y^2\right)} \qquad (6.4b)$$

and to any 2^{k-p} design with N replicates.

$$\sigma_{Loc} = \sqrt{\left(\frac{2}{2^{k-p} N}\right)^2 \left(N 2^{k-p} \sigma_y^2\right)} \qquad (6.4c)$$

6.3 THE t TEST IN A REPLICATED DESIGN

6.3.1 THE TEST

To use the test briefly described in Section 6.1 above requires the engineer to know a value for the standard deviation of all the location effects, σ_{Loc}. As equations (6.4) suggest this requires a value for σ_Y. It can only be known exactly if an infinite number of replicates are made at each test condition. This would be found by averaging the standard deviation in responses obtained at all the j test conditions when an infinite number of replications are made

$$\sigma_{jY} = \sqrt{\frac{\sum_{i}^{N}\left[Y_{ij} - \mu_{jY}\right]^2}{N}} \qquad \text{with } N = \infty \qquad (6.5a)$$

and

$$\sigma_Y = \sqrt{\frac{\sum_{j=1}^{M} \sigma_{jY}^2}{M}} \qquad (6.5b)$$

σ_{jY} is the standard deviation in the N responses (Y_{ij}) recorded at the jth test condition and where there are $M = 2^{k-p}$ different test conditions. μ_{jY} is the mean response value in the N test results at test condition j when $N = \infty$. Notice how the standard deviation at each test condition is measuring the spread in the experimental readings. At a particular test condition each response is compared with its average value and the difference squared to prevent positive deviations being offset by negative ones in the summation. The total squared deviation is then averaged out by dividing through by N to give a mean spread of responses around the average response. Finally, the square root is taken so that the resulting standard deviation is in the same units of measurement as the responses themselves.

The number of replicates needs to be very large to apply these equations. Fortunately, there is an alternative that can be used for a small number of replicates (as little as $N = 2$). S_Y will be used to show that this is an estimate made for σ_Y from a small number of replicates. The relevant formulas are

$$S_Y = \sqrt{\frac{\sum_{j=1}^{M} S_{jY}^2}{M}} \tag{6.6a}$$

M is the number of tests in the experiment (i.e. $M = 2^{k-p}$) and S_{jY} is the standard deviation for the replicated responses at each of the j test conditions. That is

$$S_{jY} = \sqrt{\frac{\sum_{i=1}^{N}\left[Y_{ij} - \overline{Y}_j\right]^2}{N-1}} \tag{6.6b}$$

where N is the number of replications made under each test condition. In equation (6.6b), $\overline{Y}_j$ is the average response recorded under the jth test condition. It differs from μ_{jY} in that it is computed from a small number of replicates

$$\overline{Y}_j = \frac{\sum_{i=1}^{N} Y_{ij}}{N} \tag{6.6c}$$

Clearly, as $N \to \infty$, $\overline{Y}_j \to \mu_{jY}$ so that $S_{jY} \to \sigma_{jY}$ [$N-1 \approx N$ when $N = \infty$]. If S_{jY} is independent of the test condition j, then this constant standard deviation, S_Y, can be estimated as a straight average of each S_{jY}, i.e. equation (6.6a).

If S_Y is used instead of σ_Y in equation (6.4b) then the sample estimate for the standard deviation of a location effect is obtained

$$S_{Loc} = \sqrt{\left(\frac{2}{2^k N}\right)^2 \left(N 2^k S_Y^2\right)} = \frac{2 S_Y}{\sqrt{N 2^k}} \tag{6.6d}$$

and for any 2^{k-p} design

$$S_{Loc} = \sqrt{\left(\frac{2}{2^{k-p} N}\right)^2 \left(N 2^{k-p} S_Y^2\right)} = \frac{2 S_Y}{\sqrt{N 2^{k-p}}} \tag{6.6e}$$

To test whether each location effect is important simply replace σ_{Loc} in equation (6.3b) with its estimate, S_{Loc}. Thus the test statistic becomes

$$t_l = \frac{(Loc_l - \mu_l)}{S_{Loc}} \qquad (6.7a)$$

This is called a **t transformation**. The replacement of σ_{Loc} by S_{Loc} unfortunately means that t_1 no longer follows the standard normal distribution. Instead it follows a Student t distribution. This distribution is generally a lot fatter, (longer), than the Z distribution but collapses onto the Z distribution when the **degrees of freedom**, V, becomes large. The degrees of freedom are simply equal to

$$V = (N - 1) \times 2^{k-p} \qquad (6.7b)$$

Table 6.3 shows areas under the t distribution for different degrees of freedom. Notice that when $V = \infty$ the t values are the same as the Z values of Table 6.2. The testing procedure is now identical to that used in Section 6.1 except we now use the table of tabulated t values together with equations (6.7). So, for the $N = 2$ replicated 2^3 design the degrees of freedom equals $(2 - 1) \times 8 = 8$. From Table 6.3 it follows that 95% of all t values must be in the range -2.31 to $+2.31$. This forms the range of plausible t values.

6.3.2 Application of the t Test to the 2^3 Ausforming Experiment

As an example of how to use the above t test, consider again the replicated 2^3 design shown in Table 6.1. Here $N = 2$ so that there are two recorded tensile strengths at each test condition. In turn there are $j = 1, 8$ tests, i.e. $M = 2^3 = 8$. Consequently, it is possible to compute eight separate estimates for S_Y.

Test (1), $j = 1$:

$$\bar{Y}_1 = \frac{2331 + 2331}{2} = 2331 \text{ and } S_{1Y} = \sqrt{\frac{(2331-2331)^2 + (2331-2331)^2}{2-1}} = 0.0$$

Test a, $j = 2$:

$$Y_2 = \frac{2100 + 2270}{2} = 2185 \text{ and } S_{2Y} = \sqrt{\frac{(2100-2185)^2 + (2270-2185)^2}{2-1}} = 120.208$$

Test b, $j = 3$:

$$Y_3 = \frac{2146 + 2270}{2} = 2208 \text{ and } S_{3Y} = \sqrt{\frac{(2146-2208)^2 + (2270-2208)^2}{2-1}} = 87.681$$

Test ab, $j = 4$:

$$Y_4 = \frac{1668 + 1559}{2} = 1613.5 \text{ and } S_{4Y} = \sqrt{\frac{(1668-1613.5)^2 + (1559-1613.5)^2}{2-1}} = 77.075$$

Table 6.3 Percentage points of the t distribution.

v \ %/200	0.25	0.2	0.15	0.10	0.05	0.025	0.01	0.005	0.0005
1	1.000	1.376	1.963	3.078	6.314	12.706	31.821	63.657	636.619
2	0.816	1.061	1.386	1.886	2.920	4.303	6.965	9.925	31.598
3	0.765	0.978	1.250	1.638	2.353	3.182	4.541	5.841	12.941
4	0.741	0.941	1.190	1.533	2.132	2.776	3.747	4.604	8.610
5	.727	.920	1.156	1.476	2.015	2.57	3.365	4.032	6.859
6	0.71	0.9	1.13	1.44	1.94	2.45	3.14	3.71	5.96
7	0.71	0.89	1.12	1.42	1.9	2.37	3.00	3.50	5.41
8	0.7	0.88	1.11	1.4	1.86	2.31	2.90	3.36	5.04
9	0.7	0.88	1.1	1.38	1.83	2.26	2.82	3.25	4.78
10	0.7	0.87	1.09	1.37	1.81	2.23	2.76	3.17	4.59
11	0.69	0.87	1.09	1.36	1.8	2.2	2.72	3.11	4.44
12	0.69	0.87	1.08	1.36	1.78	2.18	2.68	3.06	4.32
13	0.69	0.87	1.08	1.35	1.77	2.16	2.65	3.01	4.22
14	0.69	0.86	1.08	1.35	1.76	2.15	2.62	2.98	4.14
15	0.69	0.86	1.07	1.34	1.75	2.13	2.60	2.95	4.07
16	0.69	0.86	1.07	1.34	1.75	2.12	2.58	2.92	4.01
17	0.68	0.86	1.07	1.33	1.74	2.11	2.57	2.90	3.96
18	0.68	0.86	1.07	1.33	1.73	2.1	2.55	2.88	3.92
19	0.68	0.86	1.07	1.33	1.73	2.09	2.54	2.86	3.88
20	0.68	0.86	1.06	1.33	1.73	2.09	2.53	2.85	3.85
21	0.68	0.85	1.06	1.32	1.72	2.08	2.52	2.83	3.82
22	0.68	0.85	1.06	1.32	1.72	2.07	2.51	2.82	3.79
23	0.68	0.85	1.06	1.32	1.71	2.07	2.50	2.81	3.77
24	0.68	0.85	1.06	1.32	1.71	2.06	2.49	2.80	3.75
25	0.68	0.85	1.06	1.32	1.71	2.06	2.48	2.79	3.73
26	0.68	0.85	1.06	1.31	1.71	2.06	2.48	2.78	3.71
27	0.68	0.85	1.06	1.31	1.7	2.05	2.47	2.77	3.69
28	0.68	0.85	1.06	1.31	1.7	2.05	2.47	2.76	3.67
29	0.68	0.85	1.05	1.31	1.7	2.04	2.46	2.76	3.66
30	0.68	0.85	1.05	1.31	1.7	2.04	2.46	2.75	3.65
40	0.68	0.85	1.05	1.3	1.68	2.02	2.42	2.70	3.55
60	0.67	0.84	1.05	1.3	1.67	2.00	2.39	2.66	3.46
120	677	0.84	1.04	1.29	1.66	1.98	2.36	2.62	3.37
∞	0.67	0.84	1.04	1.28	1.65	1.96	2.33	2.58	3.29

Test $c, j = 5$:

$$\bar{Y}_5 = \frac{2717 + 2810}{2} = 2763.5 \text{ and } S_{5Y} = \sqrt{\frac{(2717 - 2763.5)^2 + (2810 - 2763.5)^2}{2-1}} = 65.761$$

Test $ac, j = 6$:

$$\bar{Y}_6 = \frac{2594 + 2455}{2} = 2524.5 \text{ and } S_{6Y} = \sqrt{\frac{(2594 - 2524.5)^2 + (2455 - 2524.5)^2}{2-1}} = 98.288$$

Test $bc, j = 7$:

$$\bar{Y}_7 = \frac{2717 + 2671}{2} = 2694 \text{ and } S_{7Y} = \sqrt{\frac{(2717 - 2694)^2 + (2671 - 2694)^2}{2-1}} = 32.527$$

Test $abc, j = 8$:

$$\bar{Y}_8 = \frac{2486 + 2440}{2} = 2463 \text{ and } S_{8Y} = \sqrt{\frac{(2486 - 2463)^2 + (2440 - 2463)^2}{2-1}} = 32.527$$

These can be averaged to estimate the assumed constant standard deviation in Y as

$$S_Y = \sqrt{\frac{0.0^2 + 120.208^2 + 87.681^2 + 77.075^2 + 65.761^2 + 98.288^2 + 32.527^2 + 32.527^2}{8}} = 74.31$$

Substituting this into equation (6.6d) yields the standard deviation for all of the seven location effects that can be estimated from the 2^3 design of the ausforming process.

$$S_{Loc} = \frac{2 \times 74.31}{\sqrt{2 \times 2^3}} = \frac{148.6}{4} = 37.2$$

This is a sample estimate of σ_{Loc} because S_Y has been used as an estimates of σ_Y. This t test can be applied to the 2^3 replicated design of the ausforming process introduced in Section 6.1. Table 6.4 shows the Yates algorithm for the data in Table 6.1. Notice that the response column of Table 6.4 is the average of the two tensile strengths at each test condition of Table 6.1. Shown at the bottom are the factor contrasts for the replicated 2^3 design and the location effects can be derived by diving each contrast by $2^{3-1} = 4$.

To test for important location effects in the ausforming process set $\mu_1 = 0$ in equation (6.7a) for the $l = 1, 7$ location effects. Taking each of the location effects shown in Table 6.4 in turn.

A_{Loc} effect, (hypothesis $\mu_A = 0$).

$$t = \frac{(-302.5 - 0)}{37.2} = -8.13.$$ This is an implausible t value and so the A_{Loc} effect is not zero.

B_{Loc} effect, (hypothesis $\mu_B = 0$).

Testing the Importance of Location Effects in the 2^k Design   121

Table 6.4 The Yates procedure for the replicated 2^3 design of the ausforming process.

Test	Coded Test Conditions							Mean Response	Factor Contrasts						
	A	B	AB	C	AC	BC	ABC		A	B	AB	C	AC	BC	ABC
(1)	−1	−1	1	−1	1	1	−1	2331	−2331	−2331	2331	−2331	2331	2331	−2331
a	1	−1	−1	−1	−1	1	1	2185	2185	−2185	−2185	−2185	−2185	2185	2185
b	−1	1	−1	−1	1	−1	1	2208	−2208	2208	−2208	−2208	2208	−2208	2208
ab	1	1	1	−1	−1	−1	−1	1614	1614	1614	1614	−1614	−1614	−1614	−1614
c	−1	−1	1	1	−1	−1	1	2764	−2764	−2764	2764	2764	−2764	−2764	2764
ac	1	−1	−1	1	1	−1	−1	2525	2525	−2525	−2525	2525	2525	−2525	−2525
bc	−1	1	−1	1	−1	1	−1	2694	−2694	2694	−2694	2694	−2694	2694	−2694
abc	1	1	1	1	1	1	1	2463	2463	2463	2463	2463	2463	2463	2463
							Contrasts		−1210	−826	−440	2108	270	562	456
							Effects		−302.5	−206.5	−110	527	67.5	140.5	114

$$t = \frac{(-206.5 - 0)}{37.2} = -5.55.$$ This is an implausible t value and so the B_{Loc} effect is not zero.

AB_{Loc} effect, (hypothesis $\mu_{AB} = 0$).

$$t = \frac{(-110.0 - 0)}{37.2} = -2.96.$$ This is an implausible t value and so the AB_{Loc} effect is not zero.

C_{Loc} effect, (hypothesis $\mu_C = 0$).

$$t = \frac{(+527.0 - 0)}{37.2} = 14.17.$$ This is an implausible t value and so the C_{Loc} effect is not zero.

AC_{Loc} effect, (hypothesis $\mu_{AC} = 0$).

$$t = \frac{(+67.5 - 0)}{37.2} = 1.81.$$ This is a plausible t value and so the AC_{Loc} effect is zero.

BC_{Loc} effect, (hypothesis $\mu_{BC} = 0$).

$$t = \frac{(+140.5 - 0)}{37.2} = 3.78.$$ This is an implausible t value and so the BC_{Loc} effect is not zero.

ABC_{Loc} effect, (hypothesis $\mu_{ABC} = 0$).

$$t = \frac{(+114.0 - 0)}{37.2} = 3.06.$$ This is an implausible t value and so the ABC_{Loc} effect is not zero.

Therefore the only irrelevant location effect is the AC_{Loc} effect. This means that if the above 2^3 experiment was repeated again, there is a 95% chance that all the estimated effects, except AC_{Loc}, will be non-zero.

6.4 THE t TEST WITHIN THE LEAST SQUARES PROCEDURE

The above t test statistics can be obtained in a different way. The response surface model for the 2^3 replicated design including the second order interaction is

$$Y_{ji} = \beta_0 + \beta_A A_j + \beta_B B_j + \beta_C C_j + \beta_{AB} A_j B_j + \beta_{AC} A_j C_j + \beta_{BC} B_j C_j + \beta_{ABC} A_j B_j C_j + \varepsilon_{ij} \quad (6.8)$$

As usual β_0 is the mean of all response values and each β equals half the respective location effect, e.g. β_{AB} is half the AB_{Loc} effect. ε_{ij} no longer represent higher order interactions left out from the equation because they are all included in equation (6.8). Instead it is there to represent the scatter in the replicated responses at each test condition. This term is called the **prediction error**. Substituting in the values for β shown in Table 6.4 gives

$$Y_{ji} = 2347.8 - 151.3A_j - 103.3B_j + 263.5C_j - 55.0A_jB_j + 33.8A_jC_j + 70.3B_jC_j + 57.0A_jB_jC_j + \varepsilon_{ji}$$

The two prediction errors ($i = 1, 2$) associated with each test condition ($j = 1, 8$), can now be found by rearranging this equation for ε_{ji}.

Going through each test individually the following numeric values can be attached to the two prediction errors at each test condition.

Test (1), $j = 1$. The coded test conditions in row 1 of Table 6.4 suggest $A_1 = -1$, $B_1 = -1$, $C_1 = -1$, $A_1B_1 = +1$, $A_1C_1 = +1$, $B_1C_1 = +1$, $A_1B_1C_1 = -1$. Thus,

$$\varepsilon_{11} = 2331 - \{2347.8 + 151.3 + 103.3 - 263.5 - 55.0 + 33.8 + 70.3 - 57.0\} = 0.00.$$

$$\varepsilon_{12} = 2331 - \{2347.8 + 151.3 + 103.3 - 263.5 - 55.0 + 33.8 + 70.3 - 57.0\} = 0.00.$$

Test a, $j = 2$. The coded test conditions in row 2 of Table 6.4 suggest $A_2 = +1$, $B_2 = -1$, $C_2 = -1$, $A_2B_2 = -1$, $A_2C_2 = -1$, $B_2C_2 = +1$, $A_2B_2C_2 = +1$. Thus,

$$\varepsilon_{21} = 2100 - \{2347.8 - 151.3 + 103.3 - 263.5 + 55.0 - 33.8 + 70.3 + 57.0\} = -85.0.$$

$$\varepsilon_{22} = 2270 - \{2347.8 - 151.3 + 103.3 - 263.5 + 55.0 - 33.8 + 70.3 + 57.0\} = +85.0.$$

Continuing in this way;

Test b, $j = 3$ Test ab, $j = 4$ Test c, $j = 5$ Test ac, $j = 6$ Test bc, $j = 7$ Test abc, $j = 8$

$\varepsilon_{31} = -62$ $\varepsilon_{41} = 54.5$ $\varepsilon_{51} = -46.5$ $\varepsilon_{61} = 69.5$ $\varepsilon_{71} = 23$ $\varepsilon_{81} = 23$

$\varepsilon_{32} = 62$ $\varepsilon_{42} = -54.5$ $\varepsilon_{52} = 46.5$ $\varepsilon_{62} = -69.5$ $\varepsilon_{72} = -23$ $\varepsilon_{82} = -23$

Assuming that the standard deviation of the errors is the same at all test conditions then an estimate of the standard deviation of these errors can be obtained using the formula.

$$S_\varepsilon = \sqrt{\frac{\sum_{j=1}^{M}\sum_{i=1}^{N}\left[\varepsilon_{ji} - \bar{\varepsilon}_j\right]^2}{(N \times 2^{k-p}) - V}} \quad (6.9a)$$

Notice that the average value for ε at each test condition, j, is zero. For example, when $j = 3$, $\bar{\varepsilon}_3 = (-62 + 62)/2 = 0$. Thus equation (6.9a) can be simplified.

$$S_\varepsilon = \sqrt{\frac{\sum_{j=1}^{M}\sum_{i=1}^{N}\left[\varepsilon_{ji}\right]^2}{(N \times 2^{k-p}) - V}} \quad (6.9b)$$

Substituting in the prediction errors gives

$$S_\varepsilon = \sqrt{\frac{(0)^2+(-0)^2+(-85)^2+(85)^2+(-62)^2+(62)^2+(54.5)^2+(-54.5)^2+(-46.5)^2+(46.5)^2+(69.5)^2+(-69.5)^2+(23)^2+(-23)^2+(23)^2+(-23)^2}{16-8}} = 74.31$$

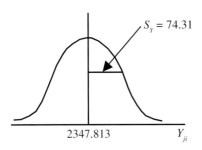 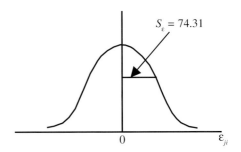

Fig. 6.2 Comparison of the distributions for Y and ε.

Notice straight away the standard deviation for the prediction errors equals the standard deviation for the responses Y, i.e.

$$S_Y = S_\varepsilon \tag{6.10}$$

Hence any assumption made about the spread in the responses is equivalent to assumptions made about the spread in the prediction errors. This means that dispersion effects (the spread in Y values depending on test conditions) can be studied through an analysis of the prediction errors. This can be visualised in Figure 6.2 where the response, Y, over all test conditions is normally distributed with a mean of 2347.8 and a standard deviation of 74.31, whilst the prediction errors are normally distributed with a mean of zero and a standard deviation of 74.31.

Consequently an alternative expression for the standard deviation of a location effect is

$$S_{Loc} = \frac{2 \times S_\varepsilon}{\sqrt{N \times 2^{k-p}}} \tag{6.11}$$

An alternative to testing the significance of each location effect is to test the significance of each β in equation (6.8). Thus testing the hypothesis that the AC_{Loc} effect is zero is equivalent to testing that β_{AC} is zero. The standard deviation for each of the βs in equation (6.8) is given by,

$$S_\beta = \frac{S_\varepsilon}{\sqrt{N \times 2^{k-p}}} \tag{6.12}$$

Hence it is clear that the standard error on β is half the standard error of its associated location effect. This is not surprising given that the value for any β is half its location effect.

To test an hypothesis about the value for any β, simply form the *t* statistic

$$t = \frac{\beta - \mu_\beta}{S_\beta} \tag{6.13}$$

where μ_β is the hypothesised mean value for β.

6.5 A GRAPHICAL TEST FOR THE IMPORTANCE OF LOCATION EFFECTS

6.5.1 Test Derivation

A simple graphical test that can be applied to replicated and unreplicated designs alike starts by assuming that all the location effects estimated from a statistically designed experiment come from a single normal distribution. This will only be so provided

i. Each recorded response comes from a normal distribution. Equations (5.1 and 5.2) show that each location effect is a linear combination of the responses (*Y* values). Any linear combination of normal variables will itself be normal. So each location effect will have an exact normal distribution provided the responses are normally distributed. In fact if the design is orthogonal each location effect is a linear combination of two average values. The central limit theorem ensures that for

$$\left(N \times \frac{2^{k-p}}{2} \right) > 30,$$

such averages are normally distributed even if the responses making up those averages are not.

For $\left(N \times \frac{2^{k-p}}{2} \right) < 30,$

the location effects will be approximately normal. Hence for 2^k, 2^{k-p} and Taguchi type designs each location effect will be close to a normal distribution, provided *k* is not too small.

ii. The statistically designed experiment is orthogonal. This is so for any 2^k and 2^{k-p} factorial design. It was shown above that for such designs each location effect has the same standard deviation. That is, $\sigma_{lLoc} = \sigma_{Loc}$ for the l = 1, *M* location effects.

iii. Each location effect in the statistically designed experiment has the same mean value of zero. That is, $\mu_{lLoc} = \mu_{Loc} = 0$. This implies that for an experiment carried out with an infinite number of replications, the estimated location effects will all come out as zero in value.

Under these conditions it is possible to show what the *M* estimated location effects should look like when presented is some graphical form. This is shown in Figure 6.3.

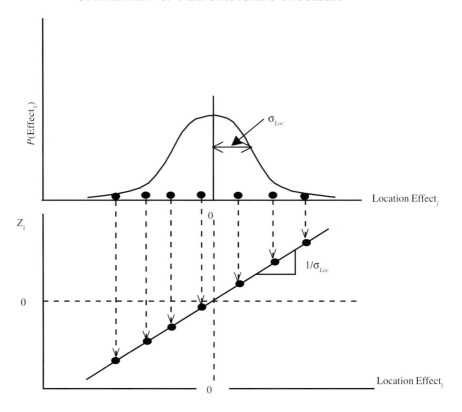

Fig. 6.3 Probability plot when there are no location effects of importance.

The top half of Figure 6.3 shows the single distribution from which all the M location effects should come from under the above conditions. Notice that the distribution has a symmetric bell shape that is characteristic of a normal distribution, has a mean of zero and a single standard deviation, σ_{Loc}. If the above assumptions are true, then all the estimated M location effects from the 2^{k-p} design should fall under this distribution in the way illustrated by the shaded dots shown on the horizontal axis for a 2^3 design with seven effects.

In the bottom half of Figure 6.3, the same information is presented as a **probability plot**. Here the estimated location effects are plotted against their standardised or Z value. Now equation (6.3b) can be rewritten as

$$Loc_{il} = \mu_l + \sigma_{lLoc} Z_{il} \tag{6.14a}$$

which if the above assumptions hold can be simplified to

$$Loc_{il} = \sigma_{Loc} Z_{il} \tag{6.14b}$$

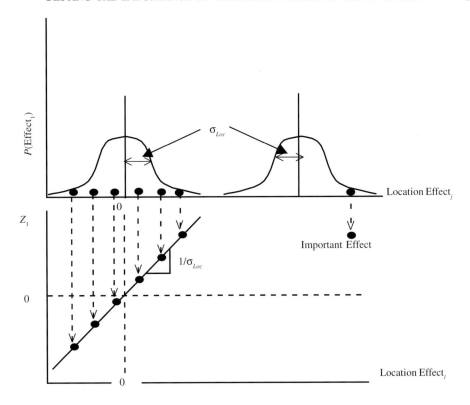

Fig. 6.4 Probability plot when there is one important location effect.

This equation is plotted as a straight line in the bottom half of Figure 6.3. If the above conditions hold true, then all the estimated M effects from the 2^k design should fall on this straight line in the way illustrated by the shaded dots shown along the drawn linear line.

The detection of an important location effect is now straight forward. If conditions i and ii above hold true, which they approximately will for a 2^{k-p} design, but that one or more of the estimated location effects falls a long way off the straight line, then this must be because assumption iii has been violated. That is, the means of these location effects are not zero. Hence they can be considered as important location effects. This is pictured in Figure 6.4. All but one of the M estimated effects fall on the straight line. Notice that this different estimated location effect still follows a normal distribution with a standard deviation identical to all the other effects. The only difference being that its distribution is shifted to the right of the others, i.e. its mean is not zero.

Hence under the conditions that all the effects are normally distributed with the same standard deviation, an important location effect (one whose mean value is not zero) shows itself as a departure from the straight line in the probability plot. For 2^{k-p} designs with

$$\left(N \times \frac{2^{k-p}}{2}\right) > 30,$$

replicates the test is a very good one in that assumption i and ii always hold so that any departure from the linear line must reflect a violation of assumption iii, i.e. that some of the effects are important.

$$\text{In practice } \left(N \times \frac{2^{k-p}}{2}\right)$$

is often a lot less than 30 so some care must be taken in the interpretation of the results. That is, the effects may not be normally distributed so that Z is the wrong transformation to use on the vertical axis of the probability plot.

6.5.2 Illustration of Graphical Test Using the High Strength Steel Case Study

To illustrate how this test can be implemented on a practical level consider the high strength steel case study discussed in Section 2.2. Car manufacturers today are imposing very stringent specifications on the sheet steel supplied to them by the steel industry for the purpose of panel forming. Designed experiments are a very efficient and effective means of ensuring that such specifications can be meet through the addition of alloying elements to a base steel. Table 6.5 summarises one such experiment designed to look at the effect of alloying content on the yield point measured in MPa. As stated in Section 2.2 the base steel has the following chemistry 0.03%C, 0.44%Mn, 0.15%Si, 0.11%Al, 0.004%P and 0.018%S. Factor A is made the quantity of nickel added to this base steel, factor B the quantity of copper added to the base steel and factor C the quantity of niobium added to the base steel. Notice also that the low value for each factor corresponds to a situation in which the steel alloy contains non of these alloying elements, i.e. the base chemistry steel.

Table 6.6 shows the Yates analysis of the data in Table 6.5. Each location effect is obtained by dividing that factor contrast by four. Clearly the addition of copper to the base steel has a large impact on the yield point of the alloyed steel, whilst the addition of niobium and nickel seems to have a smaller impact on the yield point. In each case the yield point is increased by having each alloying element at its high level, i.e. included. After this it is not clear which of the remaining interaction location effects are important, i.e. which can be assumed to have no influence on the yield point.

To identify the important location effects the Z values corresponding to each location effect needs to be quantified. The calculations required to achieve this are shown in Table 6.7. The first step is to rank the estimated location effects from most negative to

Table 6.5 Mechanical properties and alloying elements.

Test	Actual Test Conditions			Coded Test Conditions			Test Order	Yield Point, MPa
	%Ni	%Cu	%Cb	A	B	C		
(1)	0	0	0	−1	−1	−1	3	262
a	0.87	0	0	1	−1	−1	2	340
b	0	1.24	0	−1	1	−1	1	542
ab	0.87	1.24	0	1	1	−1	5	558
c	0	0	0.05	−1	−1	1	7	357
ac	0.87	0	0.05	1	−1	1	6	439
bc	0	1.24	0.05	−1	1	1	8	546
abc	0.87	1.24	0.05	1	1	1	4	612

most positive. Now under the assumptions that each location effect has a mean of zero and that each effect has the same standard deviation, it follows that each location effect has exactly the same approximate normal distribution. Thus each location effect in Table 6.7 represents a point under this distribution.

There are seven location effect estimates altogether so that each effect represents 1/7th of the information available for this location effect distribution. The most negative effect ($BC_{Loc} = -34$) characterises the lowest $(1/7) \times 100 = 14.286\%$ of the distribution. The next most negative location effect characterises the next 14.286% of the distribution and so on. Since the most negative effect represents the proportion of the distribution between 0 and 14.286% it seems reasonable to assign it a **cumulative probability** (*CP*) in between these two numbers, i.e. 7.14%. That is, $CP_1 = 7.14\%$ so that there is a 7.14% chance of estimating an effect whose value is more negative than −34. Similarly, the next smallest location effect represents the proportion of the distribution between 14.286 and 28.572% and so its cumulative probability lies in the middle of these two numbers, i.e. 21.43%. In general then the cumulative probability associated with each location effect is given by the formula

$$CP_i = \frac{100(i-0.5)}{M-1} \qquad (6.15)$$

where M is the number of test conditions, eight in this case, i is a rank index equal to 1 for the most negative location effect, 2 for the next most negative location effect and so on up to $M-1$ for the most positive location effect. The i index and *CP* values are shown in the third and fourth columns of Table 6.7.

Table 6.6 The Yates procedure for a 2^3 design of the high strength steel study.

Test	Coded Test Conditions							Mean Response	Factor Contrasts						
	A	B	AB	C	AC	BC	ABC		A	B	AB	C	AC	BC	ABC
(1)	-1	-1	1	-1	1	1	-1	262	-262	-262	262	-262	262	262	-262
a	1	-1	-1	-1	-1	1	1	340	340	-340	-340	-340	-340	340	340
b	-1	1	-1	-1	1	-1	1	542	-542	542	-542	-542	542	-542	542
ab	1	1	1	-1	-1	-1	-1	558	558	558	558	-558	-558	-558	-558
c	-1	-1	1	1	-1	-1	1	357	-357	-357	357	357	-357	-357	357
ac	1	-1	-1	1	1	-1	-1	439	439	-439	-439	439	439	-439	-439
bc	-1	1	-1	1	-1	1	-1	546	-546	546	-546	546	-546	546	-546
abc	1	1	1	1	1	1	1	612	612	612	612	612	612	612	612
								Contrasts	242	860	-78	252	54	-136	46
								Effects	60.5	215	-19.5	63	13.5	-34	11.5

Testing the Importance of Location Effects in the 2^k Design

Table 6.7 Table of calculations for a probability plot.

	Effect$_l$	Rank Index, i	CP_l	Z_l
BC	−34	1	7.14	−1.47
AB	−19.5	2	21.43	−0.79
ABC	11.5	3	35.71	−0.37
AC	13.5	4	50	0
A	60.5	5	64.29	0.37
C	63	6	78.57	0.79
B	215	7	92.86	1.47

All that remains to be done is to read off the Z values associated with each CP value from a Table of Z values (Table 6.2). The most negative location effect (−34) represents the first 7.14% of the total area under the normal distribution curve for the location effects (assuming a mean of zero and constant standard deviation for all location effects). From the Z Table, approximately 7.14% of all Z values are less than −1.47. Consequently, −1.47 is the Z value corresponding to a location effect of −34. The next most negative effect (−19.5) represents the first 21.43% of the total area under the normal distribution curve for the location effects. From the Z Table approximately 21.43% of all Z values are less than −0.79. Carrying on in this way gives the remaining Z values shown in the final column of Table 6.7.

Finally, Figure 6.5 plots the various location effect estimates against the Z values corresponding to these effect estimates. Only one effect seems to lie a long way from the linear line defining the hypothesis that all location effects have an average value of zero. This point corresponds to the B_{Loc} effect. The linear line in Figure 6.5 starts off at the origin Z_1 = effect$_1$ = 0 corresponding to the assumption that all location effects have a zero mean. The line is then extrapolated outwards from this point in such a way that seems to best fit the data points clustered around a location effect estimate of zero.

In Figure 6.5 all but one of the main and higher order interaction location effects seem to lie around the linear line that defines a location effect distribution that is normal with a mean of zero and an as yet unknown standard deviation. However, the scatter in the effects lying about this line defines the spread in the location effect distribution; all the factor effects must have this same standard deviation. The standard deviation for every factor effect can therefore be found by collapsing the location effects falling around the linear line to the horizontal axis and computing the standard deviation in such collapsed effects. Because the mean of these location effects is zero, the formula for computing the standard deviation of all the location effects is

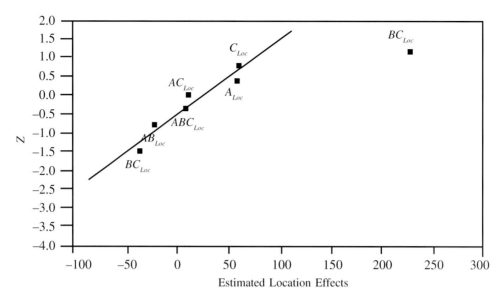

Fig. 6.5 Probability plot of estimated location effects.

$$S_{Loc} = \sqrt{\sum_{\text{All location effects around line}} \frac{[\text{effect}_l - 0]^2}{\text{Number of location effects around line}}} \qquad (6.16)$$

Here S_{Loc} is used to distinguish it as a sample estimate of the true standard deviation σ_{Loc} which could only be found if the above experiment was replicated an infinite number of times. There are six location effects around the linear line – A, C, AB, AC, BC and ABC.

$$S_{Loc} = \sqrt{\frac{[60.5-0]^2 + [63-0]^2 + [-19.5-0]^2 + [13.5-0]^2 + [-34-0]^2 + [11.5-0]^2}{6}} = 39.75$$

All seven location effects in this experiment must have a normal distribution with a standard deviation of 39.75. Thus, for example, the distribution of factor B has a standard deviation of 39.75, but the mean of this distribution is not zero, it is shifted to the right.

6.5.3 Illustration of Graphical Test Using the Ausforming Process

Probability plots are much easier to interpret in larger designs where there are likely to be more unimportant location effects that can be used to calculated the standard deviation of the distribution for location effects, i.e. that can be used to define the linear line of the probability plot. To see this return to the 2^5 design of the ausforming process shown in

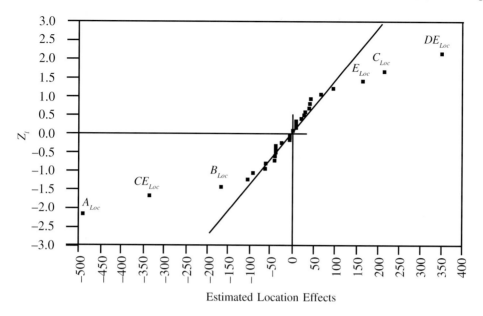

Fig. 6.6 Probability plot for the ausforming process.

Section 5.5.3. All the calculations required to build a probability plot are shown in Table 6.8. Figure 6.6 then plots each location effect against its corresponding Z value and it would appear that factors A, B, C and E are important as well as the interactions between factors C and E and D and E. The remaining factors appear to be normally distributed with a mean of zero and are scattered around the line defining such a distribution. These unimportant location effects can be used to estimate the standard deviation for each location effect. This is done using equation (6.16). In Table 6.8 the final column squares all the effects scattered around the linear line so that S_{Loc} is simply the square root of the average of this final column. That is

$$S_{Loc} = \sqrt{\frac{(-105.19)^2 + (-93.69)^2 + \ldots + (64.69)^2 + (93.69)^2}{25}} = \sqrt{2304.698} = 48.01.$$

Note that $\frac{1}{S_{Loc}} = 0.0208$ is the approximate slope of the line in Figure 6.6.

It would appear from this probability plot that a simplified first order response surface model that could be used to predict the mean tensile strength of the steel under the condition that factors F and G are held fixed at their low levels is

$$Y_j = \beta_0 + \beta_A A_j + \beta_B B_j + \beta_C C_j + \beta_E E_j + \beta_{CE} C_j E_j + \beta_{DE} D_j E_j + \varepsilon_j$$

Table 6.8 Table of a probability plot for the 2^5 design of the ausforming process.

	Effect$_i$	Rank Index, i	CP$_i$	Z$_i$	Effect$_i^2$
A	−492.94	1	1.61	−2.15	
CE	−336.69	2	4.84	−1.66	
B	−168.69	3	8.06	−1.40	
AC	−105.19	4	11.29	−1.21	11064.94
AD	−93.69	5	14.52	−1.06	8777.42
ACD	−64.69	6	17.74	−0.93	4184.80
ACE	−64.56	7	20.97	−0.81	4167.99
BCE	−45.31	8	24.19	−0.70	2053.00
BCD	−41.44	9	27.42	−0.60	1717.27
AE	−39.81	10	30.65	−0.51	1584.84
ABDE	−39.69	11	33.87	−0.42	1575.30
BE	−39.56	12	37.10	−0.33	1564.99
BDE	−26.06	13	40.32	−0.25	679.12
ABCE	−8.69	14	43.55	−0.16	75.52
CD	−8.56	15	46.77	−0.08	73.27
ABCD	−8.56	16	50.00	0	73.27
BC	−0.93	17	53.23	0.08	0.86
ABC	8.69	18	56.45	0.16	75.52
CDE	8.81	19	59.68	0.25	77.62
ABCDE	8.81	20	62.90	0.33	77.62
D	20.44	21	66.13	0.42	417.79
BD	26.06	22	69.35	0.51	679.12
AB	29.94	23	72.58	0.60	896.40
ABD	39.44	24	75.81	0.70	1555.51
ABE	39.56	25	79.03	0.81	1564.99
BCDE	41.44	26	82.26	0.93	1717.27
ACDE	64.69	27	85.48	1.06	4184.80
ADE	93.69	28	88.71	1.21	8777.82
E	163.06	29	91.94	1.40	
C	213.19	30	95.16	1.66	57617.45
DE	350.31	31	98.39	2.15	

The only terms that appear in this model are those that lie off the linear line in the probability plot of Figure 6.6. The prediction error, ε_j, picks up all the effects left out and these are scattered around the zero axis on the probability plot and so are approximately zero in value. On average then ε_j is expected to equal zero. This equation suggests that controlling the mean tensile strength, so that it equals a target requirement, can be achieved by manipulation of factor A (the ausforming temperature) and factor B (the deformation temperature). As will be seen in the next chapter, factor C can be used to minimise process variability. The way that all this comes together to minimise variability about a mean that has been placed on target is left until Chapter 8. By then process variability (Chapter 7) will have been covered in the required detail.

7. Controlling Process Variability: Dispersion Effects in Linear Designs

Up until now emphasis has been placed on designing experiments and then using the results to estimate location effects. These location effects provide information that can then be used to achieve a desired mean quality characteristic. But as discussed in Section 1.3, experimentation should not be confined to such a narrow objective. Good quality is achieved by ensuring that both the mean quality characteristic is close to a target value and that the scatter around this target is as small as possible. This chapter discusses the techniques available for minimising such scatter or process variability when the manufacturing process is linear in nature.

For linear manufacturing process three approaches to the problem of minimising process variability are developed. The first two solutions are applicable when there is no replication within a factorial or fractional factorial design. The first of these solutions involves the identification of noise~design factor interactions within a first order response surface model. Here an explicit analysis of all the noise factors has to be made. The technique assumes that process variability stems mainly from an inability to control the noise factors present within the manufacturing process. The second solution uses the prediction errors from a first order response surface model. The technique can be used when process variability is due either to the existence of noise factors and/or common cause. These two solutions are presented in Sections 7.1 to 7.5 of this chapter.

The third solution requires replication to be made within a factorial or fractional factorial experiment. The technique uses a generalised linear model that can be applied when process variability stems either from an inability to control noise factors or from common cause when a process has no noise. In the former case, noise factors are used as a means of obtaining replicate test results from experiments carried out in a laboratory. For most engineering processes, the variability in a quality characteristic is likely to be highly dependant upon the mean value for that quality characteristic. To cope with this extra complication so called PerMIA statistics, calculated from the replicates, are used to minimise variability. This forms the content of the remainder of this chapter.

Section 7.1 describes the role of noise~design factor interactions in determining process variability within the simple framework of a 2^2 design. Then in Section 7.2 these ideas are generalised to any factorial or fractional factorial design using a first order response surface model. In Section 7.3 the technique is applied to the 2^5 experiment of the ausforming process using the results of Section 6.5.3. Then in Section 7.4, the role of the prediction error of the response surface model in minimising process variability is discussed and this technique is then applied in Section 7.5 to the disk forging experiment first explained in Section 2.4. Section 7.6 introduces a simple generalised linear model which can be used to minimise process variability when replicate tests from a linear design are available. In Section 7.7 this generalised linear model is applied to the copper compact experiment of Section 2.3. Finally in Section 7.8 a comparison is made between the response surface and generalised linear model approaches to minimising process variability.

7.1 NOISE - DESIGN FACTOR INTERACTIONS AND PROCESS VARIABILITY

A **dispersion effect** is similar to a location effect. It shows what is expected to happen to process variability following a change in the level of a factor. The main dispersion effect for factor A, for example, is written as A_{Disp} and measures the change in process variability as factor A is changed from its low to its high level. Process variability can be measured in a number of ways. One obvious measure is to look at the **range** (R_j) of results. The range R_j is formally defined as the difference between the maximum and minimum response values obtained under test condition j.

$$R_j = \text{Max}(Y_j) - \text{Min}(Y_j) \tag{7.1}$$

In equation (7.1) $\text{Max}(Y_j)$ is the maximum response recorded at test condition j and $\text{Min}(Y_j)$ the minimum response recorded at test condition j. This statistic obviously requires a factorial design to be replicated so that more than one response is made at each of the j test conditions.

Alternatively, process variability can be measured as an average deviation around the mean response. The meaning and measurement of a standard deviation was discussed in detail in Section 6.2, both in relation to location effects and responses. Recall that to prevent the deviations from the mean equalling zero, the squared deviations have to be used and the square root then taken at the end of the calculation. The **standard deviation** in responses, S_{Y_j}, obtained under test condition j is therefore given by

$$S_{Y_j} = \sqrt{\frac{\sum_{i=1}^{N}\left[Y_{ij} - \bar{Y}_j\right]^2}{N-1}} \tag{7.2}$$

In equation (7.2) a total of N response values are made at each of the j test conditions making up the factorial or fractional factorial experiment. $\bar{Y}_j$ is the mean of these N responses made at test condition j. The **variance**, $S_{Y_j}^2$, is just the square of the standard deviation

$$S_{Y_j}^2 = \frac{\sum_{i=1}^{N}\left[Y_{ij} - \bar{Y}_j\right]^2}{N-1} \tag{7.3}$$

Controlling Process Variability: Dispersion Effects in Linear Designs

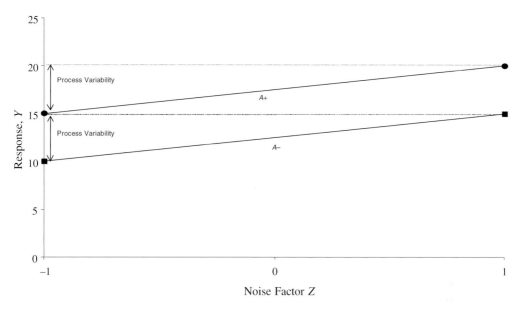

Fig. 7.1 Interaction plot for a design and noise factor as might be generated from a 2^2 design.

Both the variance and standard deviation are valid measures of process variation but the latter measure is in the same units as the response itself.

When the presence of noise factors is the major cause of process variability, it is the **noise–design factor interaction** that will determine the size of this process variability. To explain why this is so, consider the simplest possible setting. This is a 2^2 design in which one of the factors is a design factor (factor A) and one is a noise factor (factor Z). At the point of manufacture, the noise factor will vary uncontrollably between an upper and a lower limit. The size of this tolerance band is mainly determined by the cost involved in trying to control the noise factor precisely. The more costly it is to control the wider the tolerance band. The variability in the noise factor then induces the variability in a quality characteristic observed at the point of manufacture.

This process variability can be copied in the laboratory (where noise factors can be easily controlled) by setting the high and low levels for the noise factor equal to the upper and lower limits of the tolerance band. Suppose that carrying out this 2^2 design in the laboratory gave the results shown in Figure 7.1.

In Figure 7.1 the noise factor varies over time between the coded levels of -1 to $+1$. This then induces a range of response values (between 10 and 15 if factor A is set low and between 15 and 20 if factor A is set high). The size of this range determines the process variability and the important point to note is how the uncontrollable movements

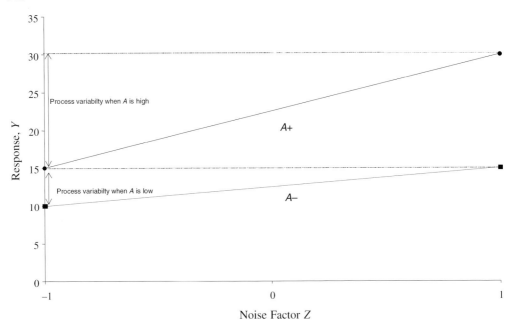

Fig. 7.2 Another interaction plot for a 2^2 design.

in the noise factor along the horizontal axis of Figure 7.1 induce this variability. Notice also that in this example it is impossible to reduce this variability by a careful setting of the design factor A. Irrespective of the level set for factor A, the range is always five.

Now consider another hypothetical situation shown in Figure 7.2. Again notice that the variance transmitted to the response comes from the noise factor. But this time the variance transmitted to the response by the noise factor is greater when factor A is set high compared to when it is set low. When factor A is set high, the range in the response is fifteen, compared with only five when factor A is set low. Variability can be minimised by setting the design factor at its low level. In this way the manufacturing process is made **robust to noise**.

Figure 7.2 depicts an interaction between factor A and factor Z, in that factor Z has a bigger influence on the response at one particular setting for factor A. Hence the key to minimising variability is to identify all the important interactions between noise and design factors. More specifically the **variance transmitted** to the response by the noise factor is proportional to the square of the slopes of the two lines shown in Figure 7.2. The transmitted variance in therefore minimised by setting the design factor to a level where the slope is flattest.

CONTROLLING PROCESS VARIABILITY: DISPERSION EFFECTS IN LINEAR DESIGNS

This can be shown more precisely using the following first order response surface model of the above 2^2 design

$$Y_j = \beta_0 + \beta_A A_j + \beta_Z Z_j + \beta_{AZ} A_j Z_j \qquad (7.4)$$

This should now be immediately recognisable as a first order response surface model. It includes both the design factor A, the noise factor Z and their interaction. As usual the β's measure half of these respective effects. Now the noise factor Z will vary randomly during actual production, and suppose that it has an average value of zero and a standard deviation of S_Z. (A variable can always be made to have a mean of zero through a suitable transformation). Then the mean response at test condition j, $\overline{Y}_j$, will be

$$\overline{Y}_j = \beta_0 + \beta_A \overline{A}_j + \beta_Z \overline{Z}_j + \beta_{AZ} \overline{A}_j \overline{Z}_j = \beta_0 + \beta_A \overline{A}_j + \beta_Z(0) + \beta_{AZ}(0) = \beta_A \overline{A}_j + \beta_0 \quad (7.5a)$$

$\overline{A}_j$ is the mean test condition for factor A and $\overline{Z}_j$, the mean test condition for factor Z. It can be shown that the variance (as defined by equation (7.3)) in a response that is governed by equation (7.4) is given by

$$S_{Yj}^2 = (\beta_Z + \beta_{AZ} A_j)^2 S_Z^2 \qquad (7.5b)$$

Notice from equation (7.5b) that the design factor can only be used to alter the variability in response if it interacts with the noise factor Z, i.e. if $\beta_{AZ} \neq 0$. That is, design factor A will only have a dispersion effect if it interacts with the noise factor. Thus analysing dispersion effects is tantamount to analysing interactions between design and noise factors. Further, selecting a value for A for which $(\beta_Z + \beta_{AZ} A_j)^2 \approx 0$ effectively neutralises the variation transmitted from the noise factor to the response and makes the process completely robust to noise.

7.2 A GENERALISED RESPONSE SURFACE APPROACH TO PROCESS VARIABILITY

The results derived above are very easy to generalise. Suppose that a manufacturing process has $v = 1$ to M_1 design factors. For ease of representation let these be labelled X_v. This breaks from the terminology used in previous chapters where design factors were labelled alphabetically as factor A, factor B etc. As before, however, Z_w will be used to represent the noise factors. Suppose the manufacturing process has $w = 1$ to M_2 such factors. In total, the process will therefore have $P = M_1 + M_2$ factors. Then, if the factorial design is unreplicated, so that a single response is obtained at each of the j test conditions, the **full first order response surface model** used to predict the responses obtained from a factorial experiment has the following form

$$Y_j = \beta_0 + \sum_{v=1}^{M_1} \beta_v X_{vj} + \sum_{v=1}^{M_1}\sum_{w=v+1}^{M_1} \beta_{vw} X_{vj} X_{wj} + \sum_{w=1}^{M_2} \lambda_w Z_{wj} + \sum_{v=1}^{M_2}\sum_{w=v+1}^{M_2} \lambda_{vw} Z_{vj} Z_{wj} + \sum_{v=1}^{M_1}\sum_{w=1}^{M_2} \delta_{vw} X_{vj} Z_{vj} + \varepsilon_j \quad (7.6)$$

Similar to previous chapters, the β_v parameters are half the main location effects for the design factors and λ_w are half the main location effects for the noise factors. The β_{vw} parameters are half the first order interaction location effects between the design factors and the λ_{vw} parameters are half the first order interaction location effects between the noise factors. The δ_{vw} parameters are half the first order interaction location effects between the design and noise factors. Y_j is the response recorded at test condition j and β_0 the average response over all test conditions. It is crucial to remember that if this model is fitted to the data from a fractional factorial design, the β, λ and δ values are actually half the aliased location effects. ε_j is included to capture all the higher order interaction terms left out of the model but which, on average, are expected to equal zero. ε_j is in effect the error in predicting the response using this model as a result of leaving out such terms, i.e. ε_j are the prediction errors.

But if the factorial design is replicated, with N responses being obtained at each of the j test conditions, the first order response surface model used to predict the responses obtained from a factorial experiment has the following form

$$Y_{ij} = \beta_0 + \sum_{v=1}^{M_1}\beta_v X_{vj} + \sum_{v=1}^{M_1}\sum_{w=v+1}^{M_1}\beta_{vw}X_{vj}X_{wj} + \sum_{w=1}^{M_2}\lambda_w Z_{wj} + \sum_{v=1}^{M_2}\sum_{w=v+1}^{M_2}\lambda_{vw}Z_{vj}Z_{wj} + \sum_{v=1}^{M_1}\sum_{w=1}^{M_2}\delta_{vw}X_{vj}Z_{vj} + \varepsilon_{ij} \quad (7.7)$$

Here ε_{ij} is included to capture all the higher order interaction terms left out of the model (which on average are expected to equal zero) and to capture the scatter in the $i = 1$ to N responses obtained at each of the j test conditions.

The **simplified first order response surface model** has the same form as equations (7.6) and (7.7) but it only includes those β, λ and δ values that are shown to be important from a probability plot. That is, the simplified model will only include a subset of the M_1 design and M_2 noise factors. In the simplified version of equation (7.6) ε_j will again represent all the higher order interaction terms left out of the model plus all the main and first order interaction effects left out because a probability plot has shown them to be unimportant (on average equal to zero). In the simplified version of equation (7.7), the prediction error ε_{ij} will again capture all the higher order interaction terms left out of the model but also the scatter in the $i = 1$ to N responses obtained at each of the j test conditions together with all the main and first order interaction effects left out because a probability plot has shown them to be unimportant.

Now if the noise factors are transformed, so that their average value is zero, and if the noise factors are independent of each other, the mean response at test condition j for a process driven by equations (7.6) or (7.7) is

$$\overline{Y}_j = \beta_0 + \sum_{v=1}^{M_1}\beta_v X_{vj} + \sum_{v=1}^{M_1}\sum_{w=v+1}^{M_1}\beta_{vw}X_{vj}X_{wj} \quad (7.8)$$

Treating the design factors as constants (because they do not vary uncontrollably) and assuming that ε_j and the noise factors are independent, the variance in the response at test condition j for a process driven by equation (7.6) is approximately equal to

Controlling Process Variability: Dispersion Effects in Linear Designs 143

$$S_{Y_j}^2 = S_{\varepsilon_j}^2 + S_{Z_v}^2 \sum_{v=M_1+1}^{M_2} \left[\frac{\partial Y}{\partial Z_v}\right]^2 \tag{7.9}$$

In equation (7.9), $S_{\varepsilon_j}^2$ is the variance in the prediction error term ε_j of equation (7.6), which may or may not depend of the jth test condition. If it is independent of the test condition then $S_{\varepsilon_j}^2$ can be replaced by the constant variance S_ε^2. $S_{Z_v}^2$ is the variability in the noise factor Z_v and the terms in the squares bracket are the partial derivates of the response with respect to each of the noise factors evaluated at the mean value for each noise factor (i.e. zero).

Steinberg and Bursztyn[17] have proposed a very simple and powerful way to interpret the results from a response surface model. The first condition that the experimental design must satisfy is that all the main location effects for the design factors are not aliased with any first order interaction effects. The second condition is that none of the first order interactions between the design and noise factors are aliased with any other first order interactions. Then under the assumptions of zero second and higher order interactions and that process variability is primarily the result of noise transmission, the following two rules apply:

i. The main design location effects (together with their first order interactions if present) can be used to push the mean of the process on target. With reference to equation (7.8), the values for β_v ($v = 1$, to M_1) and β_{vw} indicate the effect on the mean response of changing the levels for design factor X_v.
ii. The location interaction effects between the design and noise factors can be used to minimise process variability. Again with reference to equation (7.6) above the values for δ_{vw} ($v = 1$ to M_1 and $w = 1$ to M_2) show the effect on process variability (induced by the noise factor) that results from changing the levels for design factors X_v.

7.3 AN APPLICATION TO THE 2^5 DESIGN OF THE AUSFORMING

In Section 6.5.3 a probability plot (Figure 6.6) was obtained showing which of the five process variables were important in controlling the tensile strength of the steel. Six location effects showed up as being important, A_{Loc}, B_{Loc}, C_{Loc}, E_{Loc}, CE_{Loc} and DE_{Loc}. In relation to equation (7.6) there are therefore $M_1 = 3$ important design factors (A, B and C) and $M_2 = 2$ important noise factors (D and E). The simplified first order response surface model (simplified version of equation (7.6)) for the process is therefore

$$Y_j = \beta_0 + \beta_1 X_{1j} + \beta_2 X_{2j} + \beta_3 X_{3j} + \lambda_2 Z_{2j} + \lambda_{12} Z_{1j} Z_{2j} + \delta_{32} X_{3j} Z_{2j} + \varepsilon_j \tag{7.10}$$

In equation (7.10) X_1 represents design factor A (the austenitising temperature), X_2 design factor B (the deformation temperature), X_3 design factor C (the amount of deformation), Z_1 noise factor D (the quench rate) and Z_2 noise factor E (the deformation rate). β_0 is the overall average strength, β_1 half the value of A_{Loc}, β_2 half the value of B_{Loc},

β_3 half the value of C_{Loc}, λ_2 half the value of E_{Loc}, δ_{32} half the value of CE_{Loc} and λ_{12} half the value of DE_{Loc}. Substituting into the above equation the estimated values for these effects (as derived in Section 6.5) gives

$$Y_j = 2261.5 - 246.47X_{1j} - 84.34X_{2j} + 106.59X_{3j} + 81.53Z_{2j} + 175.16Z_{1j}Z_{2j} - 168.34X_{3j}Z_{2j} + \varepsilon_j \quad (7.11)$$

Assuming that the mean value for all the noise factors are zero, the mean tensile strength is, from equation (7.8), given by

$$\overline{Y}_j = \beta_0 + \beta_1 X_{1j} + \beta_2 X_{2j} + \beta_3 X_{3j} \quad (7.12a)$$

or with the estimated effects substituted in

$$\overline{Y}_j = 2261.5 - 246.47X_{1j} - 84.34X_{2j} + 106.59X_{3j} \quad (7.12b)$$

Assuming that the prediction error ε_j is independent of the test conditions, the variance in tensile strength is, from equation (7.9), given by

$$S_{Yj}^2 = S_{Z_1}^2 [\lambda_{12}Z_{2j}]^2 + S_{Z_2}^2 [\lambda_2 + \lambda_{12}Z_{1j} + \delta_{32}X_3]^2 + S_\varepsilon^2 \quad (7.13a)$$

An estimate for S_ε^2 can be made from equation (6.9b) when $N = 1$ (no replication), and when the average error is assumed to equal zero,

$$S_\varepsilon = \sqrt{\frac{\sum_{j=1}^{2^{k-p}}[\varepsilon_j - 0]^2}{(2^{k-p}) - V^*}}$$

The degrees of freedom, V^*, is equal to the number of parameters in equation (7.6) or (7.7). For the simplified version given in equation (7.11), $V^* = 7$. Values for ε_j come from rearranging equation (7.11).

$$\varepsilon_j = Y_j - \{2261.5 - 246.47X_{1j} - 84.34X_{2j} + 106.59X_{3j} + 81.53Z_{2j} + 175.16Z_{1j}Z_{2j} - 168.34X_{3j}Z_{2j}\}$$

The first test condition ($j = 1$) in the 2^5 design has all factors low. Substituting -1 in for each X_v and Z_w term gives

$$\varepsilon_1 = 2331 - \{2261.5 - 246.47(-1) - 84.34(-1) + 106.59(-1) + 81.53(-1) +$$
$$175.16(-1x - 1) - 168.34(-1x - 1)\} = -80$$

Table 7.1 shows all the other 31 values for ε_j together with their squares and the sum of these squares. Thus

$$S_\varepsilon = \sqrt{\frac{\sum_{j=1}^{2^{k-p}}[\varepsilon_j - 0]^2}{(2^{k-p}) - V^*}} = \sqrt{\frac{460926.3}{32 - 7}} = 135.8$$

CONTROLLING PROCESS VARIABILITY: DISPERSION EFFECTS IN LINEAR DESIGNS 145

Table 7.1 Calculations required for S_ε^2.

Test, j	ε_j	ε_j^2	Test, j	ε_j	ε_j^2
1	−80.01	6401.60	17	−43.43	1886.165
2	181.93	33098.52	18	16.51	272.5801
3	−96.33	9279.469	19	−106.75	11395.56
4	−81.39	6624.332	20	93.19	8684.376
5	−243.87	59472.58	21	172.07	29608.08
6	126.07	15893.64	22	−106.99	11446.86
7	−75.19	5653.536	23	16.75	280.5625
8	186.75	34875.56	24	−123.31	15205.36
9	−22.69	514.8361	25	−23.75	564.0625
10	−69.75	4865.063	26	37.19	1383.096
11	37.99	1443.24	27	−86.07	7408.045
12	129.93	16881.80	28	112.87	12739.64
13	168.45	28375.40	29	192.75	37152.56
14	−218.61	47790.33	30	−86.31	7449.416
15	152.13	23143.54	31	37.43	1401.005
16	−95.93	9202.565	32	−102.63	10532.92
				$\Sigma\varepsilon_j^2$	460926.3

Now assuming that $S_{Z_1}^2$ and $S_{Z_2}^2$ are unity (although this can of course be computed for any values of the variance in the noise factors) the variance in the tensile strength of the steel is, from equation (7.13a),

$$S_{Y_j}^2 = [175.16Z_{2j}]^2 + [81.53 + 175.16Z_{1j} - 168.34X_{3j}]^2 + 135.8^2 \qquad (7.13b)$$

From equation (7.13b) it is clear that factor X_3, (the amount of deformation), is the only factor that can be manipulated to change the variability in tensile strength. The higher the amount of deformation, the lower the variability. Equation (7.12b) then shows that changing the amount of deformation would also change the mean tensile strength. If the mean moves away from the target, factors X_1 and X_2 (austenitising and deformation temperatures) could then be used to fine-tune the mean strength so that it falls on target. More will be said about this in Chapter 8.

7.4 PREDICTION ERRORS AND PROCESS VARIABILITY

In an unreplicated design there are no repeat experiments carried out under the same test conditions from which to estimate the variability present in a manufacturing process. In such a situation an alternative to identifying the design - noise interaction effects in a first order response surface model is to use the **prediction errors** from such a model. This can be done in three simple sequential steps.

7.4.1 ESTIMATE A SIMPLIFIED RESPONSE SURFACE MODEL OF THE PROCESS

In Section 7.2 a distinction was made between the full and simplified first order response surface model for the recorded responses made during the experiment. The full model (equation 7.6) contains all the main effects and all the first order interactions between the factors. The simplified model contains only those location effects found to be important from a probability plot. Like the full model, this simplified model contains a prediction error term ε_j. Remember from Section 6.4 that this error term picks up all those effects whose mean value is zero as identified by the probability plot. This prediction error term therefore has a mean of zero and, as explained in Section 6.4, its variability equals the variability present in the response variable, Y_j. (See equation 6.10 for a proof). Thus dispersion effects can be studied through this prediction error term.

7.4.2 CALCULATE THE PREDICTION ERROR VARIABILITY AT EACH FACTOR LEVEL

If a particular factor affects process variability, then the variability in the prediction errors should be different depending on the level of that factor. The variability in the prediction errors at each factor level setting can be found from the usual variance formula. Thus when factor A is set high the **variability in the prediction error** is defined as

$$S^2_{A+} = \frac{\sum_{j^{A+}=1}^{\frac{M}{2}} \left[\varepsilon_{j^{A+}} - \overline{\varepsilon}_{A+}\right]^2}{\frac{M}{2} - 1} \tag{7.14a}$$

where M is the total number of tests defined by the experiment, j^{A+} means that the summation is made over all the prediction errors for which factor A is set high, $\overline{\varepsilon}_{A+}$ is the mean value for the prediction error calculated using only the prediction errors associated with factor A being high. When factor A is low, the prediction error variance is

$$S^2_{A-} = \frac{\sum_{j^{A-}=1}^{\frac{M}{2}} \left[\varepsilon_{j^{A-}} - \overline{\varepsilon}_{A-}\right]^2}{\frac{M}{2} - 1} \tag{7.14b}$$

where j^{A-} means that the summation is made over all the prediction errors for which factor A is low and $\bar{\varepsilon}_{A-}$ is the mean value for the prediction error calculated using only the prediction errors associated with factor A being low.

A **dispersion effect** shows up if there is a substantial difference between the prediction error variance when factor A is high compared with when it is low. That is, the dispersion effect for factor A is defined as

$$A_{Disp} = S_{A+}^2 - S_{A-}^2 \tag{7.14c}$$

Such a dispersion effect can be identified for any factor or interaction between factors. In general the effect of each factor, including all the interaction effects, on the variance in the prediction errors (and thus in the responses themselves) is found by comparing the values obtained from the following two formula

$$S_{l-}^2 = \frac{\sum_{j^{l-}=1}^{\frac{M}{2}}\left[\varepsilon_{j^{l-}} - \bar{\varepsilon}_{l-}\right]^2}{\frac{M}{2} - 1} \tag{7.15a}$$

and

$$S_{l+}^2 = \frac{\sum_{j^{l+}=1}^{\frac{M}{2}}\left[\varepsilon_{j^{l+}} - \bar{\varepsilon}_{l+}\right]^2}{\frac{M}{2} - 1} \tag{7.15b}$$

j^{l-} means that the summation is made over all the prediction errors for which factor l is low and j^{l+} means that the summation is made over all the prediction errors for which factor l is high. There are $l = 2^{k-p} - 1$ such effects in a 2^{k-p} design. l is in standard order form so that for a 2^4 design, $l = 1$ gives the dispersion effect for factor A and when $l = 15$ the $ABCD$ dispersion effect is identified.

7.4.3 Testing the Importance of Dispersion Effects

How big does the change in variability have to be for a factor to be important in contributing to process variability? The answer is found by realising that the following **F ratio** has an approximate normal distribution only when the variance in the errors are the same under the two levels for factor l.

$$F_l = \ln\left(\frac{S_{l+}^2}{S_{l-}^2}\right) \tag{7.16}$$

This ratio can obviously be calculated for each of the factor effects. From Section 6.1 it follows that there is only a 2.5% chance of observing an F value in excess

Table 7.2 Test levels for the disk forging operation.

Factor	Physical Name	Low Level	High Level
A	Height/Diameter	0.5	2
B	Friction Parameter	0.05	0.4
C	Heat Transfer Coefficient	$200/K/m^2$	$2000/J/K/m^2$
D	Emissive Power	0.1	0.8

of 1.96 or more negative than −1.96. Thus if F lies outside the range −1.96 to +1.96 we can be 95% sure that such a value can't exist. And because F only follows a normal distribution when the two variances are the same it follows that an F value outside this range implies that the two variances are different. That is, a change in the level of the factor significantly alters process variability.

7.5 THE DISK FORGING OPERATION EXPERIMENT

To illustrate the computations involved in this procedure consider again the disk forging operation discussed in Section 2.4. A current topic of interest at Rolls-Royce is the assessment of the impact of various boundary conditions of the forging operation on the energy required to forge an aero engine disk. The end objective being to minimise the mean energy required together with the variability in energy use around this mean consumption rate. As discussed in Section 2.4, four factors are considered to be of particularly importance in this area. These being the height/diameter of the aero engine disks, (factor A), the friction parameter, (factor B), the heat transfer coefficient, (factor C) and emissive power, (factor D). An obvious response variable, Y, for this problem is the energy required to forge an aero engine disk measured in kJ.

It is known that these boundary conditions have an important affect upon the response, but the size of these main location effects are not known. As well as these main location effects, a number of first order interactions are expected to be important. For example, changes in friction properties, (factor B), will have a greater effect on energy use for forgings with a large surface area, factor (A), i.e. the AB interaction effect. Also friction forces generate heat and this heat may dissipate to the dies. Since the flow properties of the forging will be temperature dependant, the friction parameter and the heat transfer coefficient are likely to be strongly interacting, i.e. the BC interaction effect.

In order to capture all possible interactions, a full 2^4 factorial experiment is used. Because the forging of an aero engine disk is expensive this was not a physical experiment, (i.e. 16 separate disks were not tested at the forge). Instead all the results shown below were obtained from a computer simulation of the forging operation. They are the results from a virtual experiment. The high and low levels for each of the four factors are shown in Table 7.2. The level settings were chosen partly because all effects were expected to

CONTROLLING PROCESS VARIABILITY: DISPERSION EFFECTS IN LINEAR DESIGNS   149

be linear over these ranges. Because all the factors are design factors, the approach used in Section 7.2 is inappropriate.

The dispersion effects associated with the forging process can be quantified by progressing through the above-mentioned steps.

7.5.1 ESTIMATE A SIMPLIFIED MODEL OF THE PROCESS

Table 7.3 contains the 2^4 design of the disk forging operation, together with the responses recorded at each of the 16 unique test conditions. Also shown is the full set of factor effects calculated using the Yates procedure. Note that each contrast is divided through by $2^{4-1} = 8$ to obtain the factor effects. Each of these location effects shows how a factor determines the mean energy requirement. The full response surface model contains all the factor effects with no error term because the experiment is unreplicated, i.e. $N = 1$, and because all the higher order interaction terms are included.

$$Y_j = \beta_0 + \beta_A A_j + \beta_B B_j + \beta_C C_j + \beta_D D_j + \beta_{AB} A_j B_j + \beta_{AC} A_j C_j + \beta_{AD} A_j D_j + \beta_{BC} B_j C_j + \beta_{BD} B_j D_j$$
$$+ \beta_{CD} C_j D_j + \beta_{ABC} A_j B_j C_j + \beta_{ABD} A_j B_j D_j + \beta_{ACD} A_j C_j D_j + \beta_{BCD} B_j C_j D_j + \beta_{ABCD} A_j B_j C_j D_j \quad (7.17)$$

β_0 is the overall mean energy requirement, β_A half the A_{Loc} effect, β_B half the B_{Loc} effect, β_{AB} half the AB_{Loc} effect and so on. The simplified response surface model is the full model with all the unimportant factors left out. The unimportant factors are those shown to be irrelevant from a probability plot. Table 7.4 shows the usual calculations required for such a probability plot, whilst Figure 7.3 shows the actual probability plot based on this Table. Factors A, B and C show up as variables that have a strong impact on the mean energy requirement. As expected, there are also strong interactions between factors A and B, factors A and C and factors B and C. The ABC second order interaction effect also appears to be important.

The simplified model for this manufacturing process is therefore

$$Y_j = \beta_0 + \beta_A A_j + \beta_B B_j + \beta_C C_j + \beta_{AB} A_j B_j + \beta_{AC} A_j C_j + \beta_{BC} B_j C_j + \beta_{ABC} A_j B_j C_j + \varepsilon_j \quad (7.18)$$

where ε_j is the prediction error. This error term picks up all those effects whose mean value is zero as identified by the probability plot, e.g. factor D amongst others. This error term therefore has a mean of zero and, as explained above, its variability equals the variability present in the response variable, Y_j. Thus we can study dispersion effects through this error term. Unusually, a second order interaction effect has shown up as being important and so is included in the response surface model. Thus equation (7.18) is not strictly a first order model.

7.5.2 CALCULATE THE ERROR VARIABILITY AT EACH FACTOR LEVEL

There are 16 tests so that 16 prediction errors need to be evaluated from equation (7.18). To do this, the simplified model is rearranged to give

Table 7.3 The Yates procedure for calculating the location effects in the disk forging experiment.

Tests	Coded Test Conditions															Response
	A	B	AB	C	AC	BC	ABC	D	AD	BD	ABD	CD	ACD	BCD	ABCD	Y_j
(1)	-1	-1	1	-1	1	1	-1	-1	1	1	-1	1	-1	-1	1	3132.6
a	1	-1	-1	-1	-1	1	1	-1	-1	1	1	1	1	-1	-1	3051.2
b	-1	1	-1	-1	1	-1	1	-1	1	-1	1	1	-1	1	-1	4115.6
ab	1	1	1	-1	-1	-1	-1	-1	-1	-1	-1	1	1	1	1	3231.1
c	-1	-1	1	1	-1	-1	1	-1	1	1	-1	-1	1	1	-1	3714.4
ac	1	-1	-1	1	1	-1	-1	-1	-1	1	1	-1	-1	1	1	3135.5
bc	-1	1	-1	1	-1	1	-1	-1	1	-1	1	-1	1	-1	1	5380.9
abc	1	1	1	1	1	1	1	-1	-1	-1	-1	-1	-1	-1	-1	3173.5
d	-1	-1	1	-1	1	1	-1	1	-1	-1	1	-1	1	1	-1	3143.7
ad	1	-1	-1	-1	-1	1	1	1	1	-1	-1	-1	-1	1	1	3099.2
bd	-1	1	-1	-1	1	-1	1	1	-1	1	-1	-1	1	-1	1	4139.8
abd	1	1	1	-1	-1	-1	-1	1	1	1	1	-1	-1	-1	-1	3111.1
cd	-1	-1	1	1	-1	-1	1	1	-1	-1	1	1	-1	-1	1	3714
acd	1	-1	-1	1	1	-1	-1	1	1	-1	-1	1	1	-1	-1	3295.9
bcd	-1	1	-1	1	-1	1	-1	1	-1	1	-1	1	-1	1	-1	5460.9
abcd	1	1	1	1	1	1	1	1	1	1	1	1	1	1	1	3330.8

Table 7.3 Continued.

	Location Contrasts and Effects							
Tests	A_{Loc} Effect A	B_{Loc} Effect B	AB_{Loc} Effect AB	C_{Loc} Effect C	AC_{Loc} Effect AC	BC_{Loc} Effect BC	ABC_{Loc} Effect ABC
(1)	−3132.6	−3132.6	3132.6	−3132.6	3132.6	3132.6	−3132.6
a	3051.2	−3051.2	−3051.2	−3051.2	−3051.2	3051.2	3051.2
b	−4115.6	4115.6	−4115.6	−4115.6	4115.6	−4115.6	4115.6
ab	3231.1	3231.1	3231.1	−3231.1	−3231.1	−3231.1	−3231.1
c	−3714.4	−3714.4	3714.4	3714.4	−3714.4	−3714.4	3714.4
ac	3135.5	−3135.5	−3135.5	3135.5	3135.5	−3135.5	−3135.5
bc	−5380.9	5380.9	−5380.9	5380.9	−5380.9	5380.9	−5380.9
abc	3173.5	3173.5	3173.5	3173.5	3173.5	3173.5	3173.5
d	−3143.7	−3143.7	3143.7	−3143.7	3143.7	3143.7	−3143.7
ad	3099.2	−3099.2	−3099.2	−3099.2	−3099.2	3099.2	3099.2
bd	−4139.8	4139.8	−4139.8	−4139.8	4139.8	−4139.8	4139.8
abd	3111.1	3111.1	3111.1	−3111.1	−3111.1	−3111.1	−3111.1
cd	−3714	−3714	3714	3714	−3714	−3714	3714
acd	3295.9	−3295.9	−3295.9	3295.9	3295.9	−3295.9	−3295.9
bcd	−5460.9	5460.9	−5460.9	5460.9	−5460.9	5460.9	−5460.9
$abcd$	3330.8	3330.8	3330.8	3330.8	3330.8	3330.8	3330.8
Loc Contrast	−7373.6	5657.2	−5127.8	4181.6	−3295.4	1315.4	−1553.2
Loc Effect	−921.7	707.15	−640.975	522.7	−411.925	164.425	−194.15
β	−460.85	353.575	−320.4875	261.35	−205.9625	82.2125	−97.075

Table 7.3 Continued.

Location Contrasts and Effects

Tests	D_{Loc} Effect D	AD_{Loc} Effect AD	BD_{Loc} Effect BD	ABD_{Loc} Effect ABD	CD_{Loc} Effect CD	ACD_{Loc} Effect ACD	BCD_{Loc} Effect BCD	$ABCD_{Loc}$ Effect ABCD
(1)	−3132.6	3132.6	3132.6	−3132.6	3132.6	−3132.6	−3132.6	3132.6
a	−3051.2	−3051.2	3051.2	3051.2	3051.2	3051.2	−3051.2	−3051.2
b	−4115.6	4115.6	−4115.6	4115.6	4115.6	−4115.6	4115.6	−4115.6
ab	−3231.1	−3231.1	−3231.1	−3231.1	3231.1	3231.1	3231.1	3231.1
c	−3714.4	3714.4	3714.4	−3714.4	−3714.4	3714.4	3714.4	−3714.4
ac	−3135.5	−3135.5	3135.5	3135.5	−3135.5	−3135.5	3135.5	3135.5
bc	−5380.9	5380.9	−5380.9	5380.9	−5380.9	5380.9	−5380.9	5380.9
abc	−3173.5	−3173.5	−3173.5	−3173.5	−3173.5	−3173.5	−3173.5	−3173.5
d	3143.7	−3143.7	−3143.7	3143.7	−3143.7	3143.7	3143.7	−3143.7
ad	3099.2	3099.2	−3099.2	−3099.2	−3099.2	−3099.2	3099.2	3099.2
bd	4139.8	−4139.8	4139.8	−4139.8	−4139.8	4139.8	−4139.8	4139.8
abd	3111.1	3111.1	3111.1	3111.1	−3111.1	−3111.1	−3111.1	−3111.1
cd	3714	−3714	−3714	3714	3714	−3714	−3714	3714
acd	3295.9	3295.9	−3295.9	−3295.9	3295.9	3295.9	−3295.9	−3295.9
bcd	5460.9	−5460.9	5460.9	−5460.9	5460.9	−5460.9	5460.9	−5460.9
abcd	3330.8	3330.8	3330.8	3330.8	3330.8	3330.8	3330.8	3330.8
Loc Contrast	360.6	130.8	−77.6	−264.6	434	345.4	232.2	97.6
Loc Effect	45.075	16.35	−9.7	−33.075	54.25	43.175	29.025	12.2
β	22.5375	8.175	−4.85	−16.5375	27.125	21.5875	14.5125	6.1

Controlling Process Variability: Dispersion Effects in Linear Designs

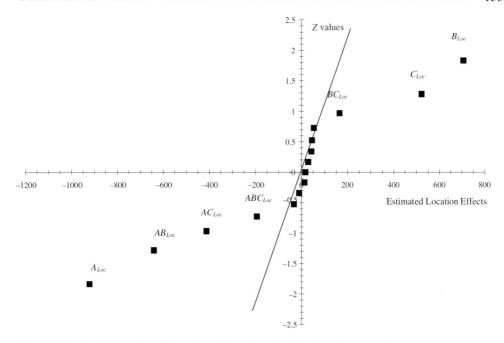

Fig. 7.3 Probability plot of location effects for the disk forging experiment.

Table 7.4 Tabulated probability plot for the disk forging operation.

Factor Effect$_l$	Estimated Effect$_l$	Rank Index, i	CP_l	Z_l
A_{Loc}	−921.70	1	3.33	−1.84
AB_{Loc}	−640.98	2	10	−1.28
AC_{Loc}	−411.93	3	16.67	−0.97
ABC_{Loc}	−194.15	4	23.33	−0.73
ABD_{Loc}	−33.08	5	30	−0.52
BD_{Loc}	−9.70	6	36.67	−0.34
$ABCD_{Loc}$	12.20	7	43.33	−0.17
AD_{Loc}	16.35	8	50	0
BCD_{Loc}	29.03	9	56.67	0.17
ACD_{Loc}	43.18	10	63.33	0.34
D_{Loc}	45.07	11	70	0.52
CD_{Loc}	54.25	12	76.67	0.73
BC_{Loc}	164.43	13	83.33	0.97
C_{Loc}	522.70	14	90	1.28
B_{Loc}	707.15	15	96.67	1.84

Table 7.5 Prediction errors from the simplified model.

Test		Prediction Error	Test		Prediction Error
(1)	$j = 1$	$\varepsilon_1 = -5.55$	d	$j = 9$	$\varepsilon_9 = 5.55$
a	$j = 2$	$\varepsilon_2 = -24.0$	ad	$j = 10$	$\varepsilon_{10} = 24.0$
b	$j = 3$	$\varepsilon_3 = -12.1$	bd	$j = 11$	$\varepsilon_{11} = 12.1$
ab	$j = 4$	$\varepsilon_4 = 60.0$	abd	$j = 12$	$\varepsilon_{12} = -60.0$
c	$j = 5$	$\varepsilon_5 = 0.2$	cd	$j = 13$	$\varepsilon_{13} = -0.2$
ac	$j = 6$	$\varepsilon_6 = -80.2$	acd	$j = 14$	$\varepsilon_{14} = 80.2$
bc	$j = 7$	$\varepsilon_7 = -40.0$	bcd	$j = 15$	$\varepsilon_{15} = 40.0$
abc	$j = 8$	$\varepsilon_8 = -78.65$	abcd	$j = 16$	$\varepsilon_{16} = 78.65$

$$\varepsilon_j = Y_j - \{\beta_0 + \beta_A A_j + \beta_B B_j + \beta_C C_j + \beta_{AB} A_j B_j + \beta_{AC} A_j C_j + \beta_{BC} B_j C_j + \beta_{ABC} A_j B_j C_j\}$$

Substituting in the values for β_0, β_A etc. from Table 7.3 yields.

$$\varepsilon_j = Y_j - \{3639.3875 - 460.85 A_j + 353.575 B_j + 261.35 C_j$$
$$- 320.49 A_j B_j - 205.965 A_j C_j + 82.215 B_j C_j - 97.075 A_j B_j C_j\}$$

For each test condition simply insert the actual responses and the positive and negative elements contained in the *A, B, C, AB, AC, BC* and *ABC* columns in the first part of Table 7.3 for factors *A, B, C, AB, AC, BC* and *ABC*. Thus for the first test condition (1), corresponding to $j = 1$,

$$\varepsilon_1 = 3132.6 - \{3639.3875 + 460.85 - 353.575 - 261.35 - 320.49 - 205.965 + 82.215 + 97.075\} = -5.55$$

For the second test condition, a, corresponding to $j = 2$,

$$\varepsilon_2 = 3051.2 - \{3639.3875 - 460.85 - 353.575 - 261.35 + 320.49 + 205.965 + 82.215 - 97.075\} = -24.0$$

This can easily be repeated for the remaining fourteen test conditions and the values for the remaining prediction errors are shown in Table 7.5.

To see if any factors affect the variability of energy usage in the forging process, each prediction error must be grouped according to whether that factor is at its high or low level. To illustrate, take factor *A*. Factor *A* is low for eight of the tests, (1), *b, c, bc, d, bd, cd* and *bcd*. In terms of *j*, factor *A* is low for tests $j = 1, j = 3, j = 5, j = 7, j = 9, j = 11, j = 13$, and $j = 15$. From Table 7.5 the errors associated with factor *A* being low are therefore

Controlling Process Variability: Dispersion Effects in Linear Designs

A− (Factor A low)

ε_1	= −5.55	ε_9	=	5.55
ε_3	= −12.1	ε_{11}	=	12.1
ε_5	= 0.2	ε_{13}	=	−0.2
ε_7	= −40.0	ε_{15}	=	40.0

Factor A is high for the other eight tests, a, ab, ac, abc, ad, abd, acd and $abcd$. In terms of j, factor A is high for tests $j = 2$, $j = 4$, $j = 6$, $j = 8$, $j = 10$, $j = 12$, $j = 14$ and $j = 16$. From Table 7.5, the errors associated with factor A being high are therefore

A+ (Factor A high)

ε_2	= −24.0	ε_{10}	=	24.0
ε_4	= 60.0	ε_{12}	=	−60.0
ε_6	= −80.2	ε_{14}	=	80.2
ε_8	= −78.65	ε_{16}	=	78.65

When factor A is low the variability in the prediction errors is given by equation (7.14b). Substituting in the numbers from the above table gives

$$S^2_{A-} = \frac{[-5.55-0]^2 + [-12.1-0]^2 + \ldots\ldots\ldots + [-0.2-0]^2 + [40.0-0]^2}{8-1} = 507.7864$$

When factor A is high the variability in the prediction errors is given by equation (7.14a). Substituting in the numbers from the above table gives

$$S^2_{A+} = \frac{[-24.0-0]^2 + [60.0-0]^2 + \ldots\ldots\ldots + [80.2-0]^2 + [78.65-0]^2}{8-1} = 4798.2464$$

It would appear from these two calculations that changes in the level for factor A results in a substantial change in the variability of the errors. Consequently, factor A appears to influence the variability present in energy usage. The dispersion effect for factor A is in fact given by equation (7.14c).

$$A_{Disp} = S^2_{A+} - S^2_{A-} = 4798.2464 - 507.7864 = 4290.46$$

Such a calculation can be made for the remaining three factors to give

$S^2_{B-} = 2011.11$ and $S^2_{B+} = 3294.92$

$S^2_{C-} = 1243.78$ and $S^2_{C+} = 4062.26$

$S^2_{D-} = 2653.07$ and $S^2_{D+} = 2653.07$

Changes in the level of factors B, C and D seem to be less important in determining variability. Just as an interaction between two factors can influence the mean of the process it is equally conceivable that an interaction between two or more factors can influence the variability in the process. For example, it may be the case that an interaction

156 OPTIMISATION OF MANUFACTURING PROCESSES

Table 7.6 The *AB* dispersion effect in the disk forging operation.

Test	*AB* Column for the 2^4 Design (Table 7.3)	Error when *AB* is Low	Error when *AB* is High
(1)	1		−5.55
a	−1	−24	
b	−1	−12.1	
ab	1		60
c	1		0.2
ac	−1	−80.2	
bc	−1	−40	
abc	1		−78.65
d	1		5.55
ad	−1	24	
bd	−1	12.1	
abd	1		−60
cd	1		−0.2
acd	−1	80.2	
bcd	−1	40	
abcd	1		78.65
		S^2_{AB-} = 2501.27	S^2_{AB+} = 2804.76

between surface area and friction influences the variability in energy requirements for disk forging. The above procedure can be used to test for such possibilities.

To see how this is done notice that equation (7.14b) works out the variance in the errors associated with the negative elements (*A* low), in column *A* of the 2^4 design shown in Table 7.3. Likewise, equation (7.14a) works out the variance in the errors associated with the positive elements (*A* high) in column *A* of Table 7.3. Thus the effect of the *AB* interaction on the process variability can be found by calculating the variance in the errors associated with the elements of the *AB* column of Table 7.3. The calculations are shown in Table 7.6. Two error variances are computed, one for all the errors associated with the negative elements of the *AB* column of Table 7.3 and one for all the errors associated with positive elements in column *AB*. There is very little difference between these two variances and so it would appear that the factors *A* and *B* do not interact to determine process variability.

Table 7.7 shows the workings behind these variance calculations for some of the other interaction dispersion effects possible in the 2^4 design. The word false is used when no error should be entered into the calculation.

Table 7.7 First and higher order dispersion effects for the disk forging operation.

	AB−	AB+	AC−	AC+	AD−	AD+	BD−	BD+	CD−	CD+
	FALSE	−5.55	FALSE	−5.55	FALSE	−5.55	FALSE	−5.55	FALSE	−5.55
	−24	FALSE	−24	FALSE	−24	FALSE	FALSE	−24	FALSE	−24
	−12.1	FALSE	FALSE	−12.1	FALSE	−12.1	−12.1	FALSE	FALSE	−12.1
	FALSE	60	60	FALSE	60	FALSE	60	FALSE	FALSE	60
	FALSE	0.2	0.2	FALSE	FALSE	0.2	FALSE	0.2	0.2	FALSE
	−80.2	FALSE	FALSE	−80.2	−80.2	FALSE	FALSE	−80.2	−80.2	FALSE
	−40	FALSE	−40	FALSE	FALSE	−40	−40	FALSE	−40	FALSE
	FALSE	−78.65	FALSE	−78.65	−78.65	FALSE	−78.65	FALSE	−78.65	FALSE
	FALSE	5.55	FALSE	5.55	5.55	FALSE	5.55	FALSE	5.55	FALSE
	24	FALSE	24	FALSE	FALSE	24	24	FALSE	24	FALSE
	12.1	FALSE	FALSE	12.1	12.1	FALSE	FALSE	12.1	12.1	FALSE
	FALSE	−60	−60	FALSE	FALSE	−60	FALSE	−60	−60	FALSE
	FALSE	−0.2	−0.2	FALSE	−0.2	FALSE	−0.2	FALSE	FALSE	−0.2
	80.2	FALSE	FALSE	80.2	FALSE	80.2	80.2	FALSE	FALSE	80.2
	40	FALSE	40	FALSE	40	FALSE	FALSE	40	FALSE	40
	FALSE	78.65	FALSE	78.65	FALSE	78.65	FALSE	78.65	FALSE	78.65
$S_l^2 =$	2501.27	2804.76	1650.3	3655.74	2576.64	2576.63	2626.13	2626.13	1812.14	1812.14
$F_l =$		0.11		0.79		-1.55×10^{-15}		2.66×10^{-15}		1.11×10^{-14}

Table 7.7 Continued.

	ABC−	ABC+	ABD−	ABD+	ACD−	ACD+	BCD−	BCD+	ABCD−	ABCD+
	−5.55	FALSE	−5.55	FALSE	−5.55	FALSE	−5.55	FALSE	FALSE	−5.55
	FALSE	−24	FALSE	−24	FALSE	−24	−24	FALSE	−24	FALSE
	FALSE	−12.1	FALSE	−12.1	−12.1	FALSE	FALSE	−12.1	−12.1	FALSE
	60	FALSE	60	FALSE	FALSE	60	FALSE	60	FALSE	60
	FALSE	0.2	0.2	FALSE	FALSE	0.2	FALSE	0.2	0.2	FALSE
	−80.2	FALSE	FALSE	−80.2	−80.2	FALSE	FALSE	−80.2	FALSE	−80.2
	−40	FALSE	FALSE	−40	FALSE	−40	−40	FALSE	FALSE	−40
	FALSE	−78.65	−78.65	FALSE	−78.65	FALSE	−78.65	FALSE	−78.65	FALSE
	5.55	FALSE	FALSE	5.55	FALSE	5.55	FALSE	5.55	5.55	FALSE
	FALSE	24	24	FALSE	24	FALSE	FALSE	24	FALSE	24
	FALSE	12.1	12.1	FALSE	FALSE	12.1	12.1	FALSE	FALSE	12.1
	−60	FALSE	FALSE	−60	−60	FALSE	−60	FALSE	−60	FALSE
	FALSE	−0.2	FALSE	−0.2	−0.2	FALSE	−0.2	FALSE	FALSE	−0.2
	80.2	FALSE	80.2	FALSE	FALSE	80.2	80.2	FALSE	80.2	FALSE
	40	FALSE	40	FALSE	40	FALSE	FALSE	40	40	FALSE
	FALSE	78.65	FALSE	78.65	FALSE	78.65	FALSE	78.65	FALSE	78.65
$S_l^2 =$	3332.24	1973.79	2340.46	2340.45	2120.42	2120.42	2412.32	2412.31	2610.49	2610.49
$F_l =$		−0.52		6.66×10^{-15}		1.62×10^{-14}		8.88×10^{-16}		5.55×10^{-15}

7.5.3 Testing the Importance of Dispersion Effects

The F_1 ratio given by equation (7.16) can be applied to factors A to D to give

$$F_A = \ln\left[\frac{4798.246}{507.7864}\right] = 2.24 \quad F_B = \ln\left[\frac{3294.924}{2011.109}\right] = 0.49$$

$$F_C = \ln\left[\frac{4062.258}{1243.775}\right] = 1.18 \quad F_D = \ln\left[\frac{2653.07}{2653.07}\right] = 0.00$$

Only factor A would appear to have an important dispersion effect. Table 7.7 shows the F ratios associated with the first, second and third order interactions. It follows from this analysis that only changes in factor A will alter the variability in energy usage and there are no interaction dispersion effects. Combining this with the information in Table 7.3 it would appear that factor A affects both the mean energy value and the variability around this mean energy requirement. More information on how to optimise this process (i.e. minimise the variability in energy use around a mean usage that is on target) will be follow in Chapter 8.

7.6 A GENERALISED LINEAR MODEL

Use of the generalised linear model requires replication so that more than one response must be available at each of the test conditions within a full or fractional factorial. If a manufacturing process has no noise factors then the variability observed in the replicated tests will be a common cause variation. Table 6.1 of Section 6.1 contains an example of a 2^3 replicated design where the variation in response at each test was solely common cause. When process variation is due primarily to noise factors, the replication must reflect the variation in the noise factors. The generalised linear model is easiest to apply when a factorial design is rearranged into an **inner** and an **outer** array.

7.6.1 Inner and Outer Arrays

Taguchi[1] has recommended assigning design factors to an inner array and noise factors to an outer array, with each array being a separate orthogonal design, e.g. a separate 2^{k-p} design. It is this outer array that induces the variability in responses at the laboratory that is supposed to mirror the inherent variability present in the manufacturing process located at the factory. Remember that both noise and design factors can be set precisely within the artificial environment of the laboratory.

The structure of the inner/outer array design is best understood by taking a very simple example. Suppose there is a manufacturing process with four important factors, two of which are design factors (factors A and B) and two of which are noise factors (factors C and D). Each factor is considered to have a linear effect on the response variable of interest so that each factor can be set at just two levels. The inner and outer array can then be made a simple 2^2 factorial design. This implies carrying out 16 tests in the laboratory. That is, four tests at each design and noise factor level combination respectively. The result is 16 response recordings as shown in Table 7.8.

In Table 7.8 each response is double subscripted. The first subscript refers to the inner array test condition and the second subscript refers to the outer array test condition. Thus $Y_{b,c}$ is the recorded response when factor A is low, B is high, C is high and D is low. There are 16 recorded responses made in total. The outer array simply induces the variability that would be observed in the responses if inner array tests were duplicated four times on the real process at the factory. This can also be shown visually as in Figure 7.4. Here a 2^2 design for the noise factors is superimposed upon a 2^2 design for the design factors.

There are two ways to visualise and analysis this experiment. The first way to visualise the above design is as an unreplicated design where one response is recorded for each of the 16 test conditions. The noise factors are then treated in the same way as design factors. Variability is minimised by identifying interactions between the noise and design factors in the way described in Sections 7.1 and 7.2 above. This can only be done if the noise factors are analysed as an integral part of the design.

Secondly, the experiment can be visualised as an $N = 4$ replicate of a 2^2 design. That is, four recorded responses are made under the same set of design factor test conditions. The noise factors can be seen as a means to induce replication and so are ignored from an analysis perspective. This leads us into the techniques associated with a generalised linear model where **PerMIA** and $(S-N)_T$ statistics are computed using the four replicates at each of the four inner array test conditions shown in Table 7.8. These statistics are designed to measure process variability at each design factor test condition. It is this second approach that is now discussed in more detail.

7.6.2 SIMPLE SUMMARY STATISTICS

To see how a **generalised linear model** can be created consider again the 2^2 design that has been replicated $N = 4$ times. This experiment is shown in Figure 7.5.

In Figure 7.5, four responses have been recorded at each of the four test conditions. The first subscript on Y refers to the replicate (1 to 4) and the second subscript to the test condition. These can be converted into two **summary statistics** for each test condition. The first summary statistic is the **mean** response at each test condition.

$$\overline{Y}_{(1)} = \frac{Y_{1(1)} + Y_{2(1)} + Y_{3(1)} + Y_{4(1)}}{4}$$

$$\overline{Y}_a = \frac{Y_{1a} + Y_{2a} + Y_{3a} + Y_{4a}}{4}$$

CONTROLLING PROCESS VARIABILITY: DISPERSION EFFECTS IN LINEAR DESIGNS

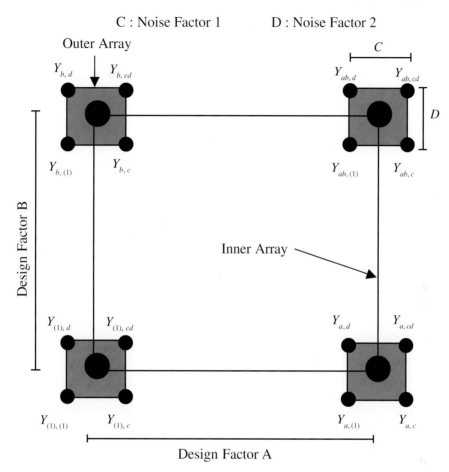

Fig. 7.4 The Taguchi inner/outer array structure.

Table 7.8 An inner/outer array representation of a hypothetical experiment.

Inner Array (2^2)			Test	Outer Array (2^2)					
				(1)	c	d	cd		
			C	−1	1	−1	1	**Summary Statistics**	
Test	**A**	**B**	D	−1	−1	1	1	PerMIA	$(S\text{-}N)_T$
				Responses					
(1)	−1	−1		$Y_{(1),(1)}$	$Y_{(1),c}$	$Y_{(1),d}$	$Y_{(1),cd}$	...	...
a	1	−1		$Y_{a,(1)}$	$Y_{a,c}$	$Y_{a,d}$	$Y_{a,cd}$	...	...
b	−1	1		$Y_{b,(1)}$	$Y_{b,c}$	$Y_{b,d}$	$Y_{b,cd}$	...	...
ab	1	1		$Y_{ab,(1)}$	$Y_{ab,c}$	$Y_{ab,d}$	$Y_{ab,cd}$	...	...

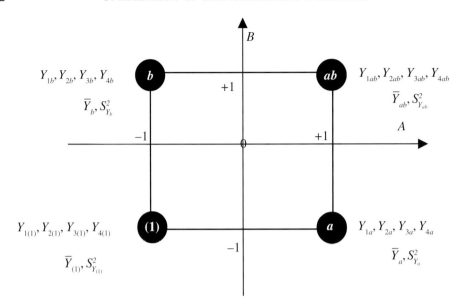

Fig. 7.5 A replicated 2^2 design.

$$\overline{Y}_b = \frac{Y_{1b} + Y_{2b} + Y_{3b} + Y_{4b}}{4}$$

and

$$\overline{Y}_{ab} = \frac{Y_{1ab} + Y_{2ab} + Y_{3ab} + Y_{4ab}}{4}$$

$\overline{Y}_{(1)}$ is the mean response from the four observations made at test condition (1) and $\overline{Y}_{ab}$ the mean response from the four observations made at test condition ab.

More important for controlling process variation is the variance in the response values at each test condition. Remember from Section 7.1 that this variance is simply the square of the standard deviation of the response variable. As such it is obtained by squaring the deviation of each response from the mean response at a particular test condition, adding up all these squares and finally averaging the resulting sum (noting that for statistical reasons $N-1$ and not N is used to average). So

$$S^2_{Y_{(1)}} = \frac{\left[(Y_{1(1)} - \overline{Y}_{(1)})^2 + (Y_{2(1)} - \overline{Y}_{(1)})^2 + (Y_{3(1)} - \overline{Y}_{(1)})^2 + (Y_{4(1)} - \overline{Y}_{(1)})^2\right]}{4-1}$$

$$S^2_{Y_a} = \frac{\left[(Y_{1a} - \overline{Y}_a)^2 + (Y_{2a} - \overline{Y}_a)^2 + (Y_{3a} - \overline{Y}_a)^2 + (Y_{4a} - \overline{Y}_a)^2\right]}{4-1}$$

$$S^2_{Y_b} = \frac{\left[(Y_{1b}-\overline{Y}_b)^2 + (Y_{2b}-\overline{Y}_b)^2 + (Y_{3b}-\overline{Y}_b)^2 + (Y_{4b}-\overline{Y}_b)^2\right]}{4-1}$$

and

$$S^2_{Y_{ab}} = \frac{\left[(Y_{1ab}-\overline{Y}_{ab})^2 + (Y_{2ab}-\overline{Y}_{ab})^2 + (Y_{3ab}-\overline{Y}_{ab})^2 + (Y_{4ab}-\overline{Y}_{ab})^2\right]}{4-1}$$

$S^2_{Y_{(1)}}$ is the variance of the four observations made at test condition (1) and $S^2_{Y_{ab}}$ is the variance of the four observations made at test condition ab.

The location effect of factor A can now be calculated in the usual way as the difference between the mean response when factor A is high compared with when it is low

$$A_{Loc} = \left[\frac{\overline{Y}_{ab} + \overline{Y}_a}{2}\right] - \left[\frac{\overline{Y}_b + \overline{Y}_{(1)}}{2}\right]$$

Remember from Section 7.1 above that a dispersion effect is present whenever changes in the level of a factor results in a change in the variability rather then the mean of the response. To be able to minimise variability, experiments will be required to help quantify these dispersion effects. The dispersion effect for factor A can be defined in more than one-way. One obvious possibility is to measure it as the difference between the average variance when factor A is high compared with when it is low.

$$A_{Disp} = \left[\frac{S^2_{Y_{ab}} + S^2_{Y_a}}{2}\right] - \left[\frac{S^2_{Y_b} + S^2_{Y_{(1)}}}{2}\right]$$

Defined in this way, the A_{Disp} effect measures the change in the average variability in response that results from a two-unit change in the level for factor A. The dispersion effect for factor B can also be defined as the difference between the average variance when factor B is high compared with when it is low.

$$B_{Disp} = \left[\frac{S^2_{Y_{ab}} + S^2_{Y_b}}{2}\right] - \left[\frac{S^2_{Y_a} + S^2_{Y_{(1)}}}{2}\right]$$

Just as there are interaction location effects there are also interaction dispersion effects. The interaction dispersion effect between factors A and B measures the extent to which changes in variability, as a consequence of a change in the level for factor A, are affected by the level for factor B,

$$AB_{Disp} = \frac{\left[S^2_{Y_{ab}} - S^2_{Y_b}\right] - \left[S^2_{Y_a} - S^2_{Y_{(1)}}\right]}{2}$$

Using $\overline{Y}_j$ to represent the mean response at the jth test condition and $S^2_{Y_j}$ to represent the variance in responses at the jth test condition, it is possible to generalise these definitions so that location and dispersion effects can be calculated from any replicated

2^{k-p} design. This can be done very simply using the Yates procedure with $\bar{Y}_j$ placed in the response column for the estimation of location effects and $S^2_{Y_j}$ placed in the response column for estimation of dispersion effects. In this generalisation,

$$\bar{Y}_j = \frac{\sum_{i=1}^{N} Y_{ji}}{N} \tag{7.19a}$$

and

$$S^2_{Y_j} = \frac{\sum_{i=1}^{N} [Y_{ji} - \bar{Y}_j]^2}{N-1} \tag{7.19b}$$

Remember that there are $j = 1, M$ test condition with $M = 2^{k-p}$.

7.6.3 THE TENDENCY FOR PROCESS MEAN AND VARIABILITY TO MOVE TOGETHER

Whilst this is a perfectly legitimate approach to identifying location and dispersion effects it does not always help achieve an optimum process. This is because for many manufacturing processes, the variance (as defined in Section 7.6.2) in response at each test condition will be dependant in some way upon the mean response at each test condition. To understand the problems that this can cause, suppose that it has been found from a 2^2 design that factor A has a positive influence on the mean response through the A_{Loc} effect but factor B has a negative influence on the variance in response through the B_{Disp} effect. There are no interaction effects. Then factor B could be increased in an attempt to minimise the variability in the response. Then having minimised variability, factor A could be increased until the mean response reached a target specification (if the current mean is below target).

But if the mean and variance are positively related a problem is immediately apparent. Raising factor A will indeed raise the mean to target, but this increase in the mean will also raise process variability so that all the efforts made in the initial change would be lost by this subsequent action. Variability will no longer be minimised. A solution to this problem can be found in the explanation as to why the mean and variance are related in many fields of engineering. There are two basic reasons for this dependency.

i. It is typical for the variance in recorded responses to increase as the magnitude of those responses increases. This occurs because measurement errors are typically a constant percentage of the size of the response being recorded. For many measuring instruments the error is a percentage of the scale reading.

ii. If the quality characteristic being measured follows a normal distribution then the mean and variance of that characteristic will be independent. Unfortunately, many of the quality characteristics that engineers wish to control do not have such a distribution. Measures of fatigue, corrosion and creep resistance, for example, all

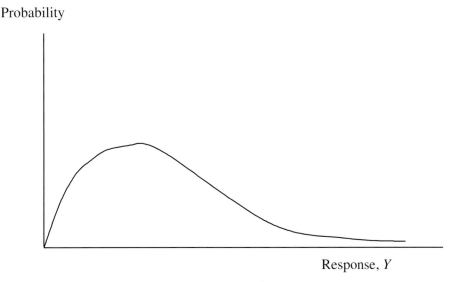

Fig. 7.6 An example of a long tailed distribution.

tend to come from skewed or long tailed distributions, such as that shown in Figure 7.6. It is this skew that induces a tendency for the mean and variance of the process to move together.

This second explanation is believed to be the major source of the problem and suggests a simple remedy. Namely, to find a transformation of the quality characteristic that will ensure that the transformed data is normally distributed. Let Y^* be such a transformation of Y. Then the process mean and variance will be independent so that variability can be minimised independent of bringing the mean on target.

Consider the specific case of a quality characteristic Y being log normally distributed. This is a particular type of skewed distribution and takes its name from the fact that $Y^* = \ln(Y)$ will follow a symmetric normal distribution. In such a situation it can be shown that the variance in Y is related to the mean value for Y in the following way.

$$S_Y^2 = \bar{Y}^2 [\exp(S_{Y^*}^2) - 1] \tag{7.20a}$$

In equation (7.20a), S_Y^2 is the variance in the untransformed response and $S_{Y^*}^2$ is the variance of the log transformed data $Y^* = \ln(Y)$. It is clear from this equation that any increase in the mean response will result in an increased variability in response. That is, the overall variability in the response data, S_Y^2, has two components. One component is dependent on the mean whilst the other is dependent on the variance in the logged data. The following two transformations give the component that is independent of the mean

$$Y^T = \frac{S_Y^2}{\overline{Y}^2} = [\exp(S_{Y*}^2) - 1] \qquad (7.20\text{b})$$

and

$$\ln(1 + Y^T) = \ln\left[1 + \frac{S_Y^2}{\overline{Y}^2}\right] = S_{Y*}^2 \qquad (7.20\text{c})$$

In equation (7.20b) the transformed response is the ratio of the variance to squared mean (in the untransformed data), whilst in equation (7.20c) it is the variance present in the natural log of Y, Y^*. The two are almost equivalent.

The important point is that the transformations Y^T and $\ln(1 + Y^T)$ are both measures of that component of overall variability that is independent of the mean. Putting the mean on target will nearly always have an affect on variability because the mean and the variance are dependant upon each other (equation (7.20a)). Little can be done about this. The aim should therefore be to minimise that component of overall variability that is independent of the mean. Once this is done, the mean can then be put on target using the estimated location effects Whilst this latter action will alter the overall variability through the dependency of the variance on the mean, the variability associated with the required mean (and target) will be minimised. Variability should therefore always be measured as that component of variability that is independent of the mean. Factors which control this measure of variability are called **control factors**. **Signal factors** are those factors that influence the mean response but not that component of variability that is independent of the mean. These signal factors can then be used to push the mean on target without altering that component of variability that is independent of the mean.

7.6.4 PerMIA Summary Statistics

Taguchi has given the name **signal to noise**, or S-N, ratio to describe the following transformation.

$$(S - N)_T = \ln\left[\frac{S_Y^2}{\overline{Y}^2}\right] \qquad (7.21)$$

He recommends that this $(S-N)_T$ ratio be used whenever the objective is to minimise variability about a required mean value. This is not completely sound advice because such a transformation will not always ensure independence between the mean and variance of the quality characteristic. Notice that equation (7.21) is similar to equation (7.20b) and so will only ensure independence if the quality characteristic is log normally distributed. But other distributions are possible and this has resulted in a variety of suggested transformations that have become known as **PerMIA** summary statistics – performance measures independent of adjustment. The name comes from the requirement that the adjustment of the mean of the process to target must not affect the variability performance of the process.

Whenever a transformation measures the variability in the process that is independent of the mean of the process, that transformation is said to be **variance stabilising**. All PerMIA summary statistics are variance stabilising. To further illustrate the restrictive nature of Taguchi's signal to noise ratio consider the following model proposed by Engel.[11] Engel has proposed that the observed variability in the responses recorded at a specific test condition j can be attributable to two causes. The first is due to the mean response value at that test condition, with the general expectation being that the variability observed at test condition j will be higher the greater the mean response recorded at that test condition. That is, $f(\overline{Y}_j)$ in equation (7.22a) can be a linear or non-linear function but its slope should always be positive. The second part, ϕ_j, is attributable to the levels of some of the factors at that test condition. This can be written as

$$S^2_{Y_j} = f(\overline{Y}_j)\phi_j \tag{7.22a}$$

where ϕ_j is that part of the variability in response that is independent of the mean. It just depends on the levels of some of the factors at test condition j. This is in fact the variance stabilising transformation. It is that part of the total variability in response that can be influenced by changing the level of the control factors themselves rather than as an indirect consequence of changing the mean through such factors.

A design factor has a **pure dispersion effect** if it influences ϕ_j. It can be defined in a similar way to a location effect. For example, consider factor A. The pure dispersion effect for factor A, A_{Disp}, is measured as the difference between the average value for ϕ when factor A is high and the average value for ϕ when factor A is low.

If $\sum_{j^{A^+}}^{\frac{M}{2}}$ reads the sum over all the $j = 1, M$ tests corresponding to factor A being at its high level and $\sum_{j^{A^-}}^{\frac{M}{2}}$ reads the sum over all the $j = 1, M$ tests corresponding to factor A being at its low level then,

$$A_{Disp} = \frac{\sum_{j^{A^+}}^{M}\phi_j}{\frac{M}{2}} - \frac{\sum_{j^{A^-}}^{M}\phi_j}{\frac{M}{2}} \tag{7.22b}$$

This is true because for any factor in a 2^{k-p} test, half of the tests will have that factor high and half will have that factor low. As usual $M = 2^{k-p}$. This formula can be applied to any factor but values for ϕ_j are needed to make it operational. This in turn requires knowledge of the exact functional form of $f(\overline{Y}_j)$. The **general linear model** is based on the **power of the mean** model.

$$f(\overline{Y}_j) = \overline{Y}_j^\theta \qquad (7.22c)$$

If this is the correct specification, then equation (7.22a) can be written as

$$S_{Y_j}^2 = \overline{Y}_j^\theta \phi_j \qquad (7.22d)$$

so that the variance stabilising transformation is

$$\phi_j = \frac{S_{Y_j}^2}{\overline{Y}_j^\theta} \qquad (7.22e)$$

Notice that if $\theta = 2$ this resembles the $(S\text{-}N)_T$ ratio in equation 7.21. Hence $\theta = 2$ suggests that the response Y is log normally distributed, which in turn suggests that $\phi_j = \exp(S^2_{y^*}) - 1$. (Compare equation (7.22e) with (7.20b) when $\theta = 2$). But θ can equally well take on other values. In this book, PerMIA summary statistics will be limited to the class of transformations given by equation (7.22e). Of course, other functional forms for $f(\overline{Y}_j)$ are feasible and there is no guarantee that the power of the means model is flexible enough to ensure variance stabilisation in all circumstances. This shortcoming needs to be borne in mind.

The following generalised linear model has stemmed from some recent work by Nelder and Pregibon,[18] Logothetis[19] and Engel.[11] The model is applied in various steps.

7.6.5 STEP 1 IDENTIFY ALL THE CONTROL FACTORS

Construct the PerMIA statistic using the power of mean model. This is done by linearising equation (7.22d) as follows:

$$\ln(S_{Y_j}^2) = \ln(\phi) + \theta \ln(\overline{Y}_j) + \varepsilon_j \qquad (7.23a)$$

An error term ε_j has been added to pick up any other factors that may influence variability and the least squares formula can be applied to minimise the sum square of this term in the estimation of θ. This can be done manually by carrying out the matrix multiplications and inversion suggested by equation (5.13). Alternatively, the regression command LINEST in Excel[20] does this automatically. As only two parameters together with their standard deviations need estimating, a block of two rows by two columns needs to be highlighted and the term LINEST(R1, R2, TRUE, TRUE) typed into the formula box at the top of the computer screen. If the first true term is replaced by the word FALSE then $\ln(\phi)$ is constrained to zero. If the second true term is replaced by the word FALSE then no standard deviation is calculated for $\ln(\phi)$ and θ.

R1 is the range of cells down a single column containing the values for the left hand side of equation (7.23a), i.e. $\ln(S_{Y_j}^2)$ and R2 is the range of cells down a single column containing values for the factors on the right hand side of equation (7.23a), i.e. $\overline{Y}_j$. Hitting the control shift and return keys simultaneously instructs Excel to return four numbers in the highlighted block. Excel returns the results in the reverse order to equation (7.23a). The first number is therefore an estimate of θ with its standard deviation shown below it, whilst the second number is the estimate for ln (φ) with its standard deviation again shown below.

Call these least squares estimates $\hat{\phi}^{(1)}$, $\hat{\theta}^{(1)}$. Notice that in obtaining $\hat{\theta}^{(1)}$, φ has replaced ϕ_j in equation (7.22b) so that it is initially assumed that any variability in Y comes solely from changes in the mean. φ does not vary with test condition j. If this is not true $\hat{\theta}^{(1)}$ will not be a completely reliable estimate of θ. A more reliable estimate is made in step 2 below. All that matters in this step is that the estimate $\hat{\theta}^{(1)}$ is good enough to achieve a reasonably reliable PerMIA statistic. This statistic can be obtained by inserting $\hat{\theta}^{(1)}$ for θ into equation (7.22e) for the variance stabilising transformation.

$$\phi_j = \frac{S^2_{Y_j}}{\overline{Y}_j^{\hat{\theta}^{(1)}}} \qquad (7.23b)$$

This PerMIA can be used to identify all the control factors by assuming, as a first guess, that all design factors are potential control factors so that all factors together with all their interactions are potential determinants of variability independent of the mean. Thus the Yates procedure needs to be implemented with either ϕ_j or $\ln(\phi_j)$ placed in the response column. It is more usual to use $\ln(\phi_j)$ to help remove any non-linearity's that might be present. The Yates procedure will give all the dispersion contrast. Then for a 2^k design each dispersion contrast needs to be divided through by 2^{k-1} and for a 2^{k-p} design by $2^{k-p}/2$.

$$Disp^{(1)} = \frac{Disp\,contrast}{2^{k-1}} \qquad for\ 2^k\ design \qquad (7.23c)$$

$$Disp^{(1)} = \frac{Disp\,contrast}{\frac{2^{k-p}}{2}} \qquad for\ 2^{k-p}\ design$$

The superscript (1) is used to indicate that these are provisional estimates of the pure dispersion effects. More reliable estimates are obtained in step 2 below.

For example, in the 2^2 design, the Yates procedure will give estimates for $A^{(1)}_{Disp}$, $B^{(1)}_{Disp}$, $AB^{(1)}_{Disp}$ and the first order response surface model for the variance independent of the mean is,

$$\ln(\phi_j) = \gamma_o + \gamma_A A_j + \gamma_B B_j + \gamma_{AB} A_j B_j$$

where $2\gamma_A = A^{(1)}_{Disp}$, $2\gamma_B = B^{(1)}_{Disp}$, $2\gamma_{AB} = AB^{(1)}_{Disp}$. Thus as an alternative, least squares could be used to estimate values for γ_o to γ_{AB} using equation (5.13). Let these estimates be $\hat{\gamma}^{(1)}_o$, $\hat{\gamma}^{(1)}_A$, $\hat{\gamma}^{(1)}_B$ and $\hat{\gamma}^{(1)}_{AB}$. Again the (1) superscript is used to indicate they are provisional estimates.

7.6.6 Step 2. Obtain Reliable Estimates of the Dispersion Effects for All the Control Factors

So far only provisional estimates of the pure dispersion effects have been looked at and they are not very reliable estimates because they are derived from an estimate of θ that assumed there were no pure dispersion effects. That is, ϕ_j was made a constant ϕ in equation (7.23a). To rectify this problem a probability plot of the provisional dispersion effects is constructed and only those pure dispersion effects shown to be important are considered in this second step. In particular, the pure dispersion effects identified from the probability plot are allowed to enter equation (7.22d).

$$\ln(\phi_j) = f(\text{important dispersion effects from probability plot}) \qquad (7.23d)$$

Thus equation (7.23a) becomes

$$\ln(S_{Y_j}^2) = \ln(\phi_j) + \theta \ln(\overline{Y}_j) + \varepsilon_j$$

and if equation (7.23d) is inserted into this equation the following model is obtained.

$$\ln(S_{Y_j}^2) = f(\text{important dispersion effects from probability plot}) + \theta \ln(\overline{Y}_j) + \varepsilon_j \qquad (7.23e)$$

As an example again consider a 2^2 design where only factor B was shown on a probability plot to affect variability independent of mean, i.e. affect ϕ_j. Then equation (7.23e) would look like,

$$\ln(S_{Y_j}^2) = \gamma_0 + \gamma_B B_j + \theta \ln(\overline{Y}_j) + \varepsilon_j$$

Least squares can be used to estimate the parameters of equation (7.23e). The superscript (2) will be used to indicate that these are the second stage estimates. These are now extremely reliable estimates of θ and γ. So in this hypothetical 2^2 design, estimation would give $\gamma_0^{(2)}$, $\lambda_B^{(2)}$ and $\hat{\theta}^{(2)}$ and a reliable estimate of the pure dispersion effect associated with the level for factor B would be given by,

$$B_{Disp}^{(2)} = 2\lambda_B^{(2)}$$

Having identified which factors have pure dispersion effects they can then be used to minimise that component of variability that is independent of the mean. So in the example above, if the $B_{Disp}^{(2)}$ effect is positive, the variability in Y that is independent of the mean can be minimised by lowering the level of B and vice versa if the $B_{Disp}^{(2)}$ effect is negative.

It is important that the reader is aware of some of the limitations of the PerMIA approach to minimising process variability. First, the approach assumes either that Y^* is log normally distributed or that the power in means model is an adequate representation of the way in which variability depends on the mean response. There is no guarantee that this will be the case. Secondly, the approach relies on there being signal factors. It may turn out to be the case that all of the factors have pure dispersion effects so that it becomes impossible to adjust the mean to target without affecting that component of variability that is independent of the mean.

Table 7.9 The copper compacts experiment.

Factor	Coded Factor	High Level	Low Level
Powder Size	A	# –52 + 100	# –200 + 300
Preform Density	B	8.03 mg m^{-3}	7.3 mg m^{-3}
Sintering Treatment	C	18 hours	0.25 hours
Forging Rate	D	2.12 mm s^{-1}	0.423 mm s^{-1}
Preheat Temperature	E	950°C	750°C
Die Temperature	F	650°C	450°C
Forging Load	G	0.0035	0.003

7.7 THE COPPER COMPACT EXPERIMENT

The above-generalised linear model can be used with any 2^{k-p} design. To demonstrate how this is done consider again the copper compact experiment of Evans and McColvin[8] given in Section 2.3. One property of interest in this type of experiment is the pore size in the centre region of the compacts, (measured in %>16.2 µ), and so this will be considered as an initial response variable. As discussed in Section 2.3 the process variables considered to be of most importance were powder size (factor A), preform density (factor B), sintering treatment (factor C), forging rate (factor D), preheat temperature (factor E), die temperature (factor F) and forging load (factor G). All these factors are design factors and so any process variability will be common cause in nature. The objective of the experiment is to ensure the minimum amount of variability around a required mean pore size.

Each test is a time consuming process so that a 2^{7-3}_{IV} fractional design was selected for the experiment. The high and low levels chosen for each factor are shown in Table 7.9.

The design generators chosen were $E = ACD$, $F = ABD$ and $G = ABC$. Notice that Evans and McColvin[8] have used none of the optimal design generators shown in Table 4.4. The resulting design is shown in Table 7.10, together with the results obtained from each of the 16 tests. Each test was replicated once so that $N = 2$. That is, a total of thirty two tests were carried out. The design was derived by following the steps outlined in Section 4.3.

As a preliminary step it is necessary to use the above data to find values for $S^2_{Y_j}$ for the $j = 1$, 16 test conditions shown in Table 7.10. Notice that for each test $N = 2$. Applying equations (7.19b) to the $j = 1$ test condition gives,

$$S^2_{Y1} = \frac{[4.4-5.55]^2 + [6.7-5.55]^2}{2-1} = 1.3225 + 1.3225 = 2.645$$

Table 7.10 A 2^{7-3}_{IV} design for the copper compacts experiment.

Coded Test Conditions							Test	Response, %	
A	B	C	D	E = ACD	F = ABD	G = ABC		Replicate 1	Replicate 2
−1	−1	−1	−1	−1	−1	−1	(1)	4.4	6.7
1	−1	−1	−1	1	1	1	aefg	6	4.3
−1	1	−1	−1	−1	1	1	bfg	8.3	8.1
1	1	−1	−1	1	−1	−1	abe	12.2	15
−1	−1	1	−1	1	−1	1	ceg	7.7	19.3
1	−1	1	−1	−1	1	−1	acf	2.6	14.1
−1	1	1	−1	1	1	−1	bcef	14.2	1
1	1	1	−1	−1	−1	1	abcg	6.4	32.5
−1	−1	−1	1	1	1	−1	def	2.3	1.9
1	−1	−1	1	−1	−1	1	adg	19.3	17.6
−1	1	−1	1	1	−1	1	bdeg	1	2.2
1	1	−1	1	−1	1	−1	abdf	8.8	9
−1	−1	1	1	−1	1	1	cdfg	2	3.5
1	−1	1	1	1	−1	−1	acde	16.2	37.3
−1	1	1	1	−1	−1	−1	bcd	32.4	11.3
1	1	1	1	1	1	1	abcdefg	5.7	13.6

Performing the same calculations for the remaining fifteen test conditions yields the results show in the fourth column of Table 7.11.

Having carried out the experiment and all provisional calculations the pure dispersion effects can be estimated by progressing through steps one and two of Sections 7.6.5 and 7.6.6.

7.7.1 STEP 1. IDENTIFY ALL THE CONTROL FACTORS FOR MAKING COPPER COMPACTS

First, the natural log of the variance values shown in the fourth column of Table 7.11 are derived. For the $j = 1$ test condition.

$$\ln(S^2_{Y_1}) = \ln(2.645) = 0.9727$$

Performing the same calculations for the remaining fifteen test conditions yields the results shown in the fifth column of Table 7.11. Next equation (7.23a) needs to be

CONTROLLING PROCESS VARIABILITY: DISPERSION EFFECTS IN LINEAR DESIGNS 173

Table 7.11 Mean, variance and stabilised variance in response at each test condition.

Test	Test Conditions	Mean, $\bar{Y}_j$	Variance, $S^2_{Y_j}$	ln Variance, $\ln(S^2_{Y_j})$	$\ln(\phi_j)$
$j=1$	(1)	5.55	2.645	0.9727	−2.8114
$j=2$	aefg	5.15	1.445	0.3681	−3.2508
$j=3$	bfg	8.2	0.02	−3.912	−8.558
$j=4$	abe	13.6	3.92	1.3661	−4.3969
$j=5$	ceg	13.5	67.28	4.2089	−1.5379
$j=6$	acf	8.35	66.125	4.1915	−0.4944
$j=7$	bef	7.6	87.12	4.4673	−0.0109
$j=8$	abcg	19.45	340.605	5.8307	−0.7223
$j=9$	def	2.1	0.08	−2.5257	−4.1639
$j=10$	adg	18.45	1.445	0.3681	−6.0684
$j=11$	bdeg	1.6	0.72	−0.3285	−1.3663
$j=12$	abdf	8.9	0.02	−3.912	−8.7388
$j=13$	cdfg	2.75	1.125	0.1177	−2.1158
$j=14$	acde	26.75	222.605	5.4054	−1.8513
$j=15$	bcd	21.85	222.605	5.4054	−1.4045
$j=16$	abcdefg	9.65	31.205	3.4406	−1.5649

estimated using the least squares procedure. As two parameters with standard deviations need to be estimated in Excel, highlight a block of two rows by two columns and type in the formula = LINEST (R1, R2, TRUE, TRUE) where R1 is the range of cells down a single column containing the values for $\ln(S^2_{Y_j})$ and R2 is the range of cells down a column containing values for $\ln(\bar{Y}_j)$. This is illustrated in the Excel image in Figure 7.7. This illustrative spreadsheet contains the two columns of data making up the least squares regression and the highlighted output range containing the results. The formula box towards the top of the sheet shows the formula lying behind the values shown in the highlighted cells. Simply hitting the control shift and return key simultaneously will return the results.

The result is

$$\ln(\hat{\phi}^{(1)}) = -3.066 \quad \text{and} \quad \hat{\theta}^{(1)} = 2.208$$

The standard deviation associated with $\hat{\theta}^{(1)}$ is 0.863. To test the hypothesis that $\theta = 2$ simply construct the usual t statistic of $t = (2.208 - 2)/0.863 = 0.24$. This is not significant

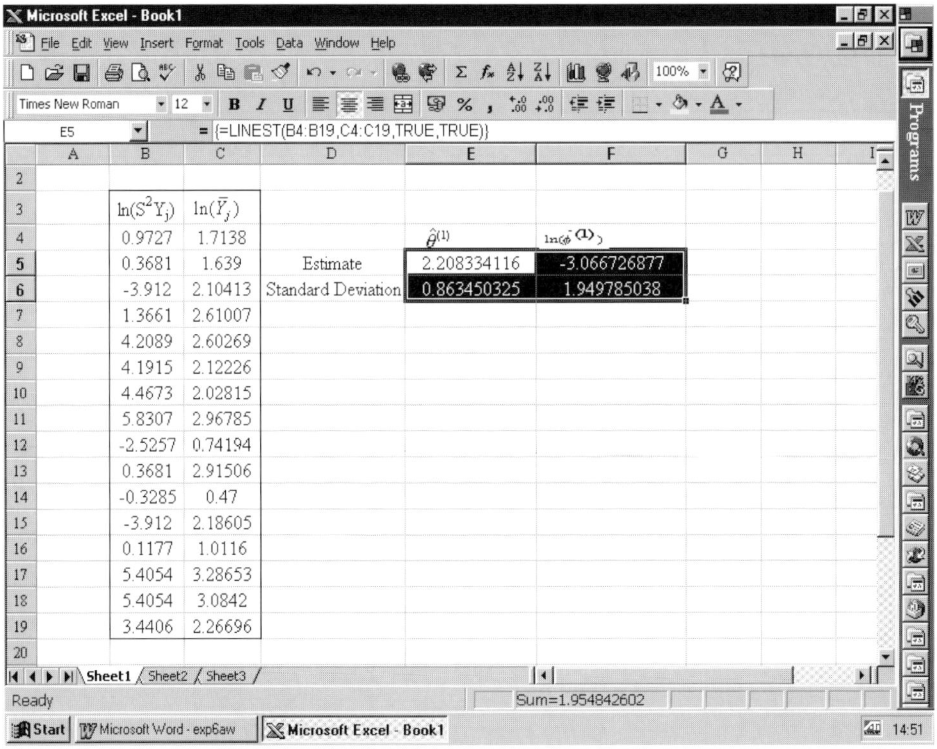

	B	C	D	E	F
3	$\ln(S^2Y_j)$	$\ln(\bar{Y}_j)$			
4	0.9727	1.7138		$\hat{\theta}^{(1)}$	$\ln(\hat{\phi}^{(1)})$
5	0.3681	1.639	Estimate	2.208334116	-3.066726877
6	-3.912	2.10413	Standard Deviation	0.863450325	1.949785038
7	1.3661	2.61007			
8	4.2089	2.60269			
9	4.1915	2.12226			
10	4.4673	2.02815			
11	5.8307	2.96785			
12	-2.5257	0.74194			
13	0.3681	2.91506			
14	-0.3285	0.47			
15	-3.912	2.18605			
16	0.1177	1.0116			
17	5.4054	3.28653			
18	5.4054	3.0842			
19	3.4406	2.26696			

Fig. 7.7 Excel spread sheet containing data making up the least squares regression and the highlighted output range containing the results.

at the 5% level so that the true value for θ could easily be two. This result suggests that the pore size data is approximately log normally distributed. It would therefore not be too much of an error to blindly select Taguchi's $(S\text{–}N)_T$ ratio (equation (7.21)) to analyse this experiment.

Notice that during the above estimation, ϕ was constrained to exp(–3.066) = 0.047 for all *j* test conditions. $\hat{\theta}^{(1)}$ is used to compute the PerMIA statistics, or values for ϕ_j, given by equation (7.23b) at each of the 16 different test conditions. So for the first test

$$\ln(\phi_1) = \ln\left[\frac{S_{Y_1}^2}{\bar{Y}_1^{\hat{\theta}^{(1)}}}\right] = \ln\left[\frac{2.645}{5.55^{2.208}}\right] = -2.8114.$$

Performing the same calculations for the remaining 15 test conditions yields the results shown in the final column of Table 7.11. These variance stabilising transformations are used for the response column of the Yates procedure shown in Table 7.12. In all, there

CONTROLLING PROCESS VARIABILITY: DISPERSION EFFECTS IN LINEAR DESIGNS   175

are $2^{7-3} - 1 = 15$ aliased pure dispersion effects that can be identified using the alias algebra described in Section 5.4.1.

A	=	CDE	=	BDF	=	BCG	=	EFG
B	=	ADF	=	ACG	=	CEF	=	DEG
C	=	ADE	=	ABG	=	BEF	=	DFG
D	=	ACE	=	ABF	=	BEG	=	CFG
E	=	ACD	=	BCF	=	BDG	=	AFG
F	=	ABD	=	BCE	=	CDG	=	AEG
G	=	ABC	=	BDE	=	CDF	=	AEF
AB	=	DF	=	CG				
AC	=	DE	=	BG				
AD	=	CE	=	BF				
AE	=	CD	=	FG				
AF	=	BD	=	EG				
AG	=	BC	=	EF				
BE	=	CF	=	DG				
ABE	=	BCD	=	DEF	=	CEG	=	ACF	=	ADG	=	BFG

These therefore form the 15 columns of coded test conditions for the Yates procedure in Table 7.12. The first seven columns of coded test conditions are as in Table 7.10. The next 8 columns contain the above first and second order interactions. The shown elements come from multiplying the required columns, so that the AB column is found by multiplying columns A and B together. The remaining columns are obtained by required multiplying these initial columns by the response column. The resulting provisional estimates for the contrasts are divided by $2^{7-3}/2 = 8$ to obtain provisional estimates for the pure dispersion effects. These pure dispersion effects could also have been obtained by applying the least squares formula to the response surface model.

$$\ln(\phi_j) = \gamma_0 + \gamma_A^{Ali} A_j + \gamma_B^{Ali} B_j + \gamma_C^{Ali} C_j + \gamma_D^{Ali} D_j + \gamma_E^{Ali} E_j + \gamma_F^{Ali} F_j + \gamma_G^{Ali} G_j + \gamma_{AB}^{Ali} A_j B_j + \gamma_{AC}^{Ali} A_j C_j$$

$$+ \gamma_{AD}^{Ali} A_j D_j + \gamma_{AE}^{Ali} A_j E_j + \gamma_{AF}^{Ali} A_j F_j + \gamma_{AG}^{Ali} A_j G_j + \gamma_{BE}^{Ali} B_j E_j + \gamma_{ABC}^{Ali} A_j B_j E_j \quad (7.24)$$

To estimate this equation in Excel, create 16 columns of data. The first column will contain the 16 values for $\ln(\phi_j)$ shown in the last column of Table 7.11. The next seven columns will contain the elements given in the first seven columns of Table 7.10. These correspond to the 16 values each for factors A_j through to G_j in the equation above. The next column will contain the product of $A_j \times B_j$, the next the product $A_j \times C_j$ and so on up to the last column that will contain the product $A_j \times B_j \times E_j$. As sixteen values for γ need to be estimated with no standard deviation (there are no replicates of $\ln(\phi_j)$ from which to compute them) highlight in Excel a block of one row by 16 columns and type in the formula = LINEST (R1, R2, TRUE) where R1 is the range of cells down a single column containing the sixteen values for $\ln(\phi_j)$ (the first column constructed in Excel) and R2 is

Table 7.12 Yates procedure for provisional estimates of the pure dispersion effects in the copper compacts experiment.

Tests	Coded Test Conditions							Response $\ln(\phi_j)$
	A	B	C	D	E = ACD	F = ABD	G = ABC	
(1)	−1	−1	−1	−1	−1	−1	−1	−2.8114
aefg	1	−1	−1	−1	1	1	1	−3.2508
bfg	−1	1	−1	−1	−1	1	1	−8.558
abe	1	1	−1	−1	1	−1	−1	−4.3969
ceg	−1	−1	1	−1	1	−1	1	−1.5379
acf	1	−1	1	−1	−1	1	−1	−0.4944
bcef	−1	1	1	−1	1	1	−1	−0.0109
abcg	1	1	1	−1	−1	−1	1	−0.7223
def	−1	−1	−1	1	1	1	−1	−4.1639
adg	1	−1	−1	1	−1	−1	1	−6.0684
bdeg	−1	1	−1	1	1	−1	1	−1.3663
abdf	1	1	−1	1	−1	1	−1	−8.7388
cdfg	−1	−1	1	1	−1	1	1	−2.1158
acde	1	−1	1	1	1	−1	−1	−1.8513
bcd	−1	1	1	1	−1	−1	−1	−1.4045
abcdefg	1	1	1	1	1	1	1	−1.5649

AB	AC	AD	AE	AF	AG
1	1	1	1	1	1
−1	−1	−1	1	1	1
−1	1	1	1	−1	−1
1	−1	−1	1	−1	−1
1	−1	1	−1	1	−1
−1	1	−1	−1	1	−1
−1	−1	1	−1	−1	1
1	1	−1	−1	−1	1
1	1	−1	−1	−1	1
−1	−1	1	−1	−1	1
−1	1	−1	−1	1	−1
1	−1	1	−1	1	−1
1	−1	−1	1	−1	−1
−1	1	1	1	−1	−1
−1	−1	−1	1	1	1
1	1	1	1	1	1

BE	ABE
1	−1
−1	−1
−1	1
1	1
−1	1
1	1
1	−1
−1	−1
−1	1
1	1
1	−1
−1	−1
1	−1
−1	−1
−1	1
1	1

Table 7.12 Continued.

	Dispersion contrasts and effects						
	A_{Disp}^{Ali} **Effect**$^{(1)}$	B_{Disp}^{Ali} **Effect**$^{(1)}$	C_{Disp}^{Ali} **Effect**$^{(1)}$	D_{Disp}^{Ali} **Effect**$^{(1)}$	E_{Disp}^{Ali} **Effect**$^{(1)}$	F_{Disp}^{Ali} **Effect**$^{(1)}$	G_{Disp}^{Ali} **Effect**$^{(1)}$
	$A = CDE =$ $BDF = BCG =$ EFG	$B = ADF =$ $ACG = CEF =$ DEG	$C = ADE =$ $ABG = BEF =$ DFG	$D = ACE =$ $ABF = BEG =$ CFG	$E = ACD =$ $BCF = BDG =$ AFG	$F = ABD =$ $BCE = CDG =$ AEG	$G = ABC =$ $BDE = CDF =$ AEF
	2.8114	2.8114	2.8114	2.8114	2.8114	2.8114	2.8114
	-3.2508	3.2508	3.2508	3.2508	-3.2508	-3.2508	-3.2508
	8.558	-8.558	8.558	8.558	8.558	-8.558	-8.558
	-4.3969	-4.3969	4.3969	4.3969	-4.3969	4.3969	4.3969
	1.5379	1.5379	-1.5379	1.5379	-1.5379	1.5379	-1.5379
	-0.4944	0.4944	-0.4944	0.4944	0.4944	-0.4944	0.4944
	0.0109	-0.0109	-0.0109	0.0109	-0.0109	-0.0109	0.0109
	-0.7223	-0.7223	-0.7223	0.7223	0.7223	0.7223	-0.7223
	4.1639	4.1639	4.1639	-4.1639	-4.1639	-4.1639	4.1639
	-6.0684	6.0684	6.0684	-6.0684	6.0684	6.0684	-6.0684
	1.3663	-1.3663	1.3663	-1.3663	-1.3663	1.3663	-1.3663
	-8.7388	-8.7388	8.7388	-8.7388	8.7388	-8.7388	8.7388
	2.1158	2.1158	-2.1158	-2.1158	2.1158	-2.1158	-2.1158
	-1.8513	1.8513	-1.8513	-1.8513	-1.8513	1.8513	1.8513
	1.4045	-1.4045	-1.4045	-1.4045	1.4045	1.4045	1.4045
	-1.5649	-1.5649	-1.5649	-1.5649	-1.5649	-1.5649	-1.5649
Disp Contrast$^{(1)}$	-5.1191	-4.4687	29.6525	-5.4913	12.7707	-8.7385	-1.3123
Disp Effect$^{(1)}$	-0.6398875	-0.5585875	3.7065625	-0.6864125	1.5963375	-1.0923125	-0.1640375

Table 7.12 Continued.

	AB_{Disp}^{Ali} Effect$^{(1)}$	AC_{Disp}^{Ali} Effect$^{(1)}$	AD_{Disp}^{Ali} Effect$^{(1)}$	AE_{Disp}^{Ali} Effect$^{(1)}$	AF_{Disp}^{Ali} Effect$^{(1)}$	AG_{Disp}^{Ali} Effect$^{(1)}$
	$AB = DF = CG$	$AC = DE = BG$	$AD = CE = BF$	$AE = CD = FG$	$AF = BD = EG$	$AG = BC = EF$
	−2.8114	−2.8114	−2.8114	−2.8114	−2.8114	−2.8114
	3.2508	3.2508	3.2508	−3.2508	−3.2508	−3.2508
	8.558	−8.558	−8.558	−8.558	8.558	8.558
	−4.3969	4.3969	4.3969	−4.3969	4.3969	4.3969
	−1.5379	1.5379	−1.5379	1.5379	−1.5379	1.5379
	0.4944	−0.4944	0.4944	0.4944	−0.4944	0.4944
	0.0109	0.0109	−0.0109	0.0109	0.0109	−0.0109
	−0.7223	−0.7223	0.7223	0.7223	0.7223	−0.7223
	−4.1639	−4.1639	4.1639	4.1639	4.1639	−4.1639
	6.0684	6.0684	−6.0684	6.0684	6.0684	−6.0684
	1.3663	−1.3663	1.3663	1.3663	−1.3663	1.3663
	−8.7388	8.7388	−8.7388	8.7388	−8.7388	8.7388
	−2.1158	2.1158	2.1158	−2.1158	2.1158	2.1158
	1.8513	−1.8513	−1.8513	−1.8513	1.8513	1.8513
	1.4045	1.4045	1.4045	−1.4045	−1.4045	−1.4045
	−1.5649	−1.5649	−1.5649	−1.5649	−1.5649	−1.5649
Disp Contrast$^{(1)}$	−3.0473	5.9915	−13.2267	−2.8507	6.7185	9.0623
Disp Effect$^{(1)}$	−0.3809125	0.7489375	−1.6533375	−0.3563375	0.8398125	1.1327875

CONTROLLING PROCESS VARIABILITY: DISPERSION EFFECTS IN LINEAR DESIGNS

Table 7.12 Continued.

	BE_{Disp}^{Ali} Effect$^{(1)}$	ABE_{Disp}^{Ali} Effect$^{(1)}$
	BE = CF = DG	ABE = BCD = DEF = CEG = ACF = ADG = BFG
	−2.8114	2.8114
	3.2508	3.2508
	8.558	−8.558
	−4.3969	−4.3969
	1.5379	−1.5379
	−0.4944	−0.4944
	−0.0109	0.0109
	0.7223	0.7223
	4.1639	−4.1639
	−6.0684	−6.0684
	−1.3663	1.3663
	8.7388	8.7388
	−2.1158	2.1158
	1.8513	1.8513
	1.4045	−1.4045
	−1.5649	−1.5649
Disp Contrast$^{(1)}$	11.3985	−7.3213
Disp Effect$^{(1)}$	1.4248125	−0.9151625

the range of cells down 15 columns containing values for A_j through to $A_jB_jE_j$ (the remaining fifteen columns in Excel).

This is illustrated in Figure 7.8. This spreadsheet image contains the 16 columns of data making up the least squares regression and the highlighted output range containing the results. The formula box towards the top of the sheet shows the formula lying behind the values shown in the highlighted cells. Simply hit the control shift and return key simultaneously to obtain these values. The result is

$\hat{\gamma}_0^{(1)} = -3.066, \hat{\gamma}_A^{(1)} = -0.319, \hat{\gamma}_B^{(1)} = -0.279, \hat{\gamma}_C^{(1)} = 1.853, \hat{\gamma}_D^{(1)} = -0.343, \hat{\gamma}_E^{(1)} = 0.798, \hat{\gamma}_F^{(1)} = -0.546$

$\hat{\gamma}_G^{(1)} = -0.082, \hat{\gamma}_{AB}^{(1)} = -0.19, \hat{\gamma}_{AC}^{(1)} = 0.374, \hat{\gamma}_{AD}^{(1)} = -0.827, \hat{\gamma}_{AE}^{(1)} = -0.178, \hat{\gamma}_{AF}^{(1)} = 0.42, \hat{\gamma}_{AG}^{(1)} = 0.566$

$\hat{\gamma}_{BE}^{(1)} = 0.712, \quad \hat{\gamma}_{ABE}^{(1)} = -0.458$

Notice how these are half the pure dispersion effects shown in Table 7.12. Provisional estimates for the pure dispersion effects are therefore found by multiplying each of these terms through by two and are shown in Table 7.13 together with the familiar construction of a probability plot for these dispersion effects.

The results in Table 7.13 are shown graphically in Figure 7.9. Factor C clearly has an important impact on dispersion or variability and possibly so too does the interaction

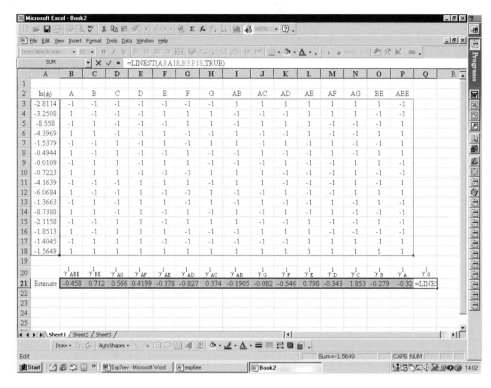

Fig. 7.8 Excel estimation of equation (7.24).

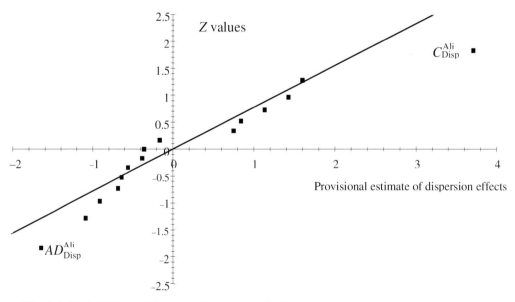

Fig. 7.9 Probability plot for pure dispersion effects in the copper compacts experiment.

Controlling Process Variability: Dispersion Effects in Linear Designs

Table 7.13 Probability plot for pure dispersion effects in the copper compact experiment.

Aliased Effect	Effect Estimate	Rank Index, i	CP_i	Z_i
AD_{Disp}^{Ali}	−1.6533	1	0.0333	−1.8344
F_{Disp}^{Ali}	−1.0923	2	0.1	−1.2816
ABE_{Disp}^{Ali}	−0.9152	3	0.1667	−0.9673
D_{Disp}^{Ali}	−0.6864	4	0.2333	−0.728
A_{Disp}^{Ali}	−0.6399	5	0.3	−0.5244
B_{Disp}^{Ali}	−0.5586	6	0.3667	−0.3406
AB_{Disp}^{Ali}	−0.3809	7	0.4333	−0.168
AE_{Disp}^{Ali}	−0.3563	8	0.5	0
G_{Disp}^{Ali}	−0.164	9	0.5667	0.168
AC_{Disp}^{Ali}	0.7489	10	0.6333	0.3406
AF_{Disp}^{Ali}	0.8398	11	0.7	0.5244
AG_{Disp}^{Ali}	1.1328	12	0.7667	0.728
BE_{Disp}^{Ali}	1.4248	13	0.8333	0.9673
E_{Disp}^{Ali}	1.5963	14	0.9	1.2816
C_{Disp}^{Ali}	3.7066	15	0.9667	1.8344

between factors A and D. These three factors are the only possible control factors that can be used to minimise variability. Remember that these are aliased dispersion effects. If second order interactions are assumed unimportant then as a provisional guess, it can be said that reducing factor C from its high to low value will reduce $\ln(S_Y^2)$ by 3.71 units independent of the mean. Further interpretation will be made once step two below has been completed.

7.2.2 STEP 2. RELIABLE ESTIMATES OF THE DISPERSION EFFECTS FOR MAKING COPPER COMPACTS

More reliable estimates of these pure dispersion effects can be obtained by estimating the dispersion effects shown above to be important simultaneously with θ, i.e. equation (7.23e). The equation to be estimated by least squares is therefore

$$\ln(S^2_{Y_j}) = \gamma_0 + \gamma_C^{Ali} C_j + \gamma_{AD}^{Ali} A_j D_j + \theta \ln(\bar{Y}_j) + \varepsilon_j$$

To estimate this equation in Excel create four columns of data. The first column will contain the 16 values for $\ln(S^2_{Y_j})$ shown in Table 7.11. The next column will contain the elements given in the C column of Table 7.12. This corresponds to the 16 values for the factor C_j in the equation above. The next column will contain the product of $A_j \times D_j$. The final column will contain the 16 values for $\ln(\bar{Y}_j)$. As three values for γ need to be estimated together with θ and their standard deviations, highlight in Excel a block of two rows by four columns and type in the formula = LINEST(R1, R2, TRUE) where R1 is the range of cells down a single column containing the values for $\ln(S^2_{Y_j})$, (the first column constructed in Excel) and R2 is the range of cells down three columns containing values for C_j, A_jD_j and $\ln(\bar{Y}_j)$, (the remaining three columns in Excel).

This is illustrated in the Excel image shown in Figure 7.10. This illustrative spreadsheet contains the four columns of data making up the least squares regression and the highlighted output range containing the results. The formula box towards the top of the sheet shows the formula lying behind the values shown in the highlighted cells. Simply hit the control shift and return key simultaneously to obtain these values.

The result is

$$\hat{\gamma}_0^{(2)} = -1.522, \hat{\gamma}_C^{(2)} = 2.082, \hat{\gamma}_{AD}^{(2)} = -0.623, \hat{\theta}^{(2)} = 1.476.$$

Hence

$$C_{Disp}^{Ali\ (2)} = 2 \times 2.082 = 4.164$$

$$AD_{Disp}^{Ali\ (2)} = 2 \times -0.623 = -1.246$$

These estimates are more reliable than those obtained in step one above. (Note that for factor C the estimate has gone up from 3.7 to 4.164 and for AD the estimate has changed from –1.65 to –1.246). The standard deviation associated with θ is 0.650. The estimate for θ has in turn changed from 2.208 to 1.476. To test the hypothesis that θ = 2 simply construct the usual t statistic of $t = (1.476–2)/0.65 = –0.81$. This is insignificant at the 5% level so that the true value for θ is likely to be two. It would therefore be advisable to use Taguchi's $(S-N)_T$ ratio (equation (7.21)) to analyse this experiment.

The standard deviation associated with $\gamma^{(2)}_C$ is 0.492. To test the hypothesis that $\gamma^{(2)}_C = 0$ simply construct the usual t statistic of $t = (2.082–0)/0.492 = 4.23$. This is significant at the 5% level so that the true value for $\gamma^{(2)}_C$ is not likely to be zero. Clearly the level of factor C is important in determining the variability in pore size that is independent of the

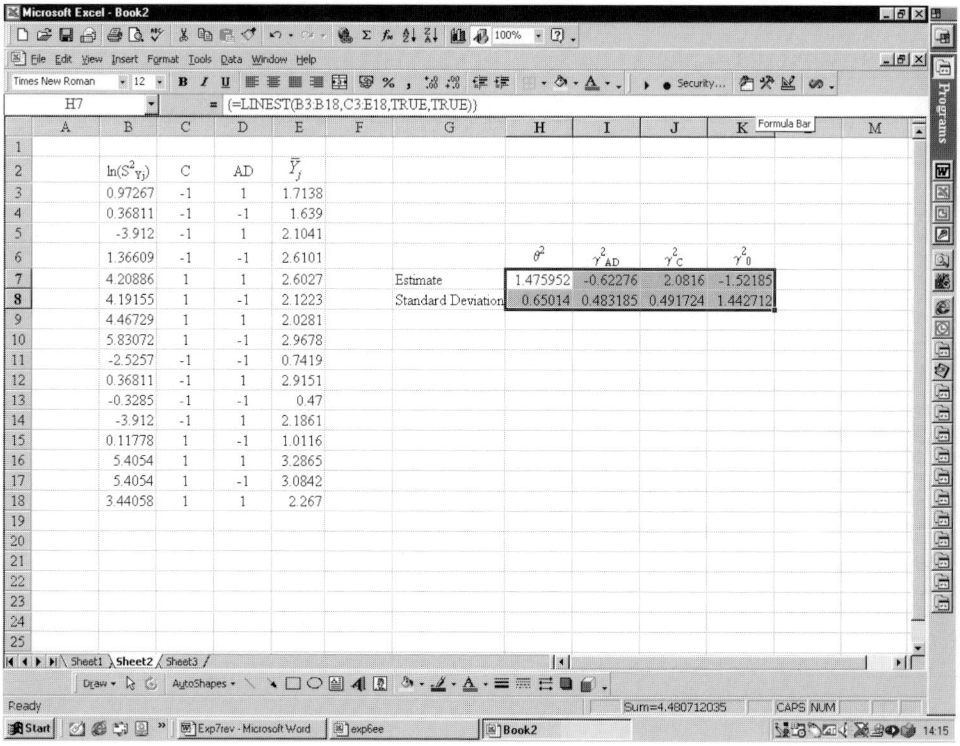

Controlling Process Variability: Dispersion Effects in Linear Designs 183

Fig. 7.10 Excel spreadsheet containing data making up least squares regression and the highlighted output range containing the results.

mean pore size. However, the standard deviation associated with $\gamma^{(2)}_{AD}$ is 0.483. To test the hypothesis that $\gamma^{(2)}_{AD} = 0$ simply construct the usual t statistic of $t = (-0.623-0)/0.483 = -1.29$. This is not significant at the 5% level so that the true value for $\gamma^{(2)}_{AD}$ is likely to be zero. Hence a simplified equation modelling process variability that is independent of the mean is given by,

$$\ln(\phi_j) = -1.522 + 2.082 C_j.$$

The only pure dispersion effect of importance is therefore $C^{Ali}_{Disp} = 2 \times 2.082 = 4.164$. This suggests that variability can be minimised independent of the mean by setting factor C low. When set low, the resulting variability is expected to be

$$\overline{Y}_j^{1.476} \phi_j = \overline{Y}_j^{1.476} \exp(-1.522 + 2.082x - 1) = \overline{Y}_j^{1.476} 0.0272$$

For a given mean, variability is minimised by holding factor C at its low level. The next step would be to ensure the mean is on target. Optimising this process in this way will be further discussed in Chapter 8.

Table 7.14 A Taguchi inner/outer array representation of the injection moulding experiment.

Inner Array for Design Factors								Outer Array for Noise Factors				
	1	2	3	4	5	6	7	Test	(1)	Z_1Z_3	Z_2Z_3	Z_1Z_2
	C	B	F	A	E	D	G	Z_1	−1	1	−1	1
								Z_2	−1	−1	1	1
Test								Z_3	−1	1	1	−1	$(S-N)_T$
(1)	−1	−1	−1	−1	−1	−1	−1		2.2	2.3	2.1	2.3	−6.2917
adeg	−1	−1	−1	1	1	1	1		0.3	2.7	2.5	0.3	−0.1721
bdfg	−1	1	1	−1	−1	1	1		0.5	0.4	3.1	2.8	−0.3193
abef	−1	1	1	1	1	−1	−1		2.0	1.8	1.9	2.0	−6.002
cefg	1	−1	1	−1	1	−1	1		3.0	3.0	3.1	3.0	−8.2053
acdf	1	−1	1	1	−1	1	−1		2.1	1.0	4.2	3.1	−1.2833
bcde	1	1	−1	−1	1	1	−1		4.0	4.6	1.9	2.2	−1.7439
abcg	1	1	−1	1	−1	−1	1		2.0	1.9	1.9	1.8	−6.2943

7.8 COMPARING THE RESPONSE SURFACE AND GENERALISED LINEAR MODELS USING THE INJECTION MOULDING EXPERIMENT

To illustrate some of the advantages and disadvantages of the response surface and generalised linear models consider the injection moulding experiment of Section 2.6. The quality control engineers conducted this experiment by assigning the seven design factors (A to G) to an inner array and the remaining three noise factors (Z_1 to Z_3) to an outer array. They used Taguchi's $L_8(2^7)$ design for the inner array. The ordering of factors to columns is not important and they assigned variable A to column four, factor B to column two, C to column one, factor D to column six all the way up to factor G in column seven. The columns of −1s and +1s in Table 7.14 are the same as the L_8 design shown in Table 4.8. A 2^{3-1} fractional factorial design was then chosen for the outer array. The total number of test carried out was therefore 32 and at each of these test conditions the response recorded was the observed percentage shrinkage. This experiment together with all the results is shown in Table 7.14. Reading along the first row, the first test (yielding a 2.2% shrinkage) involved setting all design and noise factors low. The second test (yielding a 2.3% shrinkage) involved setting all design factors low, the second noise factor low (Z_2, moisture content) and the other two noise factors (Z_1 and Z_3) high. The rest of the table can be read in this way.

CONTROLLING PROCESS VARIABILITY: DISPERSION EFFECTS IN LINEAR DESIGNS   185

There are a number of deficiencies involved with this experiment which will be looked at in the hope of installing good practice in the reader. This will be followed by a provisional analysis of the results from this experiment.

7.8.1 DESIGN PROBLEMS

The overall experiment above is a mixture of poorly selected designs. Because seven design factors are being studied, the $L_8(2^7)$ design chosen above for the inner array is actually a 2^{7-4}_{III} fractional factorial. A comparison of Table 4.8 and the inner array above will show that the two designs are identical. However, this is no ordinary 2^{7-4} design because it is a mixture of a principle and alternate fractional factorial (three design generators in Table 4.10 are for an alternate design but one is for a principle design). You should by now be able to recognise the outer array as 2^{3-1} fractional factorial with the design generator being $Z_3 = -Z_1Z_2$. It is therefore an alternate fractional factorial. The full design, with 32 tests, is then a hybrid of these two designs and so has a very poor resolution.

To prove this, the design shown in Table 7.14 can be rearranged into the more familiar format of a fractional factorial design. Replicating the inner array four times with these replicates being placed directly beneath the original inner array achieves this. This is done in Table 7.15 to give 32 rows – each corresponding to a test condition. To this is added three additional columns for the noise factors Z_1 to Z_3. Let the first eight of these rows correspond to Z_1, Z_2 and Z_3 being low so that –1s are placed in the first eight rows of the columns containing noise factors Z_1 to Z_3. Let the next eight rows correspond to Z_1 and Z_3 being at their high levels and Z_2 at its low level so columns Z_1 and Z_3 contain +1s and column Z_2 –1s. For the next eight rows only Z_1 is low so columns Z_2 and Z_3 contain +1s and column Z_1 –1s. Finally, the last eight rows of design test conditions were conducted with Z_1 and Z_2 being high but Z_3 low so that columns Z_1 and Z_2 contain +1s and Z_3 –1s. Then in Table 7.15 the responses recorded at each of the 32 test conditions are placed alongside. This is exactly the same set of test conditions as in Table 7.14. They have just been presented in a different format.

Next notice that in Table 7.15 columns A, B, C, Z_1 and Z_2 define a full 2^5 factorial in these five factors. Thus in Table 7.16 these five factors have been placed in the first five columns. Notice that the test condition column beside it is the standard order form for a 2^5 design. These five columns therefore form the base design for a 2^{10-5} fractional factorial. That is, the experiment contained in Table 7.14 is a 2^{10-5} design. To find out its resolution it is necessary to know how the remaining five factors were introduced. Notice that column D in Table 7.15 can be obtained by multiplying columns A and B together and reversing the sign, i.e. $D = -AB$. Similarly notice that $E = -AC$, $F = -BC$, $G = ABC$ and $Z_3 = -Z_1Z_2$. This 2^{10-5} design is shown in full in Table 7.16.

This design has 31 confounded interactions. The first five are obtained from the design generators in Table 7.16.

   $-ABD$ $-ACE$ $-BCF$ $ABCG$ $-Z_1Z_2Z_3$

Table 7.15 Rearrangement of the inner/outer array layout in Table 7.14.

C	B	F	A	E	D	G	Z_1	Z_2	Z_3	Y
−1	−1	−1	−1	−1	−1	−1	−1	−1	−1	2.2
−1	−1	−1	1	1	1	1	−1	−1	−1	0.3
−1	1	1	−1	−1	1	1	−1	−1	−1	0.5
−1	1	1	1	1	−1	−1	−1	−1	−1	2.0
1	−1	1	−1	1	−1	1	−1	−1	−1	3.0
1	−1	1	1	−1	1	−1	−1	−1	−1	2.1
1	1	−1	−1	1	1	−1	−1	−1	−1	4.0
1	1	−1	1	−1	−1	1	−1	−1	−1	2.0
−1	−1	−1	−1	−1	−1	−1	1	−1	1	2.3
−1	−1	−1	1	1	1	1	1	−1	1	2.7
−1	1	1	−1	−1	1	1	1	−1	1	0.4
−1	1	1	1	1	−1	−1	1	−1	1	1.8
1	−1	1	−1	1	−1	1	1	−1	1	3.0
1	−1	1	1	−1	1	−1	1	−1	1	1.0
1	1	−1	−1	1	1	−1	1	−1	1	4.6
1	1	−1	1	−1	−1	1	1	−1	1	1.9
−1	−1	−1	−1	−1	−1	−1	−1	1	1	2.1
−1	−1	−1	1	1	1	1	−1	1	1	2.5
−1	1	1	−1	−1	1	1	−1	1	1	3.1
−1	1	1	1	1	−1	−1	−1	1	1	1.9
1	−1	1	−1	1	−1	1	−1	1	1	3.1
1	−1	1	1	−1	1	−1	−1	1	1	4.2
1	1	−1	−1	1	1	−1	−1	1	1	1.9
1	1	−1	1	−1	−1	1	−1	1	1	1.9
−1	−1	−1	−1	−1	−1	−1	1	1	−1	2.3
−1	−1	−1	1	1	1	1	1	1	−1	0.3
−1	1	1	−1	−1	1	1	1	1	−1	2.8
−1	1	1	1	1	−1	−1	1	1	−1	2.0
1	−1	1	−1	1	−1	1	1	1	−1	3.0
1	−1	1	1	−1	1	−1	1	1	−1	3.1
1	1	−1	−1	1	1	−1	1	1	−1	2.2
1	1	−1	1	−1	−1	1	1	1	−1	1.8

Table 7.16 The 2^{10-5} fractional factorial representation of the injection moulding experiment.

Base Design (A full 2^5)						An Unusual 2^{10-5} Fractional Factorial Design								A 2^{10-5} Design			
Test	A	B	C	Z_1	Z_2	A	B	C	Z_1	Z_2	D = −AB	E = −AC	F = −BC	G = ABC	$Z_3 = −Z_1Z_2$	Test
(1)	−1	−1	−1	−1	−1	−1	−1	−1	−1	−1	−1	−1	−1	−1	−1	(1)
a	1	−1	−1	−1	−1	1	−1	−1	−1	−1	1	1	−1	1	−1	adeg
b	−1	1	−1	−1	−1	−1	1	−1	−1	−1	1	−1	1	1	−1	bdfg
ab	1	1	−1	−1	−1	1	1	−1	−1	−1	−1	1	1	−1	−1	abef
c	−1	−1	1	−1	−1	−1	−1	1	−1	−1	−1	1	1	1	−1	cefg
ac	1	−1	1	−1	−1	1	−1	1	−1	−1	1	−1	1	−1	−1	acdf
bc	−1	1	1	−1	−1	−1	1	1	−1	−1	1	1	−1	−1	−1	bcde
abc	1	1	1	−1	−1	1	1	1	−1	−1	−1	−1	−1	1	−1	abcg
z_1	−1	−1	−1	1	−1	−1	−1	−1	1	−1	−1	−1	−1	−1	1	z_1z_3
az_1	1	−1	−1	1	−1	1	−1	−1	1	−1	1	1	−1	1	1	$az_1deg z_3$
bz_1	−1	1	−1	1	−1	−1	1	−1	1	−1	1	−1	1	1	1	$bz_1dfg z_3$
abz_1	1	1	−1	1	−1	1	1	−1	1	−1	−1	1	1	−1	1	$abz_1efg z_3$
cz_1	−1	−1	1	1	−1	−1	−1	1	1	−1	−1	1	1	1	1	$cz_1efg z_3$
acz_1	1	−1	1	1	−1	1	−1	1	1	−1	1	−1	1	−1	1	$acz_1df z_3$
bcz_1	−1	1	1	1	−1	−1	1	1	1	−1	1	1	−1	−1	1	$bcz_1de z_3$
$abcz_1$	1	1	1	1	−1	1	1	1	1	−1	−1	−1	−1	1	1	$abcz_1g z_3$
z_2	−1	−1	−1	−1	1	−1	−1	−1	−1	1	−1	−1	−1	−1	1	z_2z_3
az_2	1	−1	−1	−1	1	1	−1	−1	−1	1	1	1	−1	1	1	$az_2deg z_3$
bz_2	−1	1	−1	−1	1	−1	1	−1	−1	1	1	−1	1	1	1	$bz_2dfg z_3$
abz_2	1	1	−1	−1	1	1	1	−1	−1	1	−1	1	1	−1	1	$abz_2ef z_3$
cz_2	−1	−1	1	−1	1	−1	−1	1	−1	1	−1	1	1	1	1	$cz_2efg z_3$
acz_2	1	−1	1	−1	1	1	−1	1	−1	1	1	−1	1	−1	1	$acz_2df z_3$
bcz_2	−1	1	1	−1	1	−1	1	1	−1	1	1	1	−1	−1	1	$bcz_2de z_3$
$abcz_2$	1	1	1	−1	1	1	1	1	−1	1	−1	−1	−1	1	1	$abcz_2g z_3$
z_1z_2	−1	−1	−1	1	1	−1	−1	−1	1	1	−1	−1	−1	−1	−1	z_1z_2
az_1z_2	1	−1	−1	1	1	1	−1	−1	1	1	1	1	−1	1	−1	az_1z_2deg
bz_1z_2	−1	1	−1	1	1	−1	1	−1	1	1	1	−1	1	1	−1	bz_1z_2dfg
abz_1z_2	1	1	−1	1	1	1	1	−1	1	1	−1	1	1	−1	−1	abz_1z_2ef
cz_1z_2	−1	−1	1	1	1	−1	−1	1	1	1	−1	1	1	1	−1	cz_1z_2efg
acz_1z_2	1	−1	1	1	1	1	−1	1	1	1	1	−1	1	−1	−1	acz_1z_2df
bcz_1z_2	−1	1	1	1	1	−1	1	1	1	1	1	1	−1	−1	−1	bcz_1z_2de
$abcz_1z_2$	1	1	1	1	1	1	1	1	1	1	−1	−1	−1	1	−1	$abcz_1z_2g$

The others are found by working out all the combinations of 2, 3, 4 and 5 of the above confounded interactions. After applying alias algebra to these products the following confounded interactions are obtained.

$BCDE$	$ACDF$	$-CDG$	$ABDZ_1Z_2Z_3$	$ABEF$
$-BEG$	$ACEZ_1Z_2Z_3$	$-AFG$	$BCFZ_1Z_2Z_3$	$-ABCGZ_1Z_2Z_3$
$-DEF$	$ADEG$	$-BCDEZ_1Z_2Z_3$	$BDFG$	$-ACDFZ_1Z_2Z_3$
$CDGZ_1Z_2Z_3$	$CEFG$	$-ABEFZ_1Z_2Z_3$	$BEGZ_1Z_2Z_3$	$AFGZ_1Z_2Z_3$
$-ABCDEFG$	$DEFZ_1Z_2Z_3$	$BDFGZ_1Z_2Z_3$	$-ADEGZ_1Z_2Z_3$	$-CEFGZ_1Z_2Z_3$
$-ABCDEFGZ_1Z_2Z_3$				

At this stage there is little point working out all the aliased effects. Note however, that when A is multiplied into all these confounded interactions, the first result is $A = -BD$. Also when AB is multiplied into all these confounded interactions, the fourth result is $AB = -CG$. Thus main location effects are aliased with first order interactions and some first order interactions are aliased with each other. This experiment is therefore a resolution III design. Yet Table 4.4 shows that with 32 tests and a different set of design generators, a resolution IV design could have been obtained. Importantly, and as will be discussed below, the first order interactions between design and noise factors are aliased with higher order interactions.

Thus it is never wise to combine different fractionated designs from an inner and an outer array and never wise to select Taguchi's orthogonal designs for either the inner or outer array. The recommended approach to deriving a response surface model is to combine design and noise factors into a single fractional factorial using the steps highlighted in Chapter 4. Having chosen the design generators from Table 4.4, the resolution of the resulting design together with the aliasing structure will be optimised and fully apparent. This is crucial because it is the interactions between design and noise factors that will be used to minimise variability and their aliasing structure is essential to an understanding of what design factors need to be changed to make a process robust.

7.8.2 Analysis of the Data

Nothing can be done here to overcome the poor resolution of the above experiment but poor data analysis can be avoided. Taguchi practitioners would carry out a row analyse of Table 7.14. That is, they would ignore the noise factors and consider them as just a means of obtaining four replicates at each of the eight design factor test conditions. In essence they would visualise the experiment as a 2^{7-4} design with four replicates made at each of the eight design factor test conditions. The signal to noise ratio would typically be used as the PerMIA summary statistic from which process variability would be minimised. This approach however can give very misleading conclusions. To see this, refer again to Table 7.14 where the $(S-N)_T$ ratio of equation (7.21) at each of the eight tests is shown. The mean response at the first set of design factor test conditions is

$$\bar{Y}_1 = \frac{2.2 + 2.3 + 2.1 + 2.3}{4} = 2.225$$

The variance at this test condition is

$$S_{Y_1}^2 = \frac{(2.2 - 2.225)^2 + \ldots + (2.3 - 2.225)^2}{4 - 1} = \frac{0.0275}{3} = 0.00917$$

The $(S\text{-}N)_T$ ratio is from equation (7.21)

$$(S-N)_{T_1} = \ln\left[\frac{0.00917}{2.225^2}\right] = -6.2917$$

Table 7.17 uses the Yates procedure to identify the pure dispersion effects with the $(S\text{-}N)_T$ ratio as the response variable. It requires a little further explanation. As noise factors are ignored it is only the inner array of design factors that are being studied and as shown in Section 4.5 this is a 2^{7-4} fractional factorial with design generators $D = -AB$, $E = -AC$, $F = -BC$ and $G = ABC$. Hence the first four confounded interactions are

$-ABD \quad -ACE \quad -BCF \quad ABCG$

The remaining confounded interactions are found from these four by multiplying together every combination of two, then every combination of three and then every combination of four. Using the rules of alias algebra this gives

$BCDE \quad ACDF \quad -CDG \quad ABEF$

$-BEG \quad -AFG \quad -DEF \quad ADEG$

$BDFG \quad CEFG \quad -ABCDEFG$

Multiplying each of these 15 confounded interactions by A gives the aliased effect for factor A

$A = -BD = -CE = -ABCF = BCG = ABCDE = CDF = -ACDG = BEF = -ABEG = -FG = -ADEF = DEG = ABDFG = ACEFG = -BCDEFG.$

Assuming that all second and higher order interactions are zero or unimportant this means that the estimated aliased A_{Disp} effect will actual equal the A_{Disp} effect minus the BD_{Disp} effect minus the CE_{Disp} effect minus the FG_{Disp} effect from the full 2^7 design

$$A_{Disp}^{Ali} = A_{Disp} - BD_{Disp} - CE_{Diso} - FG_{Disp}$$

All seven aliased main dispersion effects can be found in the same way. The following have omitted from then all the second and higher order interactions under the assumption that they equal zero

$B = -AD = -CF = -EG,$
$C = -AE = -BF = -DG,$

Table 7.17 The Yates procedure for dispersion effects in the injection moulding experiment using the $(S\text{-}N)_T$ ratio.

Coded Test Conditions								Response
Test	A	B	C	D	E	F	G	$(S\text{-}N)_T$
(1)	−1	−1	−1	−1	−1	−1	−1	−6.292
adeg	1	−1	−1	1	1	−1	1	−0.172
bdfg	−1	1	−1	1	−1	1	1	−0.319
abef	1	1	−1	−1	1	1	−1	−6.002
cefg	−1	−1	1	−1	1	1	1	−8.205
acdf	1	−1	1	1	−1	1	−1	−1.283
bcde	−1	1	1	1	1	−1	−1	−1.744
abcg	1	1	1	−1	−1	−1	1	−6.294

Dispersion Contrasts and Effects							
Test	A_{Disp}^{Ali} Effect	B_{Disp}^{Ali} Effect	C_{Disp}^{Ali} Effect	D_{Disp}^{Ali} Effect	E_{Disp}^{Ali} Effect	F_{Disp}^{Ali} Effect	G_{Disp}^{Ali} Effect
	A = −BD = −CE = −FG	B = −AD = −CF = −EG	C = −AE = −BF = −DG	D = −AB = −CG = −EF	E = −AC = −BG = −DF	F = −BC = −AG = −DE	G = −CD = −BE = −AF
(1)	6.2917	6.2917	6.2917	6.2917	6.2917	6.2917	6.2917
adeg	−0.1721	0.1721	0.1721	−0.1721	−0.1721	0.1721	−0.1721
bdfg	0.3193	−0.3193	0.3193	−0.3193	0.3193	−0.3193	−0.3193
abef	−6.002	−6.002	6.002	6.002	−6.002	−6.002	6.002
cefg	8.2053	8.2053	−8.2053	8.2053	−8.2053	−8.2053	−8.2053
acdf	−1.2833	1.2833	−1.2833	−1.2833	1.2833	−1.2833	1.2833
bcde	1.7439	−1.7439	−1.7439	−1.7439	−1.7439	1.7439	1.7439
abcg	−6.2943	−6.2943	−6.2943	6.2943	6.2943	6.2943	−6.2943
Disp Contrast	2.8085	1.5929	−4.7417	23.2747	−1.9347	−1.3079	0.3299
Disp Effect	0.702125	0.398225	−1.185425	5.818675	−0.483675	−0.326975	0.082475

D = −AB = −CG = −EF,

E = −AC = −BG = −DF,

F = −BC = −AG = −DE,

G = −CD = −BE = −AF.

Now in Table 7.17 it is these aliased effects that are used as headers for each column. From eight tests only seven aliased effects can be estimated so that there are only seven

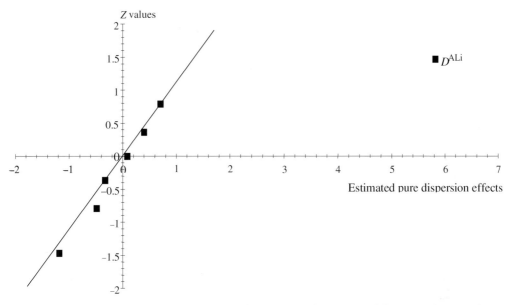

Fig. 7.11 Probability plot for dispersion effects in the injection moulding experiment using the *S-N* ratio.

columns of coded test conditions and these correspond to the inner array of Table 7.14. Each effect is found by dividing the contrasts through by $2^{7-4}/2 = 4$. Figure 7.11 shows a probability plot for these seven pure dispersion effects and only factor D (the holding time) has an impact on the variability present in the process. That is, factor D appears to be the only control factor.

In summary then, the Taguchi type analysis reveals

$$D_{Disp}^{Ali} = 5.819$$

so that variability can be minimised by setting factor D (the holding time) low. The value for this dispersion effect suggests that every two-unit code change in the level for this factor will change log variability by 5.819.

There are two fundamental shortcomings with this Taguchi type analysis. The first is to do with the choice of the PerMIA summary statistic and the second is to do with the failure to explicitly take into account the influence of the noise factors on variability. Taking each of these in turn.

7.8.2.1 The Blind Use of the $(S-N)_T$ Ratio

The above analysis used the signal to noise ratio and so explicitly forced the variability independent of the mean to be

Table 7.18 Log mean and variability in % shrinkage for the injection moulding experiment

Test	Log Mean % Shrinkage $\ln(\overline{Y}_j)$	Log Variability $\ln(S^2_{Y_j})$
(1), $j = 1$	0.7998	−4.6922
adeg $j = 2$	0.3716	0.571
bdfg $j = 3$	0.5306	0.7419
abef $j = 4$	0.6549	−4.6922
cefg $j = 5$	1.1069	−5.9915
acdf $j = 6$	0.9555	0.6277
bcde $j = 7$	1.1553	0.5667
abcg $j = 8$	0.6419	−5.0106

$$(S-N)_{T_j} = \ln(\phi_j) = \ln\left[\frac{S^2_{Y_j}}{\overline{Y}_j^{\theta=2}}\right]$$

Constraining $\theta = 2$ may not be appropriate for this experiment. In Engel's generalised linear model described above, step one would involve a test for the value for θ as follows. First the natural logs of the variance values would be calculated. For the $j = 1$ design factor test condition, $S^2_{Y_1}$ was calculated above to be 0.00917 so that

$$\ln(S^2_{Y_1}) = \ln(0.00917) = -4.6918$$

Performing the same calculation for the remaining seven test conditions yields the results shown in the first column of Table 7.18. Also shown in this table is the log mean % shrinkage's associated with the eight design factor test conditions.

These two variables enter equation (7.23a) which now needs to be estimated using the least squares procedure. As two parameters with standard deviations need to be estimated in Excel, highlight a block of two rows by two columns and type in the formula = LINEST(R1, R2, TRUE, TRUE) where R1 is the range of cells down a single column containing the values for $\ln(S^2_{Y_j})$ and R2 is the range of cells down a column containing values for $\ln(\overline{Y}_j)$. The Excel image in Figure 7.12 illustrates this procedure. As usually the spreadsheet contains the two columns of data making up the least squares regression and the highlighted output range containing the results. The formula box shows the formula lying behind the values shown in the highlighted cells. Simply hit the control shift and return key simultaneously to obtain these values.

The result is

$$\ln(\hat{\phi}^{(1)}) = -0.9165 \quad \text{and} \quad \hat{\theta}^{(1)} = -1.6966$$

Controlling Process Variability: Dispersion Effects in Linear Designs 193

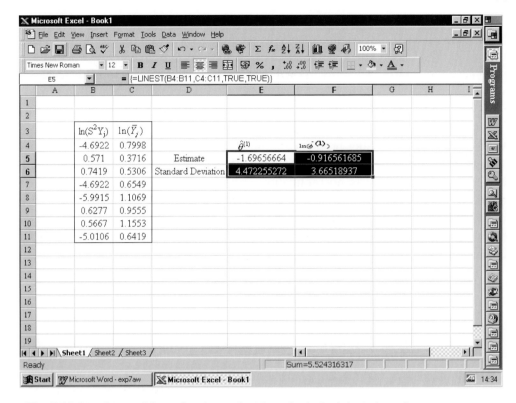

Fig. 7.12 Excel spreadsheet showing estimation of standard deviation of two parameters.

The standard deviation associated with $\hat{\theta}^{(1)}$ is 4.472. To test the hypothesis that $\theta = 0$ simply construct the usual t statistic of $t = (-1.696-0)/4.472 = -0.379$. This is not significant at the 5% level so that the true value for θ is likely to be zero and not two. It is more likely that the correct PerMIA to use is therefore

$$\phi_j = \frac{S_{Y_j}^2}{\bar{Y}_j^{\theta=0}} = (S_{Y_j}^2) \neq (S-N)_{T_j}, \text{ or } \ln(\phi_j) = \ln(S_{Y_j}^2)$$

If all the steps required for the generalised linear model were carried out the following results would be obtained.

$$\ln(\phi_j) = -5.09 + 5.72 D_j, \quad D_{Disp}^{Ali} = 2 \times 5.72 = 11.44$$

The Taguchi analysis identifies the same factor for controlling variability although the magnitude of the estimated dispersion effect is about half that shown above.

A cautionary note needs to be made about aliasing here. Both approaches identify factor D as a control factor. Assuming all second and higher order interactions are zero, the estimated dispersion effect of 11.44 is in fact an estimate of the D dispersion effect less the AB, CG and EF dispersion interaction effects. Thus it is quite possible that factors A, B, C, D, E, F and G affect variability as well and it is these factors which should be manipulated to control variability. As will be seen next, this appears to be the case.

7.8.2.2 The Lack of Analysis for Noise Factors

An insight into this aliasing problem can be obtained by using a first order response surface model that includes the noise factors in the analysis. This involves analysing all ten factors so that there is no replication at each test condition. That is, the 2^{10-5} representation of the experiment shown in Table 7.16. The 31 confounded interactions for this design were shown in Section 7.8.1 above and can be used to find the 31 aliased effects that can be estimated from this experiment. Thus multiplying all these confounded interactions by A gives the aliased A effect

$$A = -BD = -CE = -ABCF = -BCG = -AZ_1Z_2Z_3 = ABCDE = CDF = -ACDG =$$
$$BDZ_1Z_2Z_3 = BEF = -ABEG = CEZ_1Z_2Z_3 = -FG = ABCFZ_1Z_2Z_3 = -BCGZ_1Z_2Z_3 =$$
$$-ADEF = DEG = -ABCDEZ_1Z_2Z_3 = ABDFG = -CDFZ_1Z_2Z_3 = ACDGZ_1Z_2Z_3 =$$
$$ACEFG = -BEFZ_1Z_2Z_3 = ABEGZ_1Z_2Z_3 = FGZ_1Z_2Z_3 = -BCDEFG = ADEFZ_1Z_2Z_3 =$$
$$ABDFGZ_1Z_2Z_3 = -DEGZ_1Z_2Z_3 = -ACEFGZ_1Z_2Z_3 = -BCDEFGZ_1Z_2Z_3.$$

To keep things simple, and in accordance with the structure of a first order response surface model, it will be assumed from now on that all second and higher order interactions are zero. Under this assumption the above alias structure for factor A simplifies to

$$A = -BD = -CE = -FG.$$

That is, the estimated aliased A_{Loc} effect will in fact equal the A_{Loc} effect less the BD_{Loc} effect less the CE_{Loc} effect less the FG_{Loc} effect from the full 2^{10} design.

$$A_{Loc}^{Ali} = A_{Loc} - BD_{Loc} - CE_{Loc} - FG_{Loc}$$

Under the same assumptions about higher order interactions, all aliased main location effects can be found in the same way.

$$\begin{aligned}
A &= -BD = -CE = -FG, \\
B &= -AD = -CF = -EG, \\
C &= -AE = -BF = -DG, \\
D &= -AB = -CG = -EF, \\
E &= -AC = -BG = -DF, \\
F &= -BC = -AG = -DE, \\
G &= -CD = -BE = -AF, \\
Z_1 &= -Z_2Z_3, \\
Z_2 &= -Z_1Z_3, \\
Z_3 &= -Z_1Z_2.
\end{aligned}$$

Controlling Process Variability: Dispersion Effects in Linear Designs

Notice the main location effect for each design factor is aliased with first order interactions between the design factors and that the main location effect for each noise factor is aliased with first order interaction between the noise factors. But such first order interactions could be important for putting the mean on target and this experimental design is not going to yield any reliable information on how to put the mean on target using the main location effects associated with the design factors. For the same number of tests a resolution IV design could have been set up and this information obtained.

However, the remaining aliased effects are between each design and noise factor. Here all third and higher order interactions are assumed zero to give

$$
\begin{aligned}
AZ_1 &= -BDZ_1 = -CEZ_1 = -AZ_2Z_3 = -FGZ_1, \\
AZ_2 &= -BDZ_2 = -CEZ_2 = -AZ_1Z_3 = -FGZ_2, \\
AZ_3 &= -BDZ_3 = -CEZ_3 = -AZ_1Z_2 = -FGZ_3, \\
BZ_1 &= -ADZ_1 = -CFZ_1 = -BZ_2Z_3 = -EGZ_1, \\
BZ_2 &= -ADZ_2 = -CFZ_2 = -BZ_1Z_3 = -EGZ_2, \\
BZ_3 &= -ADZ_3 = -CFZ_3 = -BZ_1Z_2 = -EGZ_3, \\
CZ_1 &= -AEZ_1 = -BFZ_1 = -CZ_2Z_3 = -DGZ_1, \\
CZ_2 &= -AEZ_2 = -BFZ_2 = -CZ_1Z_3 = -DGZ_2, \\
CZ_3 &= -AEZ_3 = -BFZ_3 = -CZ_1Z_2 = -DGZ_3, \\
DZ_1 &= -ABZ_1 = -DZ_2Z_3 = -CGZ_1 = -EFZ_1, \\
DZ_2 &= -ABZ_2 = -DZ_1Z_3 = -CGZ_2 = -EFZ_2, \\
DZ_3 &= -ABZ_3 = -DZ_1Z_2 = -CGZ_3 = -EFZ_3, \\
EZ_1 &= -ACZ_1 = -EZ_2Z_3 = -BGZ_1 = -DFZ_1, \\
EZ_2 &= -ACZ_2 = -EZ_1Z_3 = -BGZ_2 = -DFZ_2, \\
EZ_3 &= -ACZ_3 = -EZ_1Z_2 = -BGZ_3 = -DFZ_3, \\
FZ_1 &= -BCZ_1 = -FZ_2Z_3 = -AGZ_1 = -DEZ_1, \\
FZ_2 &= -BCZ_2 = -FZ_1Z_3 = -AGZ_2 = -DEZ_2, \\
FZ_3 &= -BCZ_3 = -FZ_1Z_2 = -AGZ_3 = -DEZ_3, \\
GZ_1 &= -GZ_2Z_3 = -CDZ_1 = -BEZ_1 = -AFZ_1, \\
GZ_2 &= -GZ_1Z_3 = -CDZ_2 = -BEZ_2 = -AFZ_2, \\
GZ_3 &= -GZ_1Z_2 = -CZ_1Z_3 = -BEZ_3 = -AFZ_3.
\end{aligned}
$$

Notice that the interactions between the design and noise factors are aliased with second and higher order interactions only. These can reasonably be assumed to be zero. Thus this design will enable an estimate of these interactions to be made. As they are crucial to trying to minimize variability, this design should yield a strategy for variance minimisation.

Table 7.19 shows the Yates procedure for estimating all of the 31 aliased effects shown above. The first ten columns of coded test conditions are those shown in the 2^{10-5} Table 7.16 and the test conditions are shown alongside these columns. The products suggested by the aliased effects structure above then give the remaining columns. That is, column Z_3 is followed by AZ_1 whose elements are given by the product of the numbers in columns A and Z_1. And so on. Each contrast is then divided by $2^{10-5}/2 = 16$ to give the aliased location effects in Table 7.19. The β, λ and δ values of the response surface model given by equation (7.6) are simply the location effects divided through by two.

Table 7.19 The Yates procedure for location effects using the 2^{10-5} representation of the injection moulding experiment.

Test	\multicolumn{9}{c}{Coded Test Conditions}	Response
	A	B	C	Z_1	Z_2	$D = -AB$	$E = -AC$	$F = -BC$	$G = ABC$	$Z_3 = -Z_1Z_2$	Y
(1)	−1	−1	−1	−1	−1	−1	−1	−1	−1	−1	2.2
$adeg$	1	−1	−1	−1	−1	1	1	−1	1	−1	0.3
$bdfg$	−1	1	−1	−1	−1	1	−1	1	1	−1	0.5
$abef$	1	1	−1	−1	−1	−1	1	1	−1	−1	2
$cefg$	−1	−1	1	−1	−1	−1	1	1	1	−1	3
$acdf$	1	−1	1	−1	−1	1	−1	1	−1	−1	2.1
$bcde$	−1	1	1	−1	−1	1	1	−1	−1	−1	4
$abcg$	1	1	1	−1	−1	−1	−1	−1	1	−1	2
z_1z_3	−1	−1	−1	1	−1	−1	−1	−1	−1	1	2.3
az_1degz_3	1	−1	−1	1	−1	1	1	−1	1	1	2.7
bz_1dfgz_3	−1	1	−1	1	−1	1	−1	1	1	1	0.4
abz_1efgz_3	1	1	−1	1	−1	−1	1	1	−1	1	1.8
cz_1efgz_3	−1	−1	1	1	−1	−1	1	1	1	1	3
acz_1dfz_3	1	−1	1	1	−1	1	−1	1	−1	1	1
bcz_1dez_3	−1	1	1	1	−1	1	1	−1	−1	1	4.6
$abcz_1gz_3$	1	1	1	1	−1	−1	−1	−1	1	1	1.9
z_2z_3	−1	−1	−1	−1	1	−1	−1	−1	−1	1	2.1
az_2degz_3	1	−1	−1	−1	1	1	1	−1	1	1	2.5
bz_2dfgz_3	−1	1	−1	−1	1	1	−1	1	1	1	3.1
abz_2efz_3	1	1	−1	−1	1	−1	1	1	−1	1	1.9
cz_2efz_3	−1	−1	1	−1	1	−1	1	1	1	1	3.1
acz_2dfz_3	1	−1	1	−1	1	1	−1	1	−1	1	4.2
bcz_2dez_3	−1	1	1	−1	1	1	1	−1	−1	1	1.9
$abcz_2gz_3$	1	1	1	−1	1	−1	−1	−1	1	1	1.9
z_1z_2	−1	−1	−1	1	1	−1	−1	−1	−1	−1	2.3
az_1z_2deg	1	−1	−1	1	1	1	1	−1	1	−1	0.3
bz_1z_2dfg	−1	1	−1	1	1	1	−1	1	1	−1	2.8
abz_1z_2ef	1	1	−1	1	1	−1	1	1	−1	−1	2
cz_1z_2efg	−1	−1	1	1	1	−1	1	1	1	−1	3
acz_1z_2df	1	−1	1	1	1	1	−1	1	−1	−1	3.1
bcz_1z_2de	−1	1	1	1	1	1	1	−1	−1	−1	2.2
$abcz_1z_2g$	1	1	1	1	1	−1	−1	−1	1	−1	1.8

Table 7.19 Continued.

Test	\multicolumn{9}{c}{Coded Test Conditions}

Test	AZ_1	AZ_2	AZ_3	BZ_1	BZ_2	BZ_3	CZ_1	CZ_2	CZ_3
(1)	1	1	1	1	1	1	1	1	1
$adeg$	−1	−1	−1	1	1	1	1	1	1
$bdfg$	1	1	1	−1	−1	−1	1	1	1
$abef$	−1	−1	−1	−1	−1	−1	1	1	1
$cefg$	1	1	1	1	1	1	−1	−1	−1
$acdf$	−1	−1	−1	1	1	1	−1	−1	−1
$bcde$	1	1	1	−1	−1	−1	−1	−1	−1
$abcg$	−1	−1	−1	−1	−1	−1	−1	−1	−1
z_1z_3	−1	1	−1	−1	1	−1	−1	1	−1
az_1degz_3	1	−1	1	−1	1	−1	−1	1	−1
bz_1dfgz_3	−1	1	−1	1	−1	1	−1	1	−1
abz_1efgz_3	1	−1	1	1	−1	1	−1	1	−1
cz_1efgz_3	−1	1	−1	−1	1	−1	1	−1	1
acz_1dfz_3	1	−1	1	−1	1	−1	1	−1	1
bcz_1dez_3	−1	1	−1	1	−1	1	1	−1	1
$abcz_1gz_3$	1	−1	1	1	−1	1	1	−1	1
z_2z_3	1	−1	−1	1	−1	−1	1	−1	−1
az_2degz_3	−1	1	1	1	−1	−1	1	−1	−1
bz_2dfgz_3	1	−1	−1	−1	1	1	1	−1	−1
abz_2efz_3	−1	1	1	−1	1	1	1	−1	−1
cz_2efz_3	1	−1	−1	1	−1	−1	−1	1	1
acz_2dfz_3	−1	1	1	1	−1	−1	−1	1	1
bcz_2dez_3	1	−1	−1	−1	1	1	−1	1	1
$abcz_2gz_3$	−1	1	1	−1	1	1	−1	1	1
z_1z_2	−1	−1	1	−1	−1	1	−1	−1	1
az_1z_2deg	1	1	−1	−1	−1	1	−1	−1	1
bz_1z_2dfg	−1	−1	1	1	1	−1	−1	−1	1
abz_1z_2ef	1	1	−1	1	1	−1	−1	−1	1
cz_1z_2efg	−1	−1	1	−1	−1	1	1	1	−1
acz_1z_2df	1	1	−1	−1	−1	1	1	1	−1
bcz_1z_2de	−1	−1	1	1	1	−1	1	1	−1
$abcz_1z_2g$	1	1	−1	1	1	−1	1	1	−1

Table 7.19 Continued.

	Coded Test Conditions								
Test	DZ_1	DZ_2	DZ_3	EZ_1	EZ_2	EZ_3	FZ_1	FZ_2	FZ_3
(1)	1	1	1	1	1	1	1	1	1
$adeg$	−1	−1	−1	−1	−1	−1	1	1	1
$bdfg$	−1	−1	−1	1	1	1	−1	−1	−1
$abef$	1	1	1	−1	−1	−1	−1	−1	−1
$cefg$	1	1	1	−1	−1	−1	−1	−1	−1
$acdf$	−1	−1	−1	1	1	1	−1	−1	−1
$bcde$	−1	−1	−1	−1	−1	−1	1	1	1
$abcg$	1	1	1	1	1	1	1	1	1
z_1z_3	−1	1	−1	−1	1	−1	−1	1	−1
az_1degz_3	1	−1	1	1	−1	1	−1	1	−1
bz_1dfgz_3	1	−1	1	−1	1	−1	1	−1	1
abz_1efgz_3	−1	1	−1	1	−1	1	1	−1	1
cz_1efgz_3	−1	1	−1	1	−1	1	1	−1	1
acz_1dfz_3	1	−1	1	−1	1	−1	1	−1	1
bcz_1dez_3	1	−1	1	1	−1	1	−1	1	−1
$abcz_1gz_3$	−1	1	−1	−1	1	−1	−1	1	−1
z_2z_3	1	−1	−1	1	−1	−1	1	−1	−1
az_2degz_3	−1	1	1	−1	1	1	1	−1	−1
bz_2dfgz_3	−1	1	1	1	−1	−1	−1	1	1
abz_2efz_3	1	−1	−1	−1	1	1	−1	1	1
cz_2efz_3	1	−1	−1	−1	1	1	−1	1	1
acz_2dfz_3	−1	1	1	1	−1	−1	−1	1	1
bcz_2dez_3	−1	1	1	−1	1	1	1	−1	−1
$abcz_2gz_3$	1	−1	−1	1	−1	−1	1	−1	−1
z_1z_2	−1	−1	1	−1	−1	1	−1	−1	1
az_1z_2deg	1	1	−1	1	1	−1	−1	−1	1
bz_1z_2dfg	1	1	−1	−1	−1	1	1	1	−1
abz_1z_2ef	−1	−1	1	1	1	−1	1	1	−1
cz_1z_2efg	−1	−1	1	1	1	−1	1	1	−1
acz_1z_2df	1	1	−1	−1	−1	1	1	1	−1
bcz_1z_2de	1	1	−1	1	1	−1	−1	−1	1
$abcz_1z_2g$	−1	−1	1	−1	−1	1	−1	−1	1

Table 7.19 Continued.

Test	Coded Test Conditions		
	GZ_1	GZ_2	GZ_3
(1)	1	1	1
$adeg$	−1	−1	−1
$bdfg$	−1	−1	−1
$abef$	1	1	1
$cefg$	−1	−1	−1
$acdf$	1	1	1
$bcde$	1	1	1
$abcg$	−1	−1	−1
$z_1 z_3$	−1	1	−1
$az_1 deg z_3$	1	−1	1
$bz_1 dfg z_3$	1	−1	1
$abz_1 efg z_3$	−1	1	−1
$cz_1 efg z_3$	1	−1	1
$acz_1 df z_3$	−1	1	−1
$bcz_1 de z_3$	−1	1	−1
$abcz_1 g z_3$	1	−1	1
$z_2 z_3$	1	−1	−1
$az_2 deg z_3$	−1	1	1
$bz_2 dfg z_3$	−1	1	1
$abz_2 ef z_3$	1	−1	−1
$cz_2 ef z_3$	−1	1	1
$acz_2 df z_3$	1	−1	−1
$bcz_2 de z_3$	1	−1	−1
$abcz_2 g z_3$	−1	1	1
$z_1 z_2$	−1	−1	1
$az_1 z_2 deg$	1	1	−1
$bz_1 z_2 dfg$	1	1	−1
$abz_1 z_2 ef$	−1	−1	1
$cz_1 z_2 efg$	1	1	−1
$acz_1 z_2 df$	−1	−1	1
$bcz_1 z_2 de$	−1	−1	1
$abcz_1 z_2 g$	1	1	−1

Table 7.19 Continued.

Test	A_{Loc}^{Ali} $A = -BD =$ $-CE = -FG$	B_{Loc}^{Ali} $B = -AD =$ $-CF = -EG$	C_{Loc}^{Ali} $C = -AE =$ $-BF = -DG$	$Z_{1\,Loc}^{Ali}$ $Z_1 = -Z_2 Z_3$	$Z_{2\,Loc}^{Ali}$ $Z_2 = -Z_1 Z_3$	D_{Loc}^{Ali} $D = -AB =$ $-CG = -EF$	E_{Loc}^{Ali} $E = -AC =$ $-BG = -DF$	F_{Loc}^{Ali} $F = -BC =$ $-AG = -DE$	G_{Loc}^{Ali} $G = -CD =$ $-BE = -AF$	$Z_{3\,Loc}^{Ali}$ $Z_3 = -Z_1 Z_2$
(1)	-2.2	-2.2	-2.2	-2.2	-2.2	-2.2	-2.2	-2.2	-2.2	-2.2
$adeg$	0.3	-0.3	-0.3	-0.3	-0.3	0.3	0.3	-0.3	0.3	-0.3
$bdfg$	-0.5	0.5	-0.5	-0.5	-0.5	0.5	-0.5	0.5	0.5	-0.5
$abef$	2	2	-2	-2	-2	-2	2	2	-2	-2
$cefg$	-3	-3	3	-3	-3	-3	2	3	3	-3
$acdf$	2.1	-2.1	2.1	-2.1	-2.1	2.1	-2.1	2.1	-2.1	-2.1
$bcde$	-4	4	4	-4	-4	4	4	-4	-4	-4
$abcg$	2	2	2	-2	-2	-2	-2	-2	2	-2
$z_1 z_2$	-2.3	-2.3	-2.3	2.3	-2.3	-2.3	-2.3	-2.3	-2.3	2.3
$az_1 deg z_1$	2.7	-2.7	-2.7	2.7	-2.7	2.7	2.7	-2.7	2.7	2.7
$bz_1 dfg z_1$	-0.4	0.4	-0.4	0.4	-0.4	0.4	-0.4	0.4	0.4	0.4
$abz_1 efg z_1$	1.8	1.8	-1.8	1.8	-1.8	-1.8	1.8	1.8	-1.8	1.8
$cz_1 efg z_1$	-3	-3	3	3	-3	-3	3	3	3	3
$acz_1 df z_1$	1	-1	1	1	-1	1	-1	1	-1	1
$bcz_1 de z_1$	-4.6	4.6	4.6	4.6	-4.6	4.6	4.6	-4.6	-4.6	4.6
$abcz_1 g z_1$	1.9	1.9	1.9	1.9	-1.9	-1.9	-1.9	-1.9	1.9	1.9
$z_2 z_3$	-2.1	-2.1	-2.1	-2.1	2.1	-2.1	-2.1	-2.1	-2.1	2.1
$az_2 deg z_3$	2.5	-2.5	-2.5	-2.5	2.5	2.5	2.5	-2.5	2.5	2.5
$bz_2 dfg z_3$	-3.1	3.1	-3.1	-3.1	3.1	3.1	-3.1	3.1	3.1	3.1
$abz_2 ef z_3$	1.9	1.9	-1.9	-1.9	1.9	-1.9	1.9	1.9	-1.9	1.9
$cz_2 ef z_3$	-3.1	-3.1	3.1	-3.1	3.1	-3.1	3.1	3.1	3.1	3.1
$acz_2 df z_3$	4.2	-4.2	4.2	-4.2	4.2	4.2	-4.2	4.2	-4.2	4.2
$bcz_2 de z_3$	-1.9	1.9	1.9	-1.9	1.9	1.9	1.9	-1.9	-1.9	1.9
$abcz_2 g z_3$	1.9	1.9	1.9	-1.9	1.9	-1.9	-1.9	-1.9	1.9	1.9
$z_1 z_3$	-2.3	-2.3	-2.3	2.3	2.3	-2.3	-2.3	-2.3	-2.3	-2.3
$az_1 z_3 deg$	0.3	-0.3	-0.3	0.3	0.3	0.3	0.3	-0.3	0.3	-0.3
$bz_1 z_3 dfg$	-2.8	2.8	-2.8	2.8	2.8	2.8	-2.8	2.8	2.8	-2.8
$abz_1 z_3 ef$	2	2	-2	2	2	-2	2	2	-2	-2
$cz_1 z_3 efg$	-3	-3	3	3	3	-3	3	3	3	-3
$acz_1 z_3 df$	3.1	-3.1	3.1	3.1	3.1	3.1	-3.1	3.1	-3.1	-3.1
$bcz_1 z_3 de$	-2.2	2.2	2.2	2.2	2.2	2.2	2.2	-2.2	-2.2	-2.2
$abcz_1 z_3 g$	1.8	1.8	1.8	1.8	1.8	-1.8	-1.8	-1.8	1.8	-1.8
Loc Contrast	-9	-2.4	13.6	-1.6	4.4	-0.6	4.6	2	-7.4	4.8
Loc Effect	-0.5625	-0.15	0.85	-0.1	0.275	-0.0375	0.2875	0.125	-0.4625	0.3
β and λ Values	-0.28125	-0.075	0.425	-0.05	0.1375	-0.01875	0.14375	0.0625	-0.23125	0.15

Table 7.19 Continued.

Test	AZ_{1Loc}^{Ali} $AZ_1=-BDZ_1=$ $-CEZ_1=$ $-AZ_1Z_1=-FGZ_1$	AZ_{2Loc}^{Ali} $AZ_2=-BDZ_2=$ $-CEZ_2=$ $-AZ_2Z_2=-FGZ_2$	AZ_{3Loc}^{Ali} $AZ_3=-BDZ_3=$ $-CEZ_3=$ $-AZ_3Z_3=-FGZ_3$	BZ_{1Loc}^{Ali} $BZ_1=-ADZ_1=$ $-CFZ_1=$ $-BZ_1Z_1=-EGZ_1$	BZ_{2Loc}^{Ali} $BZ_2=-ADZ_2=$ $-CFZ_2=$ $-BZ_2Z_2=-EGZ_2$	BZ_{3Loc}^{Ali} $BZ_3=-ADZ_3=$ $-CFZ_3=$ $-BZ_3Z_3=-EGZ_3$	CZ_{1Loc}^{Ali} $CZ_1=-AEZ_1=$ $-BFZ_1=$ $-CZ_1Z_1=-DGZ_1$	CZ_{2Loc}^{Ali} $CZ_2=-AEZ_2=$ $-BFZ_2=$ $-CZ_2Z_2=-DGZ_2$	CZ_{3Loc}^{Ali} $CZ_3=-AEZ_3=$ $-BFZ_3=$ $-CZ_3Z_3=-DGZ_3$
(1)	2.2	2.2	2.2	2.2	2.2	2.2	2.2	2.2	2.2
adeg	−0.3	−0.3	−0.3	0.3	0.3	0.3	0.3	0.3	0.3
bdfg	0.5	0.5	0.5	−0.5	−0.5	−0.5	0.5	0.5	0.5
abef	−2	−2	−2	−2	−2	−2	2	2	2
cefg	3	3	3	3	3	3	−3	−3	−3
acdf	−2.1	−2.1	−2.1	2.1	2.1	2.1	−2.1	−2.1	−2.1
bcde	4	4	4	−4	−4	−4	−4	−4	−4
abcg	−2	−2	−2	−2	−2	−2	−2	−2	−2
z_1z_1	−2.3	2.3	−2.3	−2.3	2.3	−2.3	−2.3	2.3	−2.3
az_1degz_1	2.7	−2.7	2.7	−2.7	2.7	−2.7	−2.7	2.7	−2.7
bz_1dfgz_1	−0.4	0.4	−0.4	0.4	−0.4	0.4	−0.4	0.4	−0.4
abz_1efgz_1	1.8	−1.8	1.8	1.8	−1.8	1.8	−1.8	1.8	−1.8
cz_1efgz_1	−3	3	−3	−3	3	−3	3	−3	3
acz_1dfz_1	1	−1	1	−1	1	−1	1	−1	1
bcz_1dez_1	−4.6	4.6	−4.6	4.6	−4.6	4.6	4.6	−4.6	4.6
$abcz_1gz_1$	1.9	−1.9	1.9	1.9	−1.9	1.9	1.9	−1.9	1.9
z_2z_2	2.1	−2.1	−2.1	2.1	−2.1	−2.1	2.1	−2.1	−2.1
az_2degz_2	−2.5	2.5	2.5	2.5	−2.5	−2.5	2.5	−2.5	−2.5
bz_2dfgz_2	3.1	−3.1	−3.1	−3.1	3.1	3.1	3.1	−3.1	−3.1
abz_2efgz_2	−1.9	1.9	1.9	−1.9	1.9	1.9	1.9	−1.9	−1.9
cz_2efz_2	3.1	−3.1	−3.1	3.1	−3.1	−3.1	−3.1	3.1	3.1
acz_2dfz_2	−4.2	4.2	4.2	4.2	−4.2	−4.2	−4.2	4.2	4.2
bcz_2dez_2	1.9	−1.9	−1.9	−1.9	1.9	1.9	−1.9	1.9	1.9
$abcz_2gz_2$	−1.9	1.9	1.9	−1.9	1.9	1.9	−1.9	1.9	1.9
z_3z_3	−2.3	−2.3	2.3	−2.3	−2.3	2.3	−2.3	−2.3	2.3
az_3deg	0.3	0.3	−0.3	−0.3	−0.3	0.3	−0.3	−0.3	0.3
bz_3dfg	−2.8	−2.8	2.8	2.8	2.8	−2.8	−2.8	−2.8	2.8
abz_3ef	2	2	−2	2	2	−2	−2	−2	2
cz_3efg	−3	−3	3	−3	−3	3	3	3	−3
acz_3df	3.1	3.1	−3.1	−3.1	−3.1	3.1	3.1	3.1	−3.1
bcz_3de	−2.2	−2.2	2.2	2.2	2.2	−2.2	2.2	2.2	−2.2
$abcz_3g$	1.8	1.8	−1.8	1.8	1.8	−1.8	1.8	1.8	−1.8
Loc Contrast	−3	3.4	3.8	2	−3.6	−4.4	−1.6	−5.2	−4
Loc Effect	−0.1875	0.2125	0.2375	0.125	−0.225	−0.275	−0.1	−0.325	−0.25
δ Values	−0.09375	0.10625	0.11875	0.0625	−0.1125	−0.1375	−0.05	−0.1625	−0.125

Table 7.19 Continued.

Test	$DZ_{1\,Loc}^{Ali}$ $DZ_1=-ABZ_1=$ $-DZ_1Z_3=$ $-CGZ_1=-EFZ_1$	$DZ_{2\,Loc}^{Ali}$ $DZ_2=-ABZ_2=$ $-DZ_1Z_4=$ $-CGZ_2=-EFZ_2$	$DZ_{3\,Loc}^{Ali}$ $DZ_3=-ABZ_3=$ $-DZ_1Z_2=$ $-CGZ_3=-EFZ_3$	$EZ_{1\,Loc}^{Ali}$ $EZ_1=-ACZ_1=$ $-EZ_1Z_3=$ $-BGZ_1=-DFZ_1$	$EZ_{2\,Loc}^{Ali}$ $EZ_2=-ACZ_2=$ $-EZ_1Z_4=$ $-BGZ_2=-DFZ_2$	$EZ_{3\,Loc}^{Ali}$ $EZ_3=-ACZ_3=$ $-EZ_1Z_2=$ $-BGZ_3=-DFZ_3$	$FZ_{1\,Loc}^{Ali}$ $FZ_1=-BCZ_1=$ $-FZ_1Z_3=$ $-AGZ_1=-DEZ_1$	$FZ_{2\,Loc}^{Ali}$ $FZ_2=-BCZ_2=$ $-FZ_1Z_4=$ $-AGZ_2=-DEZ_2$	$FZ_{3\,Loc}^{Ali}$ $FZ_3=-BCZ_3=$ $-FZ_1Z_2=$ $-AGZ_3=-DEZ_3$
(1)	2.2	2.2	2.2	2.2	2.2	2.2	2.2	2.2	2.2
$adeg$	-0.3	-0.3	-0.3	-0.3	-0.3	-0.3	0.3	0.3	0.3
$bdfg$	-0.5	-0.5	-0.5	0.5	0.5	0.5	-0.5	-0.5	-0.5
$abef$	2	2	2	-2	-2	-2	-2	-2	-2
$cefg$	3	3	3	-3	-3	-3	-3	-3	-3
$acdf$	-2.1	-2.1	-2.1	2.1	2.1	2.1	-2.1	-2.1	-2.1
$bcde$	-4	-4	-4	-4	-4	-4	4	4	4
$abcg$	2	2	2	2	2	2	2	2	2
z_1z_3	-2.3	2.3	-2.3	-2.3	2.3	-2.3	-2.3	2.3	-2.3
az_3degz_3	2.7	-2.7	2.7	2.7	-2.7	2.7	-2.7	2.7	-2.7
bz_3dfgz_3	0.4	-0.4	0.4	-0.4	0.4	-0.4	0.4	-0.4	0.4
abz_3efgz_3	-1.8	1.8	-1.8	1.8	-1.8	1.8	1.8	-1.8	1.8
cz_3efgz_3	-3	3	-3	3	-3	3	3	-3	3
acz_3dfz_3	1	-1	1	-1	1	-1	1	-1	1
bcz_3dez_3	4.6	-4.6	4.6	4.6	-4.6	4.6	-4.6	4.6	-4.6
$abcz_3gz_3$	-1.9	1.9	-1.9	-1.9	1.9	-1.9	-1.9	1.9	-1.9
z_2z_3	2.1	-2.1	-2.1	2.1	-2.1	-2.1	2.1	-2.1	-2.1
az_2degz_3	-2.5	2.5	2.5	-2.5	2.5	2.5	2.5	-2.5	-2.5
bz_2dfgz_3	-3.1	3.1	3.1	3.1	-3.1	-3.1	-3.1	3.1	3.1
abz_2efz_3	1.9	-1.9	-1.9	-1.9	1.9	1.9	-1.9	1.9	1.9
$cz_2z_3efz_3$	3.1	-3.1	-3.1	-3.1	3.1	3.1	-3.1	3.1	3.1
$acz_2df z_3$	-4.2	4.2	4.2	4.2	-4.2	-4.2	-4.2	4.2	4.2
bcz_2dez_3	-1.9	1.9	1.9	-1.9	1.9	1.9	1.9	-1.9	-1.9
$abcz_2gz_3$	1.9	-1.9	-1.9	1.9	-1.9	-1.9	1.9	-1.9	-1.9
z_2z_3	-2.3	-2.3	2.3	-2.3	-2.3	2.3	-2.3	-2.3	2.3
az_2z_3deg	0.3	0.3	-0.3	0.3	0.3	-0.3	-0.3	-0.3	0.3
bz_2z_3dfg	2.8	2.8	-2.8	-2.8	-2.8	2.8	2.8	2.8	-2.8
abz_2z_3ef	-2	-2	2	2	2	-2	2	2	-2
cz_2z_3efg	-3	-3	3	3	3	-3	3	3	-3
acz_2z_3df	3.1	3.1	-3.1	-3.1	-3.1	3.1	3.1	3.1	-3.1
bcz_2z_3de	2.2	2.2	-2.2	2.2	2.2	-2.2	-2.2	-2.2	2.2
$abcz_2z_3g$	-1.8	-1.8	1.8	-1.8	-1.8	1.8	-1.8	-1.8	1.8
Loc Contrast	-1.4	4.6	5.4	3.4	-13.4	4.6	-4	14.4	-4.8
Loc Effect	-0.0875	0.2875	0.3375	0.2125	-0.8375	0.2875	-0.25	0.9	-0.3
δ values	-0.04375	0.14375	0.16875	0.10625	-0.41875	0.14375	-0.125	0.45	-0.15

Table 7.19 Continued.

Test	$GZ_{1\,Loc}^{Ali}$ $GZ_1 = -GZ_2Z_3 =$ $-CDZ_1 =$ $-BEZ_1 = -AFZ_1$	$GZ_{2\,Loc}^{Ali}$ $GZ_2 = -GZ_1Z_3 =$ $-CDZ_2 =$ $-BEZ_2 = -AFZ_2$	$GZ_{3\,Loc}^{Ali}$ $GZ_3 = -GZ_1Z_2 =$ $-CZ_1Z_3 =$ $-BEZ_3 = -AFZ_3$
(1)	2.2	2.2	2.2
$adeg$	–0.3	–0.3	–0.3
$bdfg$	–0.5	–0.5	–0.5
$abef$	2	2	2
$cefg$	–3	–3	–3
$acdf$	2.1	2.1	2.1
$bcde$	4	4	4
$abcg$	–2	–2	–2
z_1z_3	–2.3	2.3	–2.3
az_1degz_3	2.7	–2.7	2.7
bz_1dfgz_3	0.4	–0.4	0.4
abz_1efgz_3	–1.8	1.8	–1.8
cz_1efgz_3	3	–3	3
acz_1dfz_3	–1	1	–1
bcz_1dez_3	–4.6	4.6	–4.6
$abcz_1gz_3$	1.9	–1.9	1.9
z_2z_3	2.1	–2.1	–2.1
az_2degz_3	–2.5	2.5	2.5
bz_2dfgz_3	–3.1	3.1	3.1
abz_2efz_3	1.9	–1.9	–1.9
cz_2efz_3	–3.1	3.1	3.1
acz_2dfz_3	4.2	–4.2	–4.2
bcz_2dez_3	1.9	–1.9	–1.9
$abcz_2gz_3$	–1.9	1.9	1.9
z_1z_2	–2.3	–2.3	2.3
az_1z_2deg	0.3	0.3	–0.3
bz_1z_2dfg	2.8	2.8	–2.8
abz_1z_2ef	–2	–2	2
cz_1z_2efg	3	3	–3
acz_1z_2df	–3.1	–3.1	3.1
bcz_1z_2de	–2.2	–2.2	2.2
$abcz_1z_2g$	1.8	1.8	–1.8
Loc Contrast	0.6	5	5
Loc Effect	0.0375	0.3125	0.3125
δ Values	0.01875	0.15625	0.15625

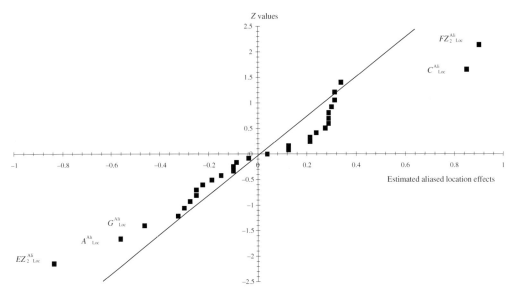

Fig. 7.13 Probability plot for the 2^{10-5} representation of the injection moulding experiment.

A probability plot for all these aliased location effects is shown in Figure 7.13. It would appear from this that factors A (holding pressure), C (cycle time) and G (gate size) are important as are the interactions between factors F and Z_2 (between cavity thickness and moisture content) and E and Z_2 (injection speed and moisture content).

The simplified first order response surface model for the injection moulding experiment is therefore (taking the β, λ and δ values from Table 7.19).

$$Y_j = \beta_0 - 0.281 A_j + 0.425 C_j - 0.231 G_j - 0.419 E_j Z_{2_j} + 0.45 F_j Z_{2_j} + \varepsilon_j$$

Guidance on how the mean can be placed on target can only be obtained if all the first order interactions between the design factors are assumed zero. Then the estimated A_{Loc}^{Ali}, C_{Loc}^{Ali} and G_{Loc}^{Ali} shown in Table 7.19 can be interpreted as the true actual A_{Loc}, C_{Loc} and G_{Loc} effects respectively. More will be said about this in Chapter 8.

The main conflict between this approach and the Taguchi/Engel approach comes when analysing variability. The Taguchi and generalised linear model analysis suggested that factor D could be used to minimise variability. In the first order response surface model estimated above, Steinberg and Bursztyn's second rule suggests that variability in shrinkage appears to be due to uncontrollable movements in noise factor Z_2, i.e. the moisture content. Because this noise factor interacts with design factors E and F the transmission of variability from moisture content to shrinkage can be minimized through a careful selection of the levels for these design factors. Factor D does not appear to be important at all in determining variability!

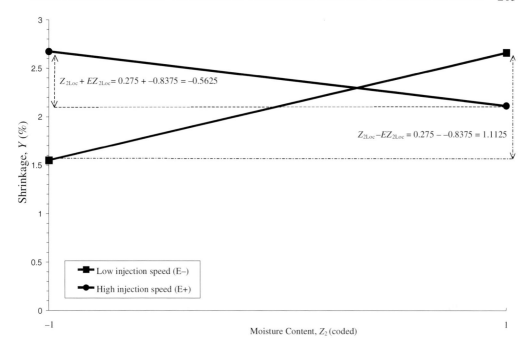

Fig. 7.14 The injection speed/moisture content, Z_2 (coded) location interaction effect, $EZ_{2\,Loc}$.

These two noise-design factor interactions highlighted in the equation above are shown in Figures 7.14 and 7.15. These are standard two way diagrams where the end point of each line is given by the average response corresponding to that test condition. For example, the point on the E low line when Z_2 is also low is given by averaging the eight responses obtained at this test condition. This corresponds to the test row numbers 1 (test(1)), 3, 6, 8, 9,11,14 and 16 (test $abcz_1z_3g$) in Table 7.16.

These two plots suggest that setting factor F low and factor E high will minimise variability in shrinkage. That is, the slopes of the two lines in Figures 7.14 and 7.15 are smallest when E is set high and F set low. Then movements in shrinkage generated by uncontrollable movements in moisture content along the horizontal axis are minimised. Most important however is the realisation that these effects are what they seem to be. As shown above, none of these interactions are aliased with any other first order interactions so complications arising from aliasing can be ignored under the realistic assumption that second order interactions are unimportant. Thus, it is factors E and F that need to be manipulated to minimise variability.

But there is an apparent contradiction in all this analysis. The Taguchi and generalised linear model approaches concluded that variability in shrinkage can be minimised by manipulating factor D. The response surface model suggests variability is minimised by

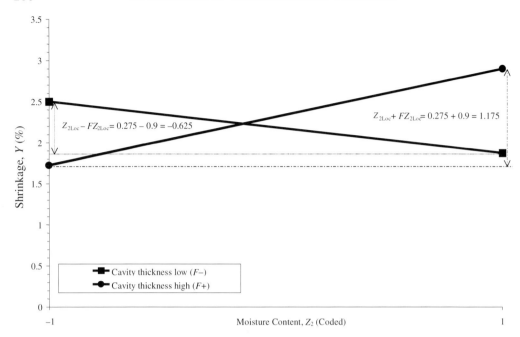

Fig. 7.15 The cavity thickness/moisture content location interaction effect, $FZ_{2\ Loc}$.

manipulation factors E and F. Factor D does not appear to enter the equation! Which is correct?

The estimated first order response surface model provides the answer. Figures 7.14 and 7.15 suggest that when factor E is set low, the variability in shrinkage resulting from changes in moisture content depends on the level for factor F. The moisture content location effect when factor E and F are set low is

$$Z_{2\ Loc} - FZ_{2\ Loc} - EZ_{2\ Loc} = 0.275 - 0.9 - -0.8375 = 0.2125$$

On the other hand the moisture content location effect when factor E is set low but F is set high is

$$Z_{2\ Loc} + FZ_{2\ Loc} - EZ_{2\ Loc} = 0.275 + 0.9 - -0.8375 = 2.01$$

So for a given level of factor E, uncontrollable movements in moisture content induce bigger changes (2.01 > 0.2125) in shrinkage when factor F is set high compared to when it is set low. That is, the variability in shrinkage, (due to uncontrollable movements in moisture content), at a given level for factor E depends on the level for factor F. Factors E and F clearly interact when determining variability so that it is to be expected that EF_{Disp} is important.

Controlling Process Variability: Dispersion Effects in Linear Designs

Bearing this in mind return to the Taguchi analysis above, which looked only at the design factors. As shown in Section 7.8.2, this inner array had the alias structure.

$D = -AB = -CG = -EF$

Hence it is very likely that the estimated $D_{Disp}^{Ali} = 5.819$ shown in Table 7.17 is in fact an estimate not of the effect of factor D on variability but of the EF location interaction.

This misleading conclusion obtained from the Taguchi and generalised linear model procedure stems from three sources. First, a failure to explicitly model the noise factors using a response surface model. Secondly, the blind application of the $(S-N)T$ ratio. But even if a response surface model had not been used and the generalised linear model used instead, the correct dispersion effect could have been identified if the experiment had been better planned. That is, if a resolution IV design with 32 tests had been carried out. Then Factor D would not have been aliased with EF and factors E and F would have shown up as having important dispersion effects in the generalised linear model. The bottom line is to try and avoid inner/outer array experiments that use Taguchi's orthogonal arrays.

8. Linear Process Optimisation

Good quality is achieved by ensuring that both the mean quality characteristic is close to a target value and that the scatter around that target is as small as possible. When this has been achieved the process is said to have been **optimised**. This dual objective can only be achieved if information is available on both location and dispersion effects. Having discussed the various ways in which location and dispersion effects can be measured, this book now goes on to describe how process optimisation can be achieved.

Section 8.1 describes a two step procedure for process optimisation when a manufacturing process is inherently linear. Implementation of this procedure using a constrained optimisation technique and a constant contour plot is then discussed. Then in Section 8.2 the ausforming, copper compact, disk forging and injection moulding processes of Chapter 2 are used to illustrate process optimisation.

8.1 A TWO STEP PROCESS OPTIMISATION PROCEDURE

8.1.1 THE PROCEDURE

No matter which technique is used to minimise process variation, optimising a process always involves a simple two-step approach. A process is said to be optimised from a quality control perspective, if the important quality characteristics of the product being manufactured are as close as possible to the customer required target specifications. Considering just the most important quality characteristic for the moment this means that process optimisation requires the mean of that quality characteristic to be on target and the variability about this mean to be as small as possible. Some examples are shown in Figure 8.1. Here a hypothetical manufacturing process is summarised using the normal distribution, which is used to describe the mean, range and frequency of occurrence of a quality characteristic for a single product manufactured over some period of time.

Suppose that T is the target specification for the response Y, where Y is the quality characteristic of interest. Figure 8.1 shows two possible states for the manufacturing process, both of which are sub optimal. In the first example, the mean of the process is on target, $\bar{Y} = T$, but any one manufactured product is likely to have a quality characteristic a long way from the target because the variability is quite high. That is, the distribution is very spread out because of a large value for the standard deviation of the process, S_Y. In the second example, the variability is much smaller ($S_Y^* < S_Y$) but the mean of the process is off target ($\bar{Y} > T$) so that again the quality characteristic of many of the products produced will be a long way from the target. Figure 8.2 shows a manufacturing process

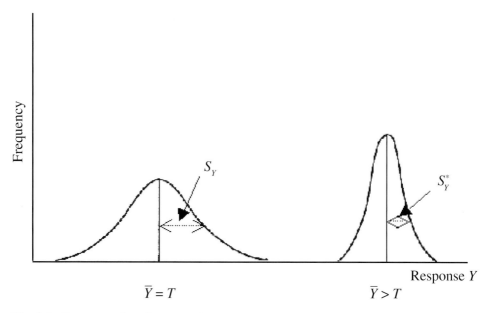

Fig. 8.1 Two examples of a sub optimal process.

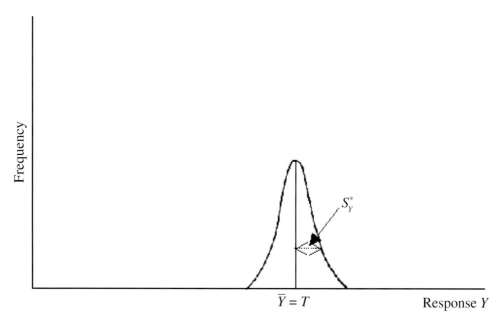

Fig. 8.2 An example of an optimal process.

Table 8.1 Types of control factors and signal factors.

Factor	Affects Mean Response?	Affects Variability in Response?	Use to
A	YES	YES	Reduce Variability
B	YES	NO	Adjust the Mean up or Down
C	NO	YES	Reduce Variability
D	NO	NO	Reduce Costs

in its optimal state. Most of the items manufactured will be very close to the target specification because the mean is on target and the variability around that mean is quite low. An **optimal process** produces a product with mean quality characteristics on target and with minimum variation about the targets.

The aim of any experimental design should be to try and achieve this optimal state. A very simple yet potentially powerful technique that can be applied to this optimisation problem is the following two step procedure. First, identify those design factors capable of compressing the distribution of the manufacturing process, i.e. those factors which will reduce S_y. Once the precise relationship between these factors and S_y is known it will then be possible to minimise the process variability by manipulating these design factors in the correct way. Remember from Chapter 7 that a dispersion effect is present whenever changes in the level of a factor results in a change in the variability rather then the mean of the response. Dispersion effects provide all the information required to minimise process variation. Any design factor, which has a dispersion effect, is termed a **control factor**.

Then, once the variability of the process has been minimised, those design factors that shift the process distribution but have no impact on process variation, can be used to place the mean of the distribution over the target. This second step involves identifying location effects, as any design factor that has a location effect, can be used to alter the mean of the process. The mean can be brought on target but only once the variability has been minimised in step one. Any design factor that has a location effect, and no dispersion effect, is termed a **signal factor**.

Some basic rules, implicit in the above two step procedure, are now summarised in Table 8.1. If only one factor affects the mean or variability of the process, use it to alter the mean or variability. No conflict or trade off between mean and variability arises in such a situation. If however one factor, such as factor A in Table 8.1, influences both the mean and variability of the process, use it to minimise the variability.

8.1.2 General Techniques

The above two-step optimisation procedure can be implemented using a **constrained optimisation** technique. Here process variability is minimised subject to the mean quality characteristic equalling some target value. Exactly how this constrained optimisation is carried out will depend upon which of the techniques discussed in Chapter 7 are used to model process variability. For example, if the generalised linear model was being used process optimisation can be achieved by minimising that component of overall variability that is independent of the mean subject to the mean equalling a target requirement. That is, find the levels for the control and signal factors that minimise

$$\phi_j = \frac{S^2_{Y_j}}{\overline{Y}_j^\theta}$$

subject to

$$\overline{Y}_j = T$$

On the other hand if the first order response surface technique is used to model process variation, then process optimisation would be achieved by minimising process variation, as given by equation (7.9), subject to the mean, given by equation (7.8), equalling a target requirement. That is, find the levels for the process variables that minimise

$$S^2_{Y_j} = S^2_{\varepsilon_j} + S^2_{Z_v} \sum_{v=M_1+1}^{M_2} \left[\frac{\partial Y}{\partial Z_v}\right]^2$$

subject to

$$\overline{Y}_j = \beta_0 + \sum_{v=1}^{M_1} \beta_v X_{v_j} + \sum_{v=1}^{M_1} \sum_{w=v+1}^{M_1} \beta_{vw} X_{v_j} X_{w_j} = T$$

For manufacturing processes that have many factors this constrained optimisation is best achieved within a package like **Excel** that has a constrained optimisation procedure built into it. **Solver** is Excel's constrained optimisation procedure. Solver will be used in some of the illustrations of Section 8.2. below. For manufacturing processes with just a few factors this optimisation can be achieved using a simple graphical technique that plots **contours** of constant mean and contours of constant process variation. As an illustration of this latter approach, suppose there is a manufacturing process that has two design factors (A and B) and no noise factors and a replicated 2^2 experiment is carried out on this process. Suppose further that the results from this experiment suggested that the variation independent of the mean, ϕ_j, is a function of both design factors

$$\ln(\phi_j) = \ln\left(\frac{S^2_{Y_j}}{\overline{Y}_j^\theta}\right) = \gamma_0 + \gamma_A A_j + \gamma_B B_j + \varepsilon_j$$

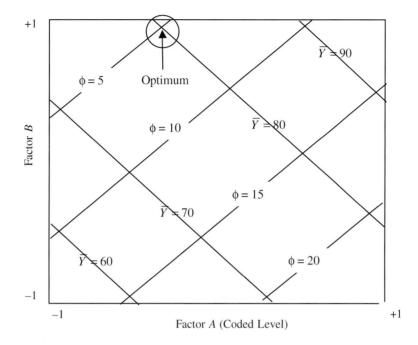

Fig. 8.3 Contours of constant mean and process variability.

Once values for the pure dispersion effects (γ values) are known and a value for the prediction error ε_j estimated, contours of constant $\ln(\phi)$ can be plotted. This is done by fixing a value for $\ln(\phi)$ and finding all those combinations of values for factors A and B in the above equation that yield this $\ln(\phi)$ value. Contour lines of constant variance independent of the mean are therefore given by

$$\frac{\ln(\phi_j) - \lambda_0 - \lambda_B B_j}{\lambda_A} = A_j$$

Figure 8.3 show hypothetical contours of constant $\ln(\phi)$ (variance independent of the mean) calculated in this way. Notice that in this example the ratio γ_A/γ_B is positive because the contours of constant variance slope up. Further, the variance independent of the mean increases as factor A increases for a given value for factor B and decreases with increases in the level for factor B given a value for factor A.

Suppose also that the results from this experiment suggested that the mean also depends on both factors without interaction

$$\overline{Y}_j = \beta_0 + \beta_A A_j + \beta_B B_j + \varepsilon_j$$

Once values for the location effects (β values) are known and a value for the prediction error ε_j estimated, contours of constant mean can be plotted. This is done by fixing a value for $\bar{Y}$ and finding all those combinations of values for factors A and B in the above equation that give this $\bar{Y}$ value. Contour lines of constant mean are therefore given by

$$\frac{\bar{Y}_j - \beta_0 - \beta_B B_j}{\beta_A} = A_j$$

Figure 8.3 shows hypothetical contours of constant $\bar{Y}$ calculated in this way. Notice that in this example β_A/β_B is negative because the contours of constant $\bar{Y}$ slope down. Further, the mean increases as factor A increases (for a given value for factor B) and increases with increases in the level for factor B (for given a value for factor A).

Suppose, finally that the main customer for the product being produced requires a quality characteristic equal to 80, i.e. $T = 80$. Figure 8.3 suggests that when the mean is placed on target, the variance independent of the mean is minimised when factor B is set very high and factor A quite low. This optimal position is circled in Figure 8.3.

8.2 ILLUSTRATIONS OF PROCESS OPTIMISATION

8.2.1 THE AUSFORMING PROCESS

Return again to the 2^5 experiment of the ausforming process (see Section 5.5.3 for a recap). Suppose the steel manufacturer intends to sell its sheet steel to a major European car manufacturer who requires the steel to have a tensile strength of 2600 MPa. (In reality the steel will also be required to have other material properties). Thus the target requirement is $T = 2600$ MPa.

In Section 7.3 the following results were obtained from a 2^5 factorial experiment that held the austenitising and isothermal incubation times fixed at their low levels

$$Y_j = 2261.5 - 246.47 X_{1j} - 84.34 X_{2j} + 106.59 X_{3j} + 81.53 Z_{2j} + 175.16 Z_{1j} Z_{2j} - 168.34 X_{3j} Z_{2j} + \varepsilon_j \quad (8.1a)$$

$$\bar{Y}_j = 2261.5 - 246.47 X_{1j} - 84.34 X_{2j} + 106.59 X_{3j} \quad (8.1b)$$

and

$$S_{Yj}^2 = [175.16 Z_{2j}]^2 + [81.53 + 175.16 Z_{1j} - 168.34 X_{3j}]^2 + 135.8^2 \quad (8.1c)$$

X_1 is the austenitising temperature, X_2 the deformation temperature, X_3 the amount of deformation, Z_1 the quench rate and Z_2 the deformation rate. X_1 to X_3 are design factors, Z_1 to Z_2 noise factors. Y_j is the actual tensile strength at test condition j, $\bar{Y}_j$ is the mean tensile strength at test condition j and S_{Yj}^2 the variance in Y at test condition j. The number 135.8 in equation (8.1c) is the estimated standard deviation in the prediction error of equation (7.13a). Equation (8.1b) shows how the mean of the process can be put on target and equation (8.1c) shows how process variability can be minimised. Equation (8.1c) assumes that the noise factors have a variance of unity and that the major source of process variation is the presence of these two noise factors.

To illustrate how these equations can be used to find the optimal values for the levels of each process variable, suppose factor X_2 is currently fixed at a coded level of -0.9. Under this condition equation (8.1b) can be written as

$$\bar{Y}_j = 2261.5 - 246.47X_{1j} - 84.34(-0.9) + 106.59X_{3j} = 2337.4 - 246.47X_{1j} + 106.59X_{3j}$$

This can be rearranged to the give the equation required to draw contours of constant mean

$$\frac{\bar{Y} - 2337.4 + 246.47X_{1j}}{106.59} = X_{3j} \tag{8.1d}$$

Equation (8.1d) shows what value X_3 must take given a value for X_1 to yield a mean equal to $\bar{Y}$, (given $X_2 = -0.9$). This equation is drawn in Figure 8.4 for values of $\bar{Y}$ ranging from 2000 to 2700 MPa.

Lines of constant variance can also be drawn under assumed values for the noise factors. For example, suppose $Z_1 = Z_2 = 0$. Then equation (8.1c) simplifies to

$$S^2_{Yj} = [81.53 - 168.34X_{3j}]^2 + 135.8^2 \tag{8.1e}$$

Clearly the smallest the variance can be is 135.8^2 obtained by setting X_3 to that value which makes the terms in brackets sum to zero. This occurs when $X_3 = 0.4843$

$$S^2_{Yj} = [81.53 - 168.34(0.4843)]^2 + 135.8^2 = [81.53 - 81.53]^2 + 135.8^2 = 135.8^2$$

The variance can't be reduced below this value because 135.8^2 is the estimated variance of the prediction error, which does not depend upon test conditions.

Thus process variability is minimised by setting $X_3 = 0.4843$. But what value must X_1 take, given $X_2 = -0.9$, to ensure the mean tensile strength equals the target of 2600 MPa? This value is found by plotting equation (8.1e) for various values of S^2_{Yj}. Because S^2_{Yj} is independent of X_1 the lines of constant variance will be horizontal in Figure 8.4. The standard deviation rather than the variance is shown in this figure. It can be seen from Figure 8.4 that the minimum variance line intersects the target constant mean line of 2600 MPa when $X_1 = -0.86$. Thus the process can be optimised by setting $X_1 = -0.86$ and $X_3 = 0.4843$ when X_2 has a value of -0.9. It may be possible to find other optimums by varying the value for X_2.

Equation (3.2) can be used to find the actual units of measurement that optimise this process. In the 2^5 experiment the high level for the austenitising temperature was

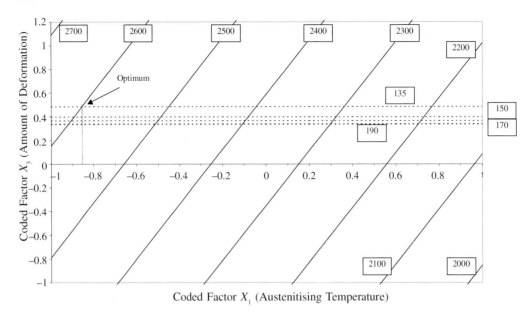

Fig. 8.4 Contours of constant mean and standard deviation.

1040°C and the low level, 930°C. Thus the mean temperature was 985°C and the range was 1040 – 930 = 110°C. Thus

Actual test condition =
mean test condition + (coded test condition × range of test condition/2)

$$\text{Actual test condition} = 985 + \left(-0.86 \times \left(\frac{110}{2}\right)\right) = 938°C$$

Performing similar calculations for X_2 and X_3 suggests that the process is optimised by running it with an austenitising temperature of 938°C, a deformation temperature of 408°C and an amount of deformation equal to 72%. At these conditions the mean tensile strength will equal the target of 2600 MPa and process variability will be minimised at a standard deviation of 135.8 MPa. (This is only 5% of the mean).

The above approach only worked because X_2 was held fixed. An optimal solution when there are many process variables can be found using Excel's Solver program. Figure 8.5 shows how to set the above optimisation problem up within Excel. Along row seven of the spreadsheet are placed the coded levels for each factor. Beneath this the actual values are then worked out. In cell C12 a formula is inserted to calculate the mean tensile strength. Notice how Excel represents equation (8.1b) by referencing the cells containing the coded levels for each process variable. Then in cell C15 a formula is inserted to calculate the variance in the tensile strength. Again notice how Excel represents equations (8.1c) by

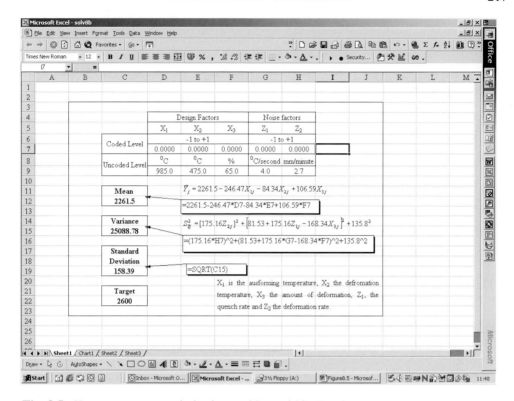

Fig. 8.5 How to setup an optimisation problem within Excel.

referencing the cells containing the coded levels for each process variable. In cell C19 the standard deviation is worked out by inserting a formula that takes the square root of the variance in Cell 15. The required target strength is inserted in cell C22. In Figure 8.5 all the factors have been set equal to zero. The resulting mean tensile strength is then 2261.5 MPa, which is a long way from the target of 2600 MPa. The variance is also a long way from the minimum value identified above. Clearly these conditions for the process variables do not optimise the process.

Excel next needs to find the values for the process variables that set the mean equal to the target and which minimises the variability around that target. By clicking on the **Tools** label on the menu bar and selecting **Solver** from the drop down menu the following sub menu shown in Figure 8.6 appears.

Starting from the top of this image the selected options read as follows. Excel will attempt to set the value in cell C15 (the variance) to a minimum. Excel will achieve this by changing the values in cells D7:F7 (the coded values for the design factors). That is, Excel will insert a set of values in cells D7:F7 and note the resulting variance. It will repeat this process until the values in cells D7:F7 give the lowest possible variance. This will then be the optimal solution.

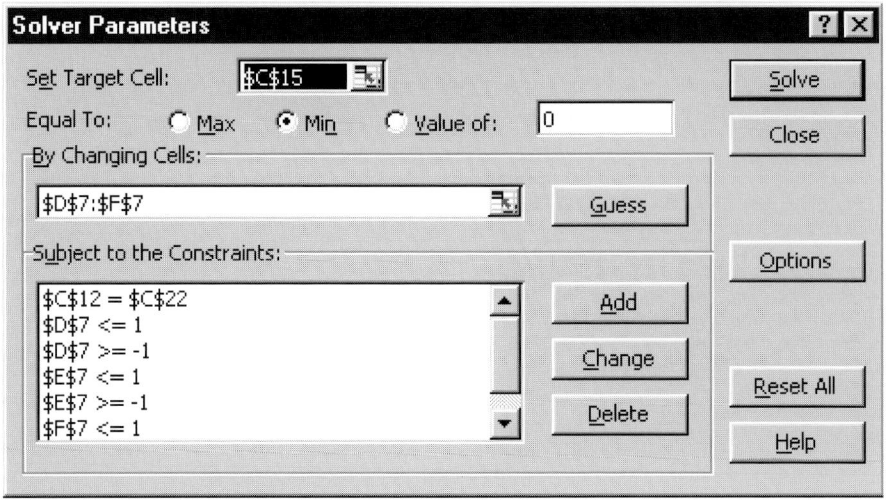

Solver Parameters

Set Target Cell: $C$15
Equal To: Min
By Changing Cells: $D$7:$F$7

Subject to the Constraints:
```
$C$12 = $C$22
$D$7 <= 1
$D$7 >= -1
$E$7 <= 1
$E$7 >= -1
$F$7 <= 1
```

Fig. 8.6 Excel's solver submenu.

But this minimum variance has to be found subject to a number of constraints that limit the search for the minimum. The first constraint shown in the **Subject to the Constraints** box is C12 = C22. That is, the minimum variance is sought subject to the condition that the mean strength in cell C12 equals the target strength in cell C22. All the other constraints instruct Excel to select values for the process variables within the limits defined by the 2^5 experiments, i.e. within the coded range -1 to $+1$. All these constraints are placed in the Subject to the Constraints box by clicking on the **Add** button and filling in the subsequent boxes. Clicking on the **Solve** button instructs Excel to search out this optimal solution. Figure 8.7 shows the optimal solution identified by Excel. Notice that it is very similar to the solution found above using the graphical approach.

Figure 8.7 can be used to find the minimum variance for other values of the noise factors by simply changing the numbers in cells G7 and H7 and running Solver again.

8.2.2 THE COPPER COMPACT EXPERIMENT

The pore size in the centre region of a copper compact is a very important quality characteristic. Suppose the 2_{IV}^{7-3} fractional experiment shown in Section 7.7 was set up with the intention that it would provide the information necessary to ensure that the copper compact manufacturing process consistently produced compacts with a pore size of 2% or less. The objective of the experiment is therefore to ensure the mean pore size from this process meets a target of $T \leq 2\%$ and that the variability around the resulting mean is at a minimum.

Linear Process Optimisation

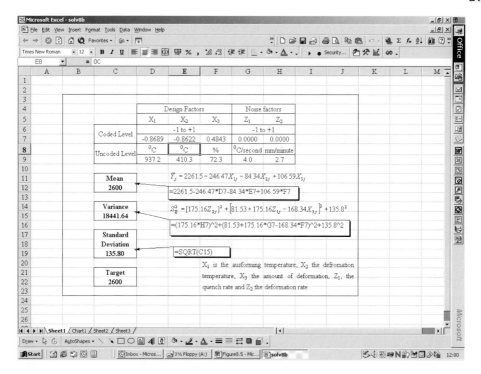

Fig. 8.7 Optimal solution identified by Excel.

In Section 7.7 use was made of a generalised linear model to analyse the data obtained from a 2^{7-3}_{IV} fractional factorial experiment. The following information on process variability was identified. First, it was found that the component of variability independent of the mean, ϕ_j, depended on factor C (the sintering treatment).

$$\ln(\phi_j) = -1.522 + 2.082C_j \tag{8.2a}$$

or

$$\phi_j = \exp[-1.522 + 2.082C_j] \tag{8.2b}$$

Total process variability, S^2_{Yj}, also depends on the mean pore size, $\overline{Y}_j$, and was found to be given by

$$\ln(S^2_{Yj}) = -1.522 + 2.082C_j + 1.476\ln(\overline{Y}_j) \tag{8.2c}$$

or

$$S^2_{Yj} = \exp[-1.522 + 2.082C_j]\overline{Y}^{1.476} \tag{8.2d}$$

Table 8.2 Probability plot for location effects in copper compact experiment.

Aliased Effect	Effect Estimate	Rank Index, I	CP_I	Z_I
F_{Loc}^{Ali}	−8.50625	1	0.0333	−1.8344
BE_{Loc}^{Ali}	−4.79375	2	0.1	−1.2816
AF_{Loc}^{Ali}	−3.04375	3	0.1667	−0.9673
AB_{Loc}^{Ali}	−2.80625	4	0.2333	−0.728
G_{Loc}^{Ali}	−1.99375	5	0.3	−0.5244
E_{Loc}^{Ali}	−1.69375	6	0.3667	−0.3406
AC_{Loc}^{Ali}	−1.26875	7	0.4333	−0.168
AG_{Loc}^{Ali}	0.76875	8	0.5	0
B_{Loc}^{Ali}	1.03125	9	0.5667	0.168
D_{Loc}^{Ali}	1.33125	10	0.6333	0.3406
AE_{Loc}^{Ali}	1.69375	11	0.7	0.5244
ABE_{Loc}^{Ali}	2.24375	12	0.7667	0.728
AD_{Loc}^{Ali}	2.96875	13	0.8333	0.9673
C_{Loc}^{Ali}	5.79375	14	0.9	1.2816
A_{Loc}^{Ali}	5.89375	15	0.9667	1.8344

However, the mean pore size was not analysed in Section 7.7 and this information is now required to optimise the process. Table 8.2 shows the results obtained by the Yates procedure for estimating the aliased location effects. Table 8.2 contains all the workings required to construct a probability plot for the location effects of this experiment. This plot is then shown in Figure 8.8. The only factor that appears to have a definite impact on the mean pore size is factor F (die temperature). Factors A (powder size) and C (sinter time) may also be important in determining the mean pore size. Hence a simplified response surface model for the mean pore size is (10.84 is the mean pore size over all test conditions).

$$\overline{Y}_j = 10.84 + \frac{5.894}{2} A_j + \frac{5.794}{2} C_j - \frac{8.506}{2} F_j \qquad (8.2e)$$

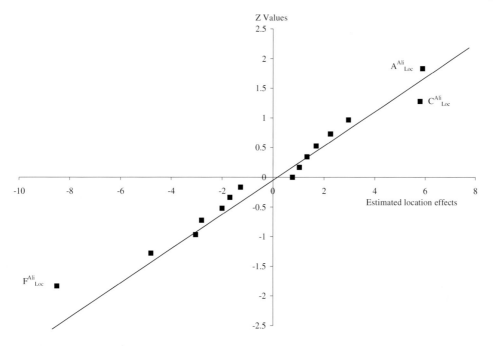

Fig. 8.8 A probability plot for location effects in the copper compact experiment.

Process optimisation turns out to be quite straightforward for this cooper compact manufacturing process. From equation (8.2b), the variance independent of the mean can be minimised by setting factor C, the sintering time, to its lowest possible level within the experiment, i.e. -1. In uncoded units this is a sintering time of 15 minutes

$$\phi_j = \exp[-1.522 + 2.082(-1)] = 0.0271$$

From equation (8.2e) the mean can be put on target by manipulating factors A, C and F. But factor C has already been fixed at its low level to minimise the variance independent of the mean. Factors A and F can then be used to put the mean on target without altering the variance independent of the mean which has been minimised to a value of 0.0271. Equation (8.2e) shows that factor F has the biggest impact on the mean and so it is best to use this factor to place the mean on target. If factor A is set at its middle level (zero in coded units), factor F must be set at the following level to put the mean on a target of 2%.

$$\overline{Y}_j = 2.0 = 10.84 + 2.947(0) + 2.897(-1) - 4.253F_j$$

$$2.0 - 10.84 + 2.897 = -4.253F_j$$

$$F_j = \frac{2.0 - 10.84 + 5.794}{-8.506} = 1.397$$

$F_j = 1.397$ in coded units corresponds to a die temperature of 690°C,

Actual test condition =
   mean test condition + (coded test condition × range of test condition/2)

$$\text{Actual test condition} = 550 + \left(1.397 \times \left(\frac{200}{2}\right)\right) = 690°C.$$

The overall level of variability associated with this mean of 2.0% is found from equation (8.2d).

$$S_{Yj}^2 = \exp[-1.522 + 2.082(-1)]\{2.0\}^{1.476} = 0.076$$

Thus the mean is put on a target pore size of 2%, by setting the die temperature to 690°C and the powder size to its middle level. The variability around this mean is then minimised by setting the sintering time to 15 minutes. This will yield a standard deviation of 0.27% (which is small faction of the mean pore size). Of course it is possible to achieve a mean below 2% if this is really required.

8.2.3 THE INJECTION MOULDING EXPERIMENT

In Section 7.8 an analysis of the injection moulding experiment was carried out and the following simplified first order response surface model was identified (taking the β, λ and δ values from Table 7.19).

$$Y_j = \beta_0 - 0.281 A_j + 0.425 C_j - 0.231 G_j - 0.419 E_j Z_{2_j} + 0.45 F_j Z_{2_j} + \varepsilon_j. \qquad (8.3a)$$

In equation (8.3a) Y_j is the percentage shrinkage in the manufactured plastic pipes during the cooling off period of the injection moulding manufacturing process. Factor A is the pressure exerted on the plastic within the mould, factor C is the cycle time, factor E the speed with which plastic is injected into the mould, factor F is the die cavity thickness and factor G the size of the safety gate on the moulding machine. These are all design factors. The noise factor Z_2 is the air moisture content. β_0 is the average percentage shrinkage over all test conditions and ε_j the error made in predicting Y_j using equation (8.3a) – the prediction error. Remember from Section 7.8.2.2 that the location effects for each factor are aliased effects because a fractional factorial has been used. Specifically, the main location effect for each design factor is aliased with first order interactions between the design factors. But such first order interactions could well be important and so this experimental design is not going to yield reliable information on how to put the mean on target using the main location effects associated with the design factors. For example, in equation (8.3a) half the aliased location effect for factor C is 0.425. This number can only be interpretated as showing what happens to the response when factor C is changed, if some first order interaction is assumed to be zero. (In particular if the interaction effects $AE = -BF = -DG = 0$. (See Section 7.8.2.2). The same is true for design factors A and G.

Proceeding under such a strong assumption, the mean percentage shrinkage and the variability around this mean are given by

$$\overline{Y}_j = 2.25 - 0.281A_j + 0.425C_j - 0.231G_j \tag{8.3b}$$

where $\beta_0 = 2.25\%$ and

$$S_{Yj}^2 = \left[-0.419E_j\right]^2 + \left[0.45F_j\right]^2 + S_\varepsilon^2 \tag{8.3c}$$

As usual equation (8.3b) assumes that the noise factors have been transformed so that their mean values are zero and equation (8.3c) assumes that the noise factors have a variance of unity and that the major source of process variation is from the existence of the two noise factors. S_ε^2 is the variance of the prediction error and is estimated from equation (6.9b) to equal 1.984. Thus

$$S_{Yj}^2 = \left[-0.419E_j\right]^2 + \left[0.45F_j\right]^2 + 1.984$$

Suppose now that our pipe manufacturer must keep the pipe shrinkage at or below 1.5% if it is to meet its major customer specifications on pipe dimensions. The objective of the experiment is therefore to ensure the mean percentage shrinkage from this process meets a target of $T \leq 1.5\%$ and that the variability around this mean is at a minimum so that the target is meet on a consistent basis.

The smallest the variance can be is 1.984 and this will be obtained when E_j and F_j take on values such that the combination shown in the square brackets of equation (8.3c) is zero. Setting factors $E_j = F_j = 0$ would be one way of minimising process variability in the injection moulding process. To see the full solution, Excel's Solver can be used in a similar way to that for the ausforming process in Section 8.2.1.

Figure 8.9 shows how to set this optimisation problem up within Excel. Along row seven of the spreadsheet are placed the coded levels for each factor. In cell C11 a formula is inserted to calculate the mean percentage shrinkage. Notice how Excel represents equation (8.3b) by referencing the cells containing the levels for each process variable. Then in cell C14 a formula is inserted to calculate the variance of the percentage shrinkage. Again notice how Excel represents equation (8.3c) by referencing the cells containing the levels for each process variable. In cell C18 the standard deviation is worked out by inserting a formula that takes the square root of the variance in Cell 14. The required target shrinkage is inserted in cell C21. In Figure 8.9 all factors that determine the mean have been set equal to zero. The resulting mean percentage shrinkage is then 2.25%, which is along way from the target of 1.5% or less. The variance is also a long way from the minimum value identified above. These conditions for the process variables do not optimise the process.

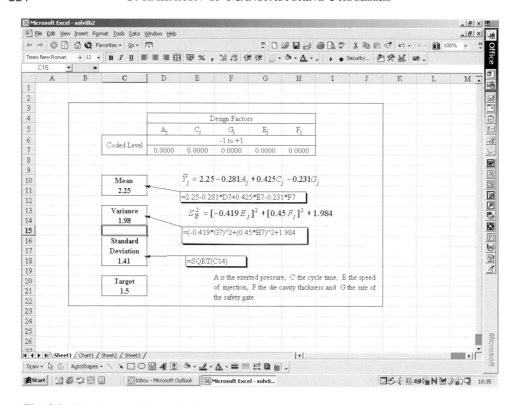

Fig. 8.9 Excel spreadsheet showing how to setup an optimisation problem.

By clicking on the **Tools** label on the menu bar in the image above and selecting **Solver** from the drop down menu the sub menu in Figure 8.10 appears.

Starting from the top of this image the selected options read as follows. Excel will attempt to set the value in cell C14 (the variance) to a minimum. Excel will achieve this by changing the values in cells D7:H7, (the values for the design factors). That is, Excel will insert a set of values in cells D7:H7 and note the resulting variance. It will repeat this process until the values in cells D7:H7 give the lowest possible variance. This will be the optimal solution.

But this minimum variance is found subject to a number of constraints that limit the search carried out by Excel. The first constraint shown in the **Subject to the Constraints** box is C11 ≤ = C21. That is the minimum variance is sought subject to the condition that the mean percentage shrinkage in cell C11 is no greater than the target shrinkage in cell C21. All the other constraints instruct Excel to use values for the process variables within the limits defined by the experiment, i.e. within the range –1, to +1. All these constraints are placed in the Subject to the Constraints box by clicking on the **Add** button and filling in the subsequent boxes. Clicking on the **Solve** button instructs Excel to find this optimal solution. Figure 8.11 shows the optimal solution.

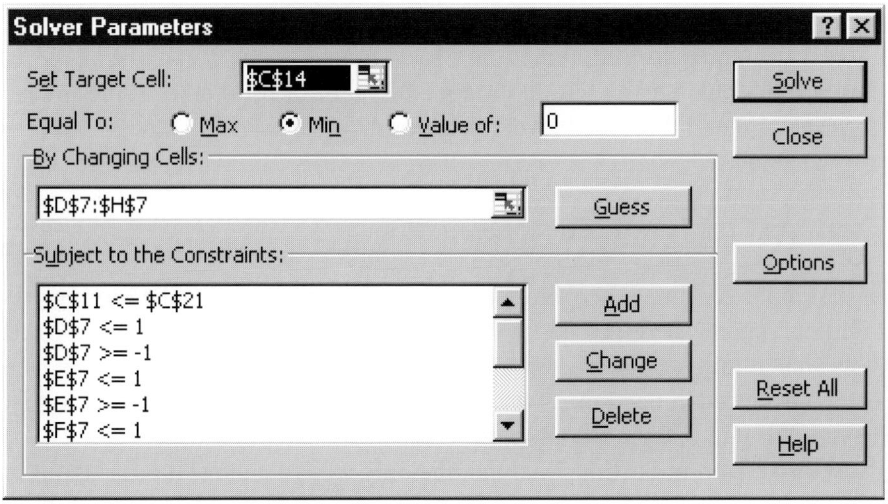

Linear Process Optimisation

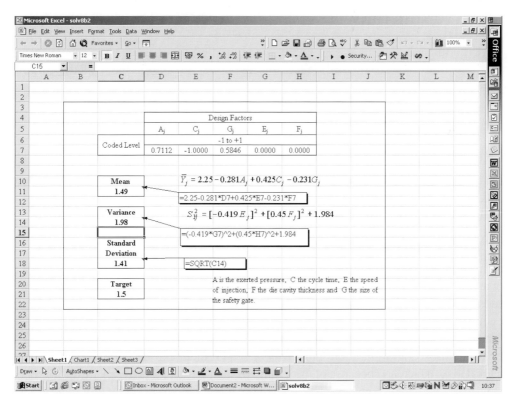

Fig. 8.10 Solves submenu in Excel.

Fig. 8.11 Excel spreadsheet showing optimal solution.

As expected the coded levels for factors E and F are zero yielding the lowest possible variance of 1.98 (equal to the prediction error variance). The mean shrinkage is just below target and is obtained by setting the coded level of factor A at 0.71, the coded level for factor C at –1.0 and the coded level for factor G at 0.58.

8.2.4 Optimising the Disk Forging Operation

A major cost in the forging of aero engine disks is the energy used in the forging process. Any forging company therefore has a strong incentive to try and find those operating conditions that minimise the energy usage. Thus the end objective of the factorial experiment introduced in Section 2.4 was to minimise the mean energy required together with the variability in energy use around this mean consumption rate.

It was discovered in Section 7.5.3 that the only way to control the variability in the energy required to forge aero engine disks was to manipulate factor A – the surface area of the aero engine disk. Factor A was the only dispersion effect. A comparison of S^2_{A-} with S^2_{A+} gave.

$$A_{Disp} = S^2_{A+} - S^2_{A-} = 4798.24 - 507.7864 = 4290.45$$

This suggests that variability can be minimised or reduced by setting factor A at its lowest level.

Unfortunately, Factor A also enters the simplified model of the process given by equation (7.18) and so also influences the mean energy requirement of the forging operation. Equation (7.18) is shown again below,

$$Y_j = 3639.3875 - 460.85A_j + 353.575B_j + 261.35C_j$$
$$- 320.49A_jB_j - 205.965A_jC_j + 82.215B_jC_j - 97.075A_jB_jC_j \qquad (8.4)$$

Furthermore, because the A_{Loc} effect is negative, (–2 × 460.85)), setting factor A low is likely to push up the mean energy requirement. Consequently, the remaining important location effects shown in equation (8.4) must be carefully analysed so as to minimise this detrimental impact on the mean energy requirement.

Equation (8.4) shows that both factors B (coefficient of friction) and C (heat transfer coefficient) can be used to manipulate the mean energy requirement. These factors interact both with each other and with factor A and so a careful consideration of these interactions needs to be made. It is not valid to just look at the sign and magnitude of the B_{Loc} and C_{Loc} effects alone. Figures 8.12 to 8.14 show the interaction effects between factors A and B, A and C and B and C respectively. Figure 8.12 suggests that given factor A must be set low to minimise variability, the mean energy requirement can be minimised by setting factor B low. In fact, and because of the strong AB interaction, there is very little difference between the B low – A high combination and the B low – A low combination. Figure 8.13 suggests that given factor A must be set low to minimise variability, the mean energy

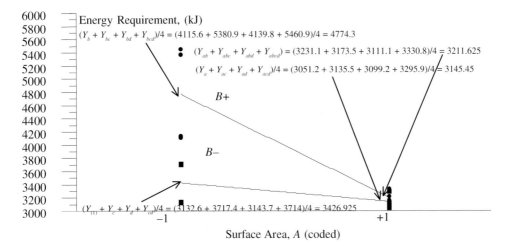

Fig. 8.12 The *AB* interaction location effect.

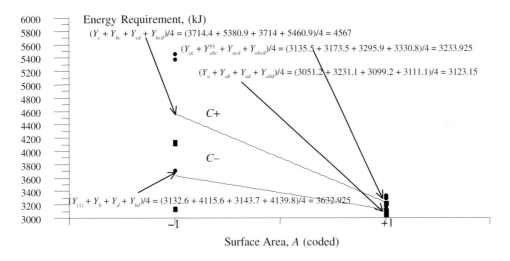

Fig. 8.13 The *AC* interaction location effect.

requirement can be minimized by setting factor *C* low. Because of the strong *AC* interaction there is very little difference between the *C* low – *A* high combination and the *C* low – *A* low combination.

The above analysis suggests that variability can be minimised by setting *A* low, and if factors *B* and *C* are also set low, the detrimental effect on the mean energy requirement

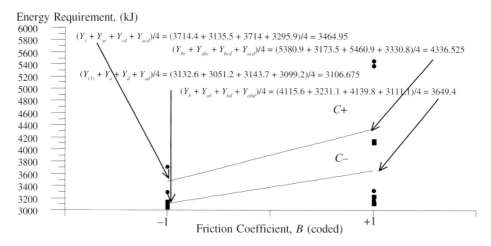

Fig. 8.14 The BC interaction location effect.

of setting A low can be minimised. Figure 8.14 shows the BC interaction and confirms the advantages of setting factors B and C low.

The predicted mean energy requirement from setting the factors at these levels can be found from equation (8.4) by inserting $A_j = -1$, $B_j = -1$, $C_j = -1$, $A_jB_j = 1$, $A_jC_j = 1$, $B_jC_j = 1$ and $A_jB_jC_j = -1$,

$Y_j = 3639.3875 + 460.85 - 353.575 - 261.35 - 320.4875 - 205.9625 + 82.2125 + 97.075 = 3138.15$ kJ

In turn the predicted variability in energy use around this average of 3138.15 kJ is $S^2_{A-} = 507.7864$. If energy use follows a normal distribution, then 95% of all the disks that are forged should use up an energy somewhere between

$3138.15 + 1.96x\sqrt{507.786} = 3182.32 kJ$ and $3138.15 - 1.96x\sqrt{507.7864} = 3093.98 kJ$

The above analysis has ignored the ABC interaction effect on mean energy requirements. Because it is a second order interaction the above conclusions are likely to be little affected by the presence of this term.

PART IV
NON–LINEAR EXPERIMENTAL DESIGNS

9. Some Non-Linear Experimental Designs

In many manufacturing processes the relationship between the quality characteristic of interest and one or more of the process variables may be non-linear in nature. This is very likely to be the case when the levels for the process variables are close to the optimum operating conditions for the process. In order to find such an optimum in a non-linear manufacturing process it is necessary to model curvature. Designs that have each factor at two levels are capable of identifying and modelling linear relationships only. To model curvature designs which have factors varying over three or more levels are required. Such designs are dealt with in this chapter. Then in Chapter 10 the concept of a non-linear effect and how it can be estimated from designs that have factors varying over more than two levels is fully explained.

There are a large variety of designs available for dealing with non-linear relationships. One obvious possibility is to extend the 2^k factorial design to the **3^k design**. That is, to study k factors each at three different levels. These designs is discussed in Section 9.1 and an illustration using the friction welding case study is given in Section 9.2. Such designs however involve a considerable amount of experimentation. For example, a study of a manufacturing process containing just four factors would involve carrying out $3^4 = 81$ tests. More efficient and statistically pleasing designs include the **central composite** design and this is explained in Sections 9.3 and 9.4 and the **Box-Behnken** design which is discussed in Section 9.5. Under certain conditions these two designs are in fact fractions of the 3^k design and can be combined to give the full 3^k design. Factorial designs with **mixed levels** are then discussed in Section 9.6.

9.1 3^k DESIGNS

As the name suggests, the **3^k factorial design** involves a factorial arrangement with k factors each at three different levels. The design consists of all the possible combinations of k factors each at three different levels. As such the number of tests required is given by $N \times 3^k$, where N is the number of replications made at each test condition. When considering four factors an unreplicated design would therefore require $3^4 = 81$ tests. For most situations this is likely to be a prohibitive number of tests. Thus this book will concentrate on explaining how to set-up and analyse the 3^2 and the 3^3 designs only.

The levels set for each factor must be equally spaced. Then it is possible to use the coded test conditions of -1, 0 and $+1$ to represent the low, middle and high levels for each factor respectively. All 3^k test conditions can be found by writing out the experiment in standard order form. To identify the standard order form, factors are introduced one at

Table 9.1 The 3^1 design.

Standard Order	Coded Test Conditions
Tests	A
0	−1
1	0
2	1

Table 9.2 The 3^2 design.

Standard Order	Coded Test Conditions	
Tests	Factor A	Factor B
0 0	−1	−1
1 0	0	−1
2 0	1	−1
0 1	−1	0
1 1	0	0
2 1	1	0
0 2	−1	1
1 2	0	1
2 2	1	1

a time with each level being combined successively with every set of three level factors above it. The best way to write out this standard order form is to use integer numbers so that, for example, 0 0 0 refers to a test with all three factors low, 1 1 1 to a test with all three factors at their middle value and 2 2 2 to a test with all three factors high. When $k = 1$ there are just 3 tests to carry out and the standard order form for this 3^1 design is shown in Table 9.1.

When $k = 2$ there are $3^2 = 9$ tests to carry out and the standard order form for this if found by first replicating the 3^1 standard order pattern until nine rows are filled up. Then a zero is added to the first $3^{k-1} = 3$ rows of numbers in the standard order column, one to the next $3^{k-1} = 3$ rows and a two to the final $3^{k-1} = 3$ rows. This pattern is shown in the standard order test column of Table 9.2 which is then used to derive the coded test condition

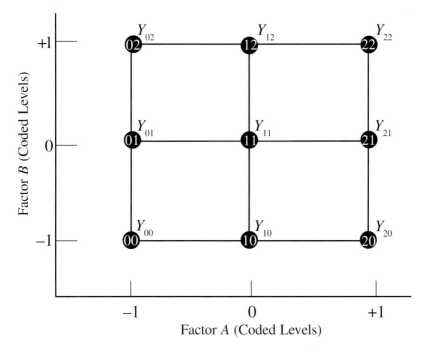

Fig. 9.1 Geometric representation of the 3^2 factorial design.

levels for factors A and B. Test 0 0 involves setting factors A and B low and so a minus one is placed in the coded test condition columns. Test 1 1 involves setting factors A and B at their middle levels and so a zero is placed in the coded test condition columns. Test 2 2 involves setting factors A and B at their high levels and so a one is placed in the coded test condition columns.

When $k = 3$ there are $3^3 = 27$ tests to carry out. The standard order form for this is found by first replicating the standard order pattern for the 3^2 design until 27 rows are filled up. Then a zero is added to the first $3^{k-1} = 9$ rows of numbers in the standard order column, one to the next $3^{k-1} = 9$ rows and a two to the last $3^{k-1} = 9$ rows. This pattern is shown in the standard order test column of Table 9.3 which is then used to derive the coded test condition levels for factors A, B and C. Test 0 0 0 involves setting factors A, B and C low and so a minus one is placed in the coded test condition columns. Test 1 1 1 involves setting factors A, B and C at their middle levels and so a zero is placed in the coded test condition columns. Test 2 2 2 involves setting factors A, B and C at their high levels and so a one is placed in the coded test condition columns.

As with 2^k designs, these 3^k designs can also be represented geometrically. Figure 9.1 represents the 3^2 design as a square. Notice that unlike the 2^2 design, the centre of the

Table 9.3 The 3^3 design.

Standard Order	Coded Test Conditions		
Tests	A	B	C
0 0 0	−1	−1	−1
1 0 0	0	−1	−1
2 0 0	1	−1	−1
0 1 0	−1	0	−1
1 1 0	0	0	−1
2 1 0	1	0	−1
0 2 0	−1	1	−1
1 2 0	0	1	−1
2 2 0	1	1	−1
0 0 1	−1	−1	0
1 0 1	0	−1	0
2 0 1	1	−1	0
0 1 1	−1	0	0
1 1 1	0	0	0
2 1 1	1	0	0
0 2 1	−1	1	0
1 2 1	0	1	0
2 2 1	1	1	0
0 0 2	−1	−1	1
1 0 2	0	−1	1
2 0 2	1	−1	1
0 1 2	−1	0	1
1 1 2	0	0	1
2 1 2	1	0	1
0 2 2	−1	1	1
1 2 2	0	1	1
2 2 2	1	1	1

square and the centre points on the sides of the square become a set of test conditions. For 3^k designs it is useful to replace the letter subscripts of the 2^k designs on the response variable Y with the number subscripts discussed above. Thus Y_{02} in Figure 9.1 refers to the value for the response recorded when factor A is at its lowest level but factor B is at its highest level. Y_{11} on the other hand is the response obtained when both factors are at their middle levels.

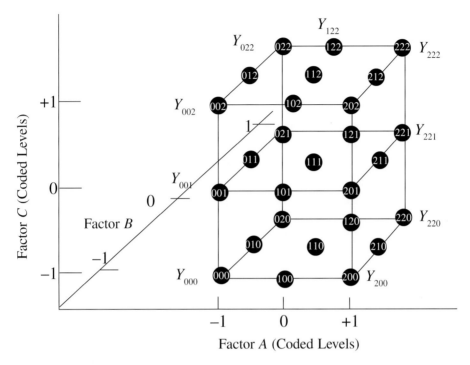

Fig. 9.2 Geometric representation of the 3^3 factorial design.

Figure 9.2 represents the 3^3 design as a cube. In this figure, Y_{002} refers to the value for the response recorded when factors A and B are at their lowest levels but factor C is at its highest level. Y_{222}, on the other hand, is the response obtained when all three factors are at their highest levels.

9.2 A 3^2 DESIGN FOR THE FRICTION WELDING CASE STUDY

To illustrate designs which have each factor at three levels, consider again the friction welding case study of Section 2.5. Let factor A be the axial force, factor B the frequency of vibration and factor C the amplitude of vibration. In this section, the amplitude of vibration is held constant at 2 mm. The high levels for factors A and B are 60 kN and 35 Hz respectively. The low levels for factors A and B are 40 kN and 25 Hz respectively. The third level in a 3^2 design must be in the middle of these high and low values. Thus the middle levels for factors A and B are 50 kN and 30 Hz respectively. The response recorded is the degree of upset in mm. As discussed in Section 2.5, the quality and strength of the

Table 9.4 A 3^2 design for part of the friction welding case study.

Test	Axial Force, A (kN)	Frequency of Vibration, B (Hz)	A (Coded)	B (Coded)	Degree of Upset, mm
00	40	25	−1	−1	0.4378
10	50	25	0	−1	1.6496
20	60	25	1	−1	4.0982
01	40	30	−1	0	1.7164
11	50	30	0	0	4.641
21	60	30	1	0	7.165
02	40	35	−1	1	3.9696
12	50	35	0	1	6.4726
22	60	35	1	1	9.7698

finished weld is strongly dependant upon this factor with strength increasing with the degree of upset. A clear objective is therefore to try and maximise the degree of upset.

Table 9.4 shows a 3^2 design for this process using the levels stated above. Notice that it is written out in the standard order form described above. There are $3^2 = 9$ tests to carry out and the standard order form shown in Table 9.4 was found by first replicating the 3^1 design until nine rows had been filled up. Then a zero was added to the first 3^{k-1} rows of numbers in the standard order column, one to the next 3^{k-1} rows and a two to the final 3^{k-1} rows. This pattern is shown in the test column of Table 9.4 and was then used to derive the coded and uncoded test condition levels for factors A and B. Thus, the first test in standard order form is test 0 0. That is, a low level for both the axial force and the frequency of vibration. Thus a minus one is placed in the coded test condition columns for factors A and B. The final test in standard order form is test 2 2. That is, a high level for both the axial force and the frequency of vibration. Hence the plus ones in coded test condition columns for factors A and B. The results obtained from this experiment are shown in the last column of Table 9.4.

This design can be represented geometrically as a block of four squares as in Figure 9.3. Each corner and mid side of the square represents a test condition and so a response is shown alongside it. In Figure 9.3 coded units are used for the levels of factors A and B. These results will be analysed in Chapter 10.

9.3 CENTRAL COMPOSITE DESIGNS

A 3^3 design involves $3^3 = 27$ tests. If a non-linear manufacturing process has five factors then the required 3^5 experiment would involve $3^5 = 243$ tests. In most situations this

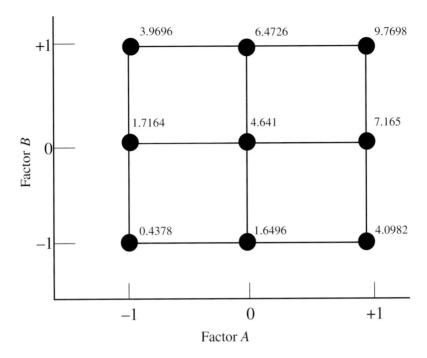

Fig. 9.3 Geometric representation of the friction welding experiment.

would be just too time consuming or too costly to consider. It is a characteristic of a non-linear design that substantial experimentation will be required. To try and minimise the amount of experimentation it is suggested that the following **sequential** experimental programme be carried out.

i. First set up a two level factorial, or if possible a two level fractional factorial design, i.e. set up a 2^{k-p} design. From this design carry out a series of t tests or probability plots to identify which location effects are important and which are not.

ii. Carry out a simple test for curvature. If this tests suggests that non-linearity's may be present carry out an efficient non-linear design for the important factors only.

The **central composite design** is ideal for such sequential testing. This is because such designs have a number of useful properties. First, they will involve less experimentation than the alternative 3^k design. Secondly, central composite designs have at their core a 2^k or a 2^{k-p} design. Thirdly, such designs are orthogonal and so the analysis of the data remains as straightforward as in the 2^k designs. Fourthly, when it comes to optimising a process, the optimum point on the response surface is not known before hand and so it is important that the design chosen to locate this optimum provides equal

Table 9.5 Rotatable central composite designs.

k	2 $p=0$	3 $p=0$	4 $p=0$	5 $p=0$	5 $p=1$	6 $p=0$	6 $p=1$	7 $p=1$	8 $p=1$
2^{k-p}	4	8	16	32	16	64	32	64	128
N_a	4	6	8	10	10	12	12	14	16
N_c	2–4	2–4	2–4	0–4	0–4	0–4	0–4	2–4	2–4
No. of Tests	10–12	16–18	26–28	42–46	26–30	76–80	44–48	80–82	146–148
α	1.414	1.682	2	2.378	2	2.828	2.378	2.828	3.364

precision of estimation from all directions on the surface. Such a design is said to be **rotatable**. All 2^{k-p} designs are rotatable and central composite designs can be made rotatable. 3^k designs can not be made rotatable.

Any central composite design can be built up from an initial 2^k or 2^{k-p} design by adding **axial points** and **centre points** to the two level design. These axial points are located in the coded test condition space through the parameter α. For the design to remain rotatable, α must be determined through the equation

$$\alpha = (2^{k-p})^{\frac{1}{4}} \qquad (9.1)$$

α is a number in the same units as the coded test conditions from a 2^{k-p} design. For all the test points of the central composite design to lie on a sphere α must take on the value.

$$\alpha = \sqrt{k} \qquad (9.2)$$

The design is then said to be **spherical**. The number of axial points, N_a, for a central composite design is always equal to

$$N_a = 2 \times k \qquad (9.3)$$

For the design to be orthogonal, the number of centre points, N_c, must be set correctly. Unfortunately, there is no unique expression for the value of N_c, but such N_c values for a number of factors are shown in Table 9.5.

Central composite designs can be visualised as a star, with the test conditions tracing out a spherical region of the response surface. To illustrate, consider an experiment involving two factors. For this design to be orthogonal two to four centre points must be added to a standard 2^2 design. From equation (9.3) an additional $2 \times 2 = 4$ axial points must be added. Thus a central composite design involving just two factors requires at most $2^k + N_c + N_a = 12$ tests. (This is a little more than a 3^2 design, but for $k > 2$ the central

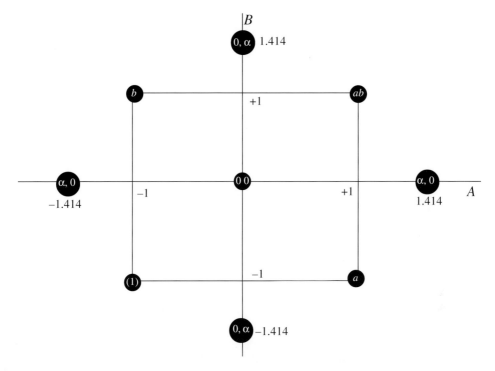

Fig. 9.4 A central composite design for $k = 2$.

composite design always involves less tests than the alternative 3^k design). The four axial points are determined through the value for α given by equation (9.1).

$$\alpha = (2^{k-p})^{\frac{1}{4}} = (2^{2-0})^{0.25} = 1.414$$

This design can be represented by a two-dimensional star as shown in Figure 9.4.

Notice that at the core of this design is the familiar 2^2 design involving the tests (1), a, b and ab that define the corners of the square. To this is added a further two to four replicated tests conducted at the centre point of the 2^2 design. Thus a test is carried out at levels for factors A and B that are exactly half way between their high and low levels defined in the 2^2 design. Lastly, four axial points are added. The first axial test is conducted with factor A at level 1.414 and factor B at its centre level. Remember that 1.414 is the coded test level for factors A and B. The second axial test is conducted with factor A at its centre level and factor B at level 1.414. The third axial test is conducted with factor A at level −1.414 and factor B at its centre level and the fourth axial test has factor A at its centre level and factor B at level −1.414. When $\alpha = k^{0.5}$, the axial and 2^k factorial tests

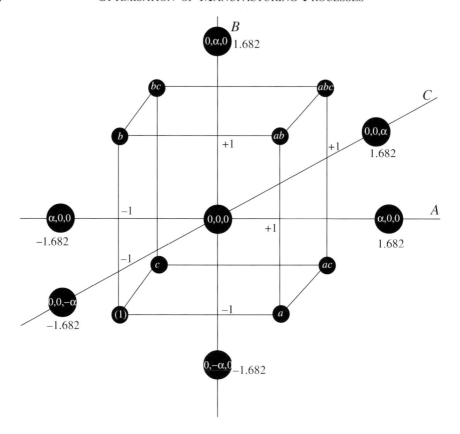

Fig. 9.5 A central composite design for $k = 3$.

define points that trace out a circle on the response surface, hence the name spherical design.

Notice how useful the design is in terms of the sequential approach to experimentation. Because the design has at its core the 2^2 design, this two level experiment can be carried out first and important effects identified. Next centre points can be added and used to test for curvature (see Section 9.4 for details). If such curvature is present, the axial point tests can be carried out and the totality of results used to analyse the non-linear manufacturing process.

This central composite design is easily generalised to any number of factors. For example, Figure 9.5 shows a central composite design for $k = 3$ factors.

For the design to be rotatable $\alpha = (2^3)^{0.25} = 1.682$. The design involves $2^3 = 8$ tests from the two level factorial part of the experiment. Test (1), a, b, ab, c, ac, bc and abc define the corners of the cube in Figure 9.5. To this is added two to four centre point tests for orthogonality and $2 \times 3 = 6$ axial points. Again these axial tests define points that

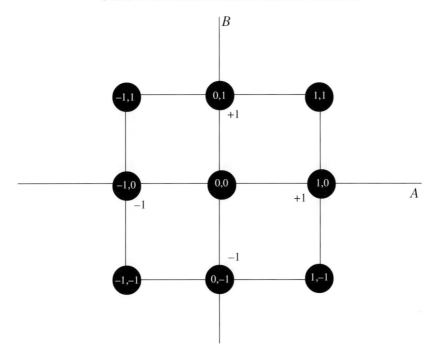

Fig. 9.6 A face centred central composite design for $k = 2$ factors.

trace out an approximate sphere on the response surface (an exact sphere when $\alpha = k^{0.5}$). This represents at most a total of 18 tests which is less than the $3^3 = 27$ tests required for a design having three factors all at three levels. Table 9.5 gives the recommended number of centre points for a central composite design involving $k = 2$ to $k = 8$ factors.

When constructing a central composite design from a two level fractional factorial it is advisable that the fractional factorial be at least resolution V. The reason for this will become clearer in the next chapter.

A **face centred central composite** design is a special case of the central composite design. It is not rotatable because the axial point α is always set equal to one, no matter how many factors are included in the experiment. Because of this the axial points will trace out a cuboidal region on the response surface. They are therefore sometime called **cuboidal designs**. Such designs are found by forming a 2^k or 2^{k-p} design and then adding on an additional 2^k axial points and a number of replicated centre points. If $p > 0$ then the 2^{k-p} part of the design must be at least resolution V. Each of the k factors will have two axial points. The first axial point has factor k set low with all other factors at their centre values. The second axial point has factor k high with all other factors at their centre values. Finally, the centre point of the design is replicated between one and four times.

Figure 9.6 shows a geometric representation of a face centred central composite design for $k = 2$ factors. Notice that it happens to be the same as the 3^2 design shown in Figure 9.1.

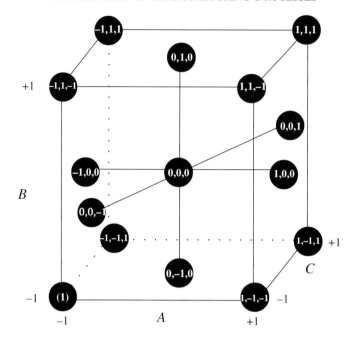

Fig. 9.7 A face centred central composite design for $k = 3$ factors.

Figure 9.7 shows a geometric representation of a face centred central composite design for $k = 3$ factors. This time it is not the same as the 3^3 design shown in Figure 9.2– it involves less tests.

The standard order form for the central and face centred central composite designs is as follows. The first 2^{k-p} rows for such designs in standard order form are the tests corresponding to the 2^{k-p} design (in its standard order form). The next 2^k rows of such designs in standard order form are the tests corresponding to the axial points in alphabetical order. The final rows of such designs in standard order form are the tests corresponding to the centre point.

Whilst face centred central composite designs are not rotatable, the test levels required for the axial points are not so extreme. Hence such designs are advantageous when the axial points in the central composite design are test combinations that are prohibitively expensive or impossible to test because of physical process constraints.

9.4 A CENTRAL COMPOSITE DESIGN FOR THE LINEAR FRICTION WELDING CASE STUDY

To illustrate the sequential approach to experimentation discussed above, consider all three factors in the friction welding experiment of Section 2.5. Without any prior

Table 9.6 A 2^3 factorial design for the friction welding experiment.

Test	Axial Force, A (kN)	Frequency of Vibration, B (Hz)	Amplitude of Vibration, C (mm)	A, (Coded)	B, (Coded)	C, (Coded)	Degree of Upset, (mm)
(1)	40	25	2	−1	−1	−1	0.4378
a	60	25	2	1	−1	−1	4.0982
b	40	35	2	−1	1	−1	3.9696
ab	60	35	2	1	1	−1	9.7698
c	40	25	3	−1	−1	1	4.9792
ac	60	25	3	1	−1	1	10.7212
bc	40	35	3	−1	1	1	9.1712
abc	60	35	3	1	1	1	12.6532

knowledge on the extent of curvature, the first stage of experimentation should be to set up a simple 2^3 factorial design. Table 9.6 shows the high and low levels chosen for the three factors of the friction welding process. The degree of upset recorded at each of these test conditions is shown in the last column of Table 9.6.

The next step is see whether there are any non-linear effects present in the friction welding process. A simple test for this is

$$\text{Degree of curvature} = \bar{Y} - \bar{Y}_{cp} \tag{9.4}$$

where $\bar{Y}$ is the mean response from the results obtained from all the tests excluding the centre point tests, and $\bar{Y}_{cp}$ is the average over all the responses from the centre point test conditions.

The next sequential step is therefore to duplicate a centre point test so that $\bar{Y}_{cp}$ can be calculated. Table 9.5 suggests that the centre point test should be duplicated between two and four times. Three duplicates were made in the friction welding experiment. The uncoded centre point test condition is easy to identify. The centre level for factor A is simply half way between 40 and 60 kN, i.e. 50 kN. The centre level for factor B is half way between the high and low levels for factor B, i.e. 30 Hz and the centre level for factor C is 2.5 mm. The average degree of upset obtained by replicating tests at this centre point condition was 7.8286 mm. The average of the eight degrees of upset shown in Table 9.6 is 6.97503 mm. Thus the degree of curvature is estimated at

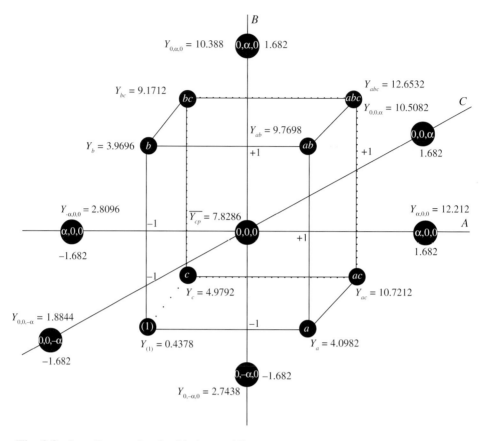

Fig. 9.8 Star diagram for the friction welding process.

Degree of curvature = 6.97503 − 7.8286 = −0.85357 mm.

This is clearly different from zero so that it is necessary to proceed to a non-linear experiment. At least three levels for each factor are going to be required to identify a model that can accurately predict the degree of upset. An efficient option is to carry out just six new tests and add them to the previous 2^3 design with three centre points to create a full central composite design. The six new tests are of course the axial points of the central composite design. Table 9.7 shows the results from the full central composite design. Notice that the experiment in Table 9.7 is written out in the standard order form described above. Figure 9.8 shows this experiment in graphical form and these results will be analysed in Chapter 10.

Table 9.7 Central composite design for the friction welding experiment.

Central Composite Design	Factor A	Factor B	Factor C	Degree of Upset, mm
2³ Factorial Design	−1	−1	−1	0.4378
	1	−1	−1	4.0982
	−1	1	−1	3.9696
	1	1	−1	9.7698
	−1	−1	1	4.9792
	1	−1	1	10.7212
	−1	1	1	9.1712
	1	1	1	12.6532
Axial Points	1.682	0	0	12.2122
	−1.682	0	0	2.8096
	0	1.682	0	10.388
	0	−1.682	0	2.7438
	0	0	1.682	10.5082
	0	0	−1.682	1.8844
Centre Points	0	0	0	7.0845
	0	0	0	7.8286
	0	0	0	8.5727

9.5 THE BOX-BEHNKEN DESIGN

The **Box-Behnken** design is a very efficient three level design with evenly spaced levels. The design is an important alternative to the central composite designs. The design is nearly rotatable and when studying four or seven factors the designs are exactly rotatable. A Box-Behnken design is found by pairing together two variables in a 2^2 design while holding all the remaining factors fixed at their centre levels. This is best achieved in three stages.

9.5.1 FIND ALL COMBINATIONS OF TWO

For an experiment containing k factors the first step involves finding out all possible combinations of two factors, i.e. the pairings. For example, consider $k = 3$ and $k = 4$

Table 9.8 Pairings for $k = 3$.

	A	B	C
Pairing 1	+	+	
Pairing 2	+		+
Pairing 3		+	+

Table 9.9 Pairings for $k = 4$ factors.

	A	B	C	D
Pairing 1	+	+		
Pairing 2	+		+	
Pairing 3	+			+
Pairing 4		+	+	
Pairing 5		+		+
Pairing 6			+	+

factors. Table 9.8 shows that there are just three different combinations of two factors when $k = 3$. Factor A can be paired with factor B and factor C, leaving one more pairing between factors B and C.

Likewise, Table 9.9 shows that there are just six different combinations of two factors when $k = 4$. Factor A can be paired with factors B, C and D, factor B can also be paired with factors C and D, leaving one more pairing between factors C and D.

9.5.2 Form 2^2 Designs for all Pairings

The second step involves taking each identified pairing and forming a 2^2 experiment with them, whilst setting the remaining factors at their middle levels. So the first four tests of a Box-Behnken design with three factors is a 2^2 design in factors A and B (the first pairing in Table 9.8) with factor C set at its middle level. Using coded values for each test condition gives.

A	B	C
−1	−1	0
1	−1	0
−1	1	0
1	1	0

2^2 for Pairing 1: other factor set to center level

The next four tests of a Box-Behnken design with three factors is a 2^2 design in factors A and C (the second pairing in Table 9.8) with factor B set at its middle level.

A	B	C
−1	−1	0
1	−1	0
−1	1	0
1	1	0

2^2 for Pairing 1: other factor set to centre level

A	B	C
−1	0	−1
1	0	−1
−1	0	1
1	0	1

2^2 for Pairing 2: other factor set to centre level

The next four tests of a Box-Behnken design with three factors is a 2^2 design in factors B and C (the third pairing in Table 9.8) with factor A set at its middle level.

A	B	C
−1	−1	0
1	−1	0
−1	1	0
1	1	0

2^2 for Pairing 1: other factor set to centre level

A	B	C
−1	0	−1
1	0	−1
−1	0	1
1	0	1

2^2 for Pairing 2: other factor set to centre level

A	B	C
0	−1	−1
0	1	−1
0	−1	1
0	1	1

2^2 for Pairing 3: other factor set to centre level

The first 24 tests for a Box-Behnken design with $k = 4$ factors are identified in the same way as above using the pairings in Table 9.9.

A	B	C	D
−1	−1	0	0
1	−1	0	0
−1	1	0	0
1	1	0	0

2^2 for Pairing 1: other factor set to centre level

A	B	C	D
−1	0	−1	0
1	0	−1	0
−1	0	1	0
1	0	1	0

2^2 for Pairing 2: other factor set to centre level

A	B	C	D
−1	0	0	−1
1	0	0	−1
−1	0	0	1
1	0	0	1

2^2 for Pairing 3: other factor set to centre level

A	B	C	D
0	−1	−1	0
0	1	−1	0
0	−1	1	0
0	1	1	0

2^2 for Pairing 4: other factor set to centre level

A	B	C	D
0	−1	0	−1
0	1	0	−1
0	−1	0	1
0	1	0	1

2^2 for Pairing 5: other factor set to centre level

A	B	C	D
0	0	−1	−1
0	0	1	−1
0	0	−1	1
0	0	1	1

2^2 for Pairing 6: other factor set to centre level

9.5.3 Replication of Centre Points

Finally, the complete Box-Behnken design is obtained by adding one to three replications of a centre point test. Table 9.10 shows the complete Box-Behnken design for $k = 3$

Table 9.10 A Box-Behnken design for $k = 3$ factors.

Factor A	Factor B	Factor C
−1	−1	0
1	−1	0
−1	1	0
1	1	0
−1	0	−1
1	0	−1
−1	0	1
1	0	1
0	−1	−1
0	1	−1
0	−1	1
0	1	1
0	0	0
0	0	0
0	0	0

factors with three replications at the centre point of the design. All test conditions are in coded units (−1, 0 and +1, for low middle and high).

Table 9.11 shows the complete Box-Behnken design for $k = 4$ factors with one test carried out at the centre point of the design. All test conditions are in coded units (−1, 0 and +1, for low middle and high).

The experiments shown in Tables 9.10 and 9.11 are written in the standard order form for such designs. Proceeding through the three steps above will always identify a Box-Behnken design in its standard order form. The Box-Behnken design for $k = 3$ factors shown in Table 9.10 can be shown graphically using a simple cube. The circles on the cube in Figure 9.9 represent all the test conditions shown in Table 9.10. Unlike in a two level factorial or a central composite design, all the test points fall on the edges rather than the corners of the cube. There are therefore no factorial or face cantered points in these designs. Such designs should therefore be used when the extreme points at the corners of the cube represent factor level combinations that are prohibitively expensive or impossible to test because of physical process constraints.

The Box-Behnken design should not be used if predictions of the response at the extremes are required, that is at the corners of the cube. If this is the requirement a central composite design would be better.

Notice what happens when the Box-Behnken design is combined with the face centred central composite design for $k = 3$ factors. Imagine superimposing Figure 9.9

Table 9.11 A Box-Behnken design for $k = 4$ factors.

Factor A	Factor B	Factor C	Factor D
−1	−1	0	0
1	−1	0	0
−1	1	0	0
1	1	0	0
−1	0	−1	0
1	0	−1	0
−1	0	1	0
1	0	1	0
−1	0	0	−1
1	0	0	−1
−1	0	0	1
1	0	0	1
0	−1	−1	0
0	1	−1	0
0	−1	1	0
0	1	1	0
0	−1	0	−1
0	1	0	−1
0	−1	0	1
0	1	0	1
0	0	−1	−1
0	0	1	−1
0	0	−1	1
0	0	1	1
0	0	0	0

onto Figure 9.7. All the corners of the cube in Figure 9.9 would also become test conditions as would the face centres of the sides of the cube in Figure 9.9. In fact the resulting cube would look identical to the 3^3 design shown in Figure 9.2. Combining a face centred central composite design with a Box-Behnken design gives a 3^k design.

9.6 MIXED LEVEL FACTORIAL DESIGNS

Two level factorial and fractional factorial designs should form the foundation of any initial industrial experimentation. If non-linearities need to be considered, one option is

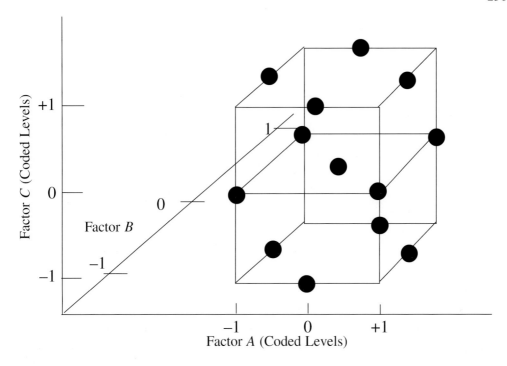

Fig. 9.9 Geometric representation of the Box-Behnken design ($k = 3$).

to carry out a central composite design or a 3^k design for just those factors identified as being important from a 2^{k-p} design. In some situations it may be necessary to consider just a sub set of the factors at more than two levels. In such a situation, **mixed level factorials** can be used. The most popular designs in this area are those that mix factors at two and three levels and those that mix factors at two and four levels.

9.6.1 FACTORS AT TWO AND THREE LEVELS

Consider a situation in which k factors need to be studied, but that a fraction of these need to be at three levels. Let f stand for the number of factors that are at three levels. The remaining, $k-f$ factors being of course at two levels. To determine what the levels are for each factor, a 2^{k+f} factorial design must be set up. The $k-f$ two level factors must be assigned to the first $k-f$ columns of this design and the remaining columns then used for the f three level factors. Two columns are assigned to each three level factor and whenever these two columns have two plus ones on the same row then that three level factor should be set high. When the two columns have two minus signs on the same row then that three

Table 9.12 Design for two factors at two levels and one factor at three levels.

No. of Tests	A	B	C	D	Actual Test Conditions		
					A	B	X
1	−1	−1	−1	−1	−1 (Low)	−1 (Low)	−1 (Low)
2	1	−1	−1	−1	1 (High)	−1 (Low)	−1 (Low)
3	−1	1	−1	−1	−1 (Low)	1 (High)	−1 (Low)
4	1	1	−1	−1	1 (High)	1 (High)	−1 (Low)
5	−1	−1	1	−1	−1 (Low)	−1 (Low)	0 (Centre)
6	1	−1	1	−1	1 (High)	−1 (Low)	0 (Centre)
7	−1	1	1	−1	−1 (Low)	1 (High)	0 (Centre)
8	1	1	1	−1	1 (High)	1 (High)	0 (Centre)
9	−1	−1	−1	1	−1 (Low)	−1 (Low)	0 (Centre)
10	1	−1	−1	1	1 (High)	−1 (Low)	0 (Centre)
11	−1	1	−1	1	−1 (Low)	1 (High)	0 (Centre)
12	1	1	−1	1	1 (High)	1 (High)	0 (Centre)
13	−1	−1	1	1	−1 (Low)	−1 (Low)	1 (High)
14	1	−1	1	1	1 (High)	−1 (Low)	1 (High)
15	−1	1	1	1	−1 (Low)	1 (High)	1 (High)
16	1	1	1	1	1 (High)	1 (High)	1 (High)

level factor should be set low and when the signs alternate, that three level factor should be set at its intermediate level. This intermediate level must be half way between the high and low levels.

To illustrate consider an experiment involving three factors, two at two levels and one at three levels. Here $k = 3$ and $f = 1$. Consequently a $2^{3+1} = 2^4$ design must be set up in the usual way. Let A and B represent the two level factors and X the one three level factor. Then the test levels are derived in the way shown in Table 9.12.

In Table 9.12, a zero is used to signify a factor set at its middle or centre value. C and D do not represent real factors as such, but these two columns are used to identify the levels for the three level factor, X. In all, 16 tests need to be carried out and the levels for the three factors at each of these tests is shown in the last three columns of Table 9.12. For the first test, columns A and B contain a negative element so these two factors are set at their low levels. Columns C and D both contain negative elements so factor X is also set at its low level. For test eight, columns A and B both contain positive elements and so are set high. Column C has a positive element but D has a negative element so that factor X must be set at its intermediate level for this test. Finally, in test 16 columns A to D all have positive elements and so factors A, B and X are all set high.

Table 9.13 Design for two factors at two levels and one factor at four levels.

No. of Tests	A	B	C	D	Actual Test Conditions		
					X	C	D
1	−1	−1	−1	−1	−1 (Low)	−1 (Low)	−1 (Low)
2	1	−1	−1	−1	−0.33	−1 (Low)	−1 (Low)
3	−1	1	−1	−1	0.33	−1 (Low)	−1 (Low)
4	1	1	−1	−1	1 (High)	−1 (Low)	−1 (Low)
5	−1	−1	1	−1	−1 (Low)	1 (High)	−1 (Low)
6	1	−1	1	−1	−0.33	1 (High)	−1 (Low)
7	−1	1	1	−1	0.33	1 (High)	−1 (Low)
8	1	1	1	−1	1 (High)	1 (High)	−1 (Low)
9	−1	−1	−1	1	−1 (Low)	−1 (Low)	1 (High)
10	1	−1	−1	1	−0.33	−1 (Low)	1 (High)
11	−1	1	−1	1	0.33	−1 (Low)	1 (High)
12	1	1	−1	1	1 (High)	−1 (Low)	1 (High)
13	−1	−1	1	1	−1 (Low)	1 (High)	1 (High)
14	1	−1	1	1	−0.33	1 (High)	1 (High)
15	−1	1	1	1	0.33	1 (High)	1 (High)
16	1	1	1	1	1 (High)	1 (High)	1 (High)

9.6.2 Factors at Two and Four levels

Consider a situation in which k factors need to be studied, but that a fraction of these need to be at four levels. Let f stand for the number of factors that are at four levels. The remaining, $k-f$ factors being of course at two levels. To determine what the levels are for each factor, a 2^{k+f} factorial design must be set up. The $k-f$ two level factors must be assigned to the last $k-f$ columns of this design and the remaining columns are used for the f four level factors. Two columns are assigned to each four level factor and whenever these two columns have two plus ones in the same row then that four level factor should be set at its highest level. When the two columns have two minus signs in the same row then that four level factor should be set its lowest level. When the first column contains a plus one and the second a minus one then that four level factor should be set at its lowest but one level. Then when the first column contains a minus one and the second a plus one that four level factor should be set at its highest but one level. The spacing between each of the four levels must be equal both in the uncoded and coded measures.

So if a plus one is used to represent a factor at its high level and a minus one for its low level, −0.33 and +0.33 represent the two intermediate coded levels.

To illustrate consider an experiment involving three factors, two at two levels and one at four levels. Here $k = 3$ and $f = 1$. Consequently a $2^{3+1} = 2^4$ design must be set up in the usual way. Let C and D represent the two level factors and X the one four level factor. Then the test levels are derived in the way shown in Table 9.13.

In Table 9.13, a −0.33 is used to signify a factor set at its lowest but one level and a 0.33 for its highest but one level. A and B do not represent real factors as such. Instead these two columns are used to identify the levels for the four level factor, X. In all 16 tests need to be carried out and the levels for the three factors at each of these tests are shown in the last three columns of Table 9.13. For the first test, columns A and B contain a negative element so factor X is set at its lowest level. Columns C and D both contain negative elements so these two level factors are both set at their low levels. For test eight, columns A and B both contain positive elements and so factor X is set at its highest level. Column C has a positive element but D has a negative element so that factor C is set high and D low. In test 15 column A contains a negative element and column B contains a positive element and so factor X is set at its one but highest level. Columns C and D have positive elements and so factors C and D are set high. Finally, in test 14 Column A contains a positive element and column B contains a negative element and so factor X is set at its one-but-lowest level. Columns C and D have positive elements and so factors C and D are set high.

10. Linear and Non-linear Effects

Designs that have each factor at two levels are capable of identifying and modelling linear relationships only. To model curvature designs that have factors varying over three or more levels are required and such designs were dealt with in great detail in the previous chapter. It is now necessary to distinguish between the **linear effects** studied so far and the following more general **non-linear effects** that can be derived from the non-linear designs of the previous chapter.

Designs that have factors at three levels can be used to estimated a **quadratic** relationship between the quality characteristics of a product and the process variables of the manufacturing process used to make that product. Designs that have factors at four levels are capable of identifying a **cubic** relationship. In general the estimation of an rth order polynomial requires a design with each factor at $r + 1$ levels. The concept of a non-linear effect is fully explained in Section 10.1. Then in Section 10.2 the structure of a **second order response surface model** is given. Some more general models for analysing data from a 3^k design are also presented. The second order model is an extension of the first order response surface model discussed in Chapter 7 and can be used to analyse the data from non-linear experimental designs. Section 10.3 then describes how least squares can be used to estimate the parameters of this second order model and how the least squares procedure can be implemented within Excel. This section also describes how least squares can be used to estimate models that are more general than the second order model. In Section 10.4 the friction welding experiments of Sections 9.2 and 9.4 are analysed.

10.1 A NON-LINEAR EFFECT

It is important at this point to be clear about the meaning of **non-linearity**. The term non-linear is not just used to refer to a response surface that is curved. It is used to describe a situation in which a plot of the response against the level of a factor, whilst holding the levels for all the other factors constant, traces out a curve rather than a line. To make this distinction clear, consider a manufacturing process that has just one important design factor, factor A. If, in an experiment, a single factor is given just two levels and a single response is recorded at each level, then all that can be identified is a linear relationship between the response and that factor. In Figure 10.1 there is only enough information to connect the two experimental points with a straight line. This is not to say that the underlying relationship is linear, simply that the design can only detect a linear relationship. The line in Figure 10.1 is given by the following equation

$$Y_j = \beta_0 + \beta_A A_j \qquad (10.1)$$

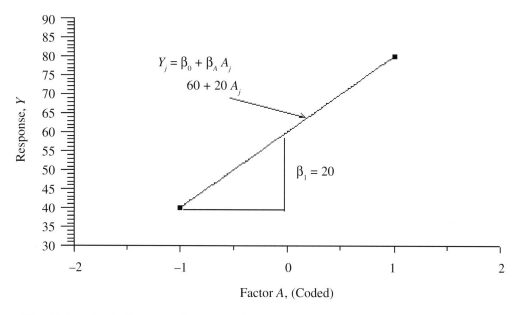

Fig. 10.1 A simple linear one factor model.

where the levels for A_j are in the usual coded units. In Figure 10.1, $\beta_0 = 60$ is the average of the two responses, (or the value for Y_j when $A_j = 0$). $\beta_A = 20$ is half the A_{Loc} effect as defined in Chapter 5. Speaking more generally, β_A is half the **linear effect** and factor A is said to impact upon the response in a linear fashion. Equation (10.1) is therefore an example of a **linear one factor** model. The linear A_{Loc} effect has a very simple interpretation. This being that every unit change in the coded level of factor A will bring forth a $\beta_A = 20$ unit change in the response variable irrespective of the level for factor A.

Now suppose that a further response reading is made at a third level for factor A with this new level being exactly half way between the high and the low levels shown in Figure 10.1 for factor A. If the relationship between the response and factor A is truly linear this point should fall on that part of the linear line corresponding to $A_j = 0$. That is, a response of $Y_j = 60 + 20 \times 0 = 60$ should be recorded. The degree of curvature or non-linearity can then be measured as the distance between the linear line and the observed response when $A_j = 0$. Suppose the result of this additional test is as shown in Figure 10.2.

When $A_j = 0$, the response takes on a value of 45 and not 60 so that a considerable amount of curvature is present. So instead of having a single straight line in Figure 10.2, there is a segmented line that is kinked at the centre point observation. The slope of each of these segmented lines measures two distinct linear location effects for factor A. The first measures the effect that a unit change in factor A has on the response

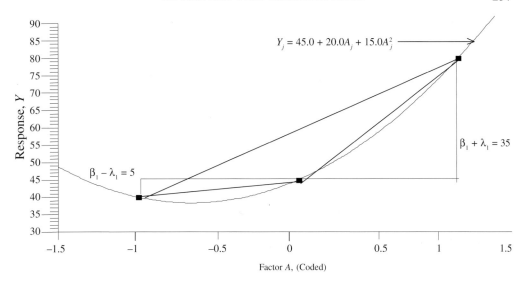

Fig. 10.2 A simple quadratic one factor model.

when factor A is changed from its lowest level. The second shows the effect that a unit change in factor A has on the response when factor A is changed from its middle value. The two together add up to give the location effect of a two unit change in factor A on the response. This is the familiar definition of a location effect derived in Chapter 5. So in a three level design each of the l effects, whether they are location or dispersion effects, are made up of two components.

$$l_{Loc} = l_{Loc1} + l_{Loc2} \tag{10.2}$$

This generalises to an rth level design where each effect, defined as the impact on the average response or variability following a two unit change in the level of the factor, is made up of $r-1$ individual components given by the slope of the $r-1$ segmented lines

$$l_{Loc} = l_{Loc1} + l_{Loc2} + \ldots\ldots + l_{Locr-1} \tag{10.3}$$

Each component of an effect can be found by realising that the segmented line in Figure 10.2 is just an approximation to a smooth quadratic curve, or in general to a smooth polynomial of order r. Thus a curve going through the three data points in Figure 10.2 can be described by the quadratic equation

$$Y_j = \beta_0 + \beta_A A_j + \beta_{AA} A_j^2 \tag{10.4}$$

In Figure 10.2 the smooth curve is given by equation (10.4) with $\beta_0 = 45$, $\beta_A = 20$ and $\beta_{AA} = 15$. Equation (10.4) is an example of a **quadratic one factor** model. The model is non-linear because factor A is raised to a power in excess of one. The meaning of the A_{Loc} effect is no longer straight forward because a unit change in the level of factor A will have a different impact on the response depending on the level for factor A. Thus when $A_j = -1$ in Figure 10.2, a unit increase in factor A will increase the response by 5 units, but when $A_j = 0$, the same unit increase in the level of A will increase the response by 35 units. That is

$$A_{Loc1} = \beta_A - \beta_{AA} = 20 - 15 = 5$$

$$A_{Loc2} = \beta_A + \beta_{AA} = 20 + 15 = 35$$

$$A_{Loc} = A_{Loc1} + A_{Loc2} = 5 + 35 = 40$$

Consequently, a two unit change in factor A will bring forth a 40 unit change in the response. Unlike the linear model however, this change is not evenly split over unit changes in the level of factor A. In the linear model of Figure 10.1 each unit change in the level of factor A lead to a $\beta_A = 20$ unit change in response. In the quadratic model of Figure 10.2 this is split into a $\beta_A - \beta_{AA} = 5$ unit change and a $\beta_A + \beta_{AA} = 35$ unit change in response.

For this reason it is often more practical to redefine the meaning of an effect. Instead of it measuring the change in average response or variability following a two unit coded change in the level of a factor, (as is the case in the linear designs), it is more compact to define an effect in terms a very small change in the level of that factor. That is, an effect is measured by the slope of a curve like that shown in Figure 10.2. The symbols A'_{Loc}, B'_{Loc} etc will be used to represent this new definition of a location effect and A'_{Disp}, B'_{Disp} etc. will be used to represent this new definition of a dispersion effect. Such an effect will no longer be a single number because the slope of the curve in Figure 10.2 changes with the level of factor A. Given equation (10.4) the A'_{Loc} effect is given by the derivative of Y with respect to factor A.

$$A'_{Loc} = \frac{\partial Y_j}{\partial A_j} = \beta_A + 2\beta_{AA} A_j \quad (10.5a)$$

In equation (10.5a) the value for the location effect of factor A depends on the constant β_A and on the level from which factor A is changed through the parameter β_{AA}. Notice that if there are no non-linearity's present, so that $\beta_{AA} = 0$, this definition of an effect collapses to the one given in Chapter 5, i.e. to β_A which is half the location effect in a linear design.

Similarly, if the variance independent of the mean, ϕ_j, is given by the following quadratic one factor model

$$\phi_j = \gamma_0 + \gamma_A A_j + \gamma_{AA} A_j^2 \quad (10.5b)$$

then the pure dispersion effect for factor A, A'_{Disp}, is given by the derivative of ϕ_j with respect to factor A

$$A'_{Disp} = \frac{\partial \phi_j}{\partial A_j} = \gamma_A + 2\gamma_{AA} A_j \tag{10.5c}$$

This definition of an effect is easily extended to designs with more than one variable. In equation (10.6a) below factor B also has a quadratic impact on the response and factors A and B interact in a linear fashion

$$Y_j = \beta_0 + \beta_A A_j + \beta_B B_j + \beta_{AB} A_j B_j + \beta_{AA} A_j^2 + \beta_{BB} B_j^2 \tag{10.6a}$$

Now the location effect for factor A is given by

$$A'_{Loc} = \frac{\partial Y_j}{\partial A_j} = \beta_A + 2\beta_{AA} A_j + \beta_{AB} B_j \tag{10.6b}$$

The value for the location effect of factor A now depends on the constant β_A, on the level from which factor A is changed through the parameter β_{AA} and on the level at which factor B is set when A is changed (through parameter β_{AB}). Complexity therefore increases with the number of factors and the number of levels for these factors.

The values for β_0, β_A and β_{AA} in equation (10.4) are easily obtained. β_0 is simply the value for the response obtained at the middle level for factor A, whilst β_{AA} is obtained by averaging the response at the high and low levels for factor A and subtracting from this average the response obtained at the middle level for factor A. $\beta_{AA} = (40 + 80)/2 - 45 = 15$. Finally, β_A is found by halving the slope of a linear line joining the responses obtained at the high and low levels for factor A, $\beta_{AA} = (80 - 40)/2 = 20$.

A non-linear relationship is therefore defined as one in which the model used to predict the response contains quadratic or cubic terms in the factor levels and not just one in which the response surface is curved. There is a subtle difference between these two definitions which shows up from a comparison of equation (10.6a) with equation (10.6c).

$$Y_j = \beta_0 + \beta_A A_j + \beta_B B_j + \beta_{AB} A_j B_j \tag{10.6c}$$

All the effects in equation (10.6c) are linear in nature so that factor A impacts on the response in a linear fashion, (albeit that the size of this linear impact depends on the level of factor B). The same is true for factor B. None of the effects are raised to a power higher than one and so the model is indeed linear. However, the response surface associated with this equation is actually twisted because of the presence of the AB_{Loc} interaction term. Curved surfaces can be modelled by introducing either interaction effects or quadratic (and cubic) effects. Only when the latter are present in the model is the relationship between the response and a factor described as non-linear.

10.2 THE SECOND ORDER RESPONSE SURFACE MODEL

10.2.1 Structure of the Second Order Response Surface Model

Equation (10.6c) above is an example of a first order response surface model that can be used to analyse a linear process having just two factors, A and B. A first order model contains the levels for each factor together with their interactions. It can be estimated from any experiment that sets each factor at just two different levels. Equation (7.6) of Section 7.2 gave the general form for a first order response surface model. This model included the levels for each factor and their first order interactions. Each factor is sub classified as a design or noise factor and each first order interaction is classified as that between two design factors, two noise factors or between a noise and design factor.

Equation (10.6a) is an example of a **second order response surface model** that can be used to analyse a non-linear process having two factors. Notice (by comparing equations (10.6a) and (10.6c)) that a second order model is constructed by adding onto the first order response surface model terms containing the squares of the levels of each factor. It can be estimated from an experiment that sets the level for each factor at more than two different levels. In fact all of the designs shown in the previous chapter will allows a second order response surface model to be identified.

Generalising to any number of factors (including design and noise factors) the second order response surface model for analysing the $i = 1$ to N responses made at test condition j is

$$Y_{ij} = \beta_0 + \sum_{v=1}^{M_1} \beta_v X_{vj} + \sum_{v=1}^{M_1}\sum_{w=v+1}^{M_1} \beta_{vw} X_{vj} X_{wj} + \sum_{w=1}^{M_2} \lambda_w Z_{wj} + \sum_{v=1}^{M_2}\sum_{w=v+1}^{M_2} \lambda_{vw} Z_{vj} Z_{wj} + \sum_{v=1}^{M_1}\sum_{w=1}^{M_2} \delta_{vw} X_{vj} Z_{vj}$$

$$+ \sum_{v=1}^{M_1} \beta_{vv} X_{vj}^2 + \sum_{w=1}^{M_2} \lambda_{ww} Z_{wj}^2 + \varepsilon_{ij} \qquad (10.7a)$$

β, λ and δ are the parameters of the model which help define all the location effects as defined in Section 10.1. For unreplicated designs, $N = 1$ and the ith subscript drops out of equation (10.7a). Comparing this equation to equation (7.7) it becomes clear that a new type of variable is added to the first order model – the squares of the levels for the design factors (X_{vj}^2) and the squares of the levels of the noise factors (Z_{wj}^2).

As in the first order model, the prediction error ε_{ij} will again capture the scatter in the $i = 1$ to N responses obtained at each of the j test conditions, together with all the higher order interaction terms left out of the model. Unlike the first order model, the higher order terms left out are not just the linear interactions between three or more factors (e.g. the interaction between X_1, X_2 and X_3) but also the non-linear interactions such as those between the squares of two or more factors (e.g. the interaction between X_1^2 and X_2^2) and the interactions between squared and non squared factors (e.g. X_1 and X_2^2).

In all other respects these two equations are the same. Thus it is clear from the properties of the prediction error that a second order response surface model is not the only functional form that can be used to analyse experimental data. It is in itself a simplified model and this will become clear in the sub-sections below.

Now, if the noise factors are transformed so that their average value is zero, and the noise factors are independent of each other, the mean response at test condition j for a process driven by equation (10.7a) is

$$\overline{Y}_j = \beta_0 + \sum_{v=1}^{M_1} \beta_v X_{vj} + \sum_{v=1}^{M_1} \beta_{vv} X_{vj}^2 + \sum_{v=1}^{M_1} \sum_{w=v+1}^{M_1} \beta_{vw} X_{vj} X_{wj} \qquad (10.7b)$$

Treating the design factors as constants (because they do not vary uncontrollably) and assuming that ε_j and the noise factors are independent, the variance in the response at test condition j for a process driven by equation (10.7a) is approximately equal to

$$S_{Yj}^2 = S_{\varepsilon j}^2 + S_{Zv}^2 \sum_{v=M_1+1}^{M_2} \left[\frac{\partial Y}{\partial Z_v}\right]^2 \qquad (10.7c)$$

The total number of variables, P, in the second order response surface model given by equation (10.7a) is

$$P = 2[M_1 + M_2] + [M_1 M_2] + \left[\frac{M_1!}{2!(M_1-2)!}\right] + \left[\frac{M_2!}{2!(M_2-2)!}\right] \qquad (10.7d)$$

where $M_1!$ reads the factorial of M_1 and $M_2!$ reads the factorial of M_2.

10.2.2 Some Models for the 3^k Design

Any 3^k design involves 3^k tests and so at most 3^k parameters in a response surface model can be estimated from this type of design. When studying two factors (A and B) using a 3^2 design the most general model that can be used to represent the results from this design is therefore

$$Y_j = \beta_0 + \beta_A A_j + \beta_B B_j + \beta_{AB} A_j B_j + \beta_{AA} A_j^2 + \beta_{BB} B_j^2 + \beta_{A^2B^2} A_j^2 B_j^2$$
$$+ \beta_{A^2B} A_j^2 B_j + \beta_{AB^2} A_j B_j^2 \qquad (10.8a)$$

A_j and B_j are the levels for factors A and B at test condition j and $A_j B_j$ the product of these levels at test condition j. A_j^2 is the squared level for factor A, B_j^2 is the squared level for factor B, $A_j^2 B_j$ is the squared level for factor A multiplied by the level for factor B, $A_j B_j^2$ is the squared level for factor B multiplied by the level for factor A and $A_j^2 B_j^2$ is the squared level for factor A multiplied by the squared level for factor B. The values for β in equation (10.8a) combine to form the location effect for factors A and B as defined in Section 10.1. For example, if equation (10.8a) is a realistic model for predicting the response Y_j, then the location effect for factor A is defined as

$$A'_{Loc} = \frac{\partial Y_j}{\partial A_j} = \beta_A + \beta_{AB}B_j + 2\beta_{AA}A_j + 2\beta_{A^2B^2}A_jB_j^2 + 2\beta_{A^2B}A_jB_j + \beta_{AB^2}B_j^2$$

So when factor A is changed by a small amount, the resulting change in response depends on the initial level from which factor A is changed through the values for $2\beta_{AA}$, $2\beta_{A^2B^2}$ and $2\beta_{A^2B}$. It also depends on the level at which factor B is set when factor A is changed through the terms β_{AB}, $2\beta_{A^2B^2}$, $2\beta_{A^2B}$ and β_{AB^2}. So when factor A is at its middle level ($A_j = 0$), and factor B is at its low level ($B_j = -1$), a small change in the level of factor A will change the response by

$$\beta_A - \beta_{AB} + \beta_{AB^2}$$

In a similar fashion the location effect for factor B is given by differentiating equation (10.8a) with respect to B_j

$$B'_{Loc} = \beta_B + \beta_{AB}A_j + 2\beta_{BB}B_j + 2\beta_{A^2B^2}A_j^2B_j + \beta_{A^2B}A_j^2 + 2\beta_{AB^2}B_j$$

The general model given by equation (10.8a) for analysing the results from a 3^2 experiment contains some terms which on their own are very difficult to interpret. The first of these is the interaction between the square of factor A and the square of factor B, as measured by $2\beta_{A^2B^2}$. Then there is the interaction between factor A and the square of factor B, as measured by $2\beta_{AB^2}$. Also there is the interaction between factor B and square of factor A, as measured by $2\beta_{A^2B}$. For this reason research engineers often assume that these non-linear interactions are unimportant. That is, they all come from the same distribution of factor effects that has a mean of zero and some standard deviation. If ε_{ij} is used to represent the sum of these effects, together with the scatter in the N responses recorded at each test condition (if there is replication), then the second order model for analysing data from a 3^2 design is of the form

$$Y_{ij} = \beta_0 + \beta_A A_j + \beta_B B_j + \beta_{AB}A_jB_j + \beta_{AA}A_j^2 + \beta_{BB}B_j^2 + \varepsilon_{ij} \tag{10.8b}$$

The prediction error, ε_{ij}, takes on the distribution for the three omitted interactions and so has a mean of zero. This is a lot easier to interpret. But clearly the second order response surface model given by equation (10.8b) is a simplified version of the most general from given by equation (10.8a). For an 3^3 design the second order model takes the form

$$Y_{ij} = \beta_0 + \beta_A A_j + \beta_B B_j + \beta_C C_j + \beta_{AB}A_jB_j + \beta_{AC}A_jC_j + \beta_{BC}B_jC_j + \beta_{AA}A_j^2 + \beta_{BB}B_j^2 + \beta_{CC}C_j^2 + \varepsilon_{ij} \tag{10.8c}$$

This model now includes four extra terms in comparison to its equivalent for a 3^2 design (i.e. equation (10.8b)). These being the level for factor C, the interaction between factors A and C, B and C and the square of the level for factor C. This design has 27 tests and so 27 parameters could be included in the most general model for this design. These terms are assigned to the variable ε_{ij} (together with the scatter in the N responses recorded at each test condition if there is replication), and include the variables $A_j B_j C_j$, $A_j^2 B_j$, $A_j^2 C_j$, $B_j^2 C_j$, $A_j^2 B_j^2$, $A_j^2 C_j^2$, $B_j^2 C_j^2$, $A_j^2 B_j^2 C_j^2$ etc. So again the second order response surface model is a simplified model of the most general model that could be used.

The results from a 3^k design will be enough to allow the estimation of the most general type of model, i.e. to identify the β parameter in front of every conceivable variable. It follows that 3^k designs will also allow the identification of the simpler second order response surface model. For example in the 3^2 design, models given by either equation (10.8a) or (10.8b) can be estimated from the results obtained in the design.

10.2.3 SOME MODELS FOR THE CENTRAL COMPOSITE AND BOX-BEHNKEN DESIGNS

The results from both these designs are enough to enable a second order response surface model, given by equation (10.7a), to be estimated. This is irrespective of the number of factors being studied provided that the fractional component of the central composite design is at least resolution V. If the resolution is lower than this, then the obtained values for the β parameters in equation (10.8c), for example, may not be showing what they appear to be showing. In such low resolution designs first order interactions will be alaised with each other so that β_{AB} in equation (10.8c) for example may not be showing the interaction between factors A and B but that between two other factors in the alaising structure (or even their sums). This is why resolution V was stipulated in Chapter 9 when describing the central composite design. However, the most general types of model can't be estimated from central composite or Box-Behnken designs. For example, the results from a central composite or Box-Behnken design for $k = 2$ factors will allow the parameters in equation (10.8b) to be estimated but not all the parameters in equation (10.8a).

10.2.4 A MODEL FOR MIXED FACTORIAL DESIGNS

The results from mixed level designs that are not fractional or have a resolution V or more are enough to enable a second order response surface model, given by equation (10.7a), to be estimated. For example, an experiment that has two factors at two levels (A and B) and one factor at three levels (X) can be analysed using the following second order model

$$Y_{ij} = \beta_0 + \beta_A A_j + \beta_B B_j + \beta_X X_j + \beta_{AB} A_j B_j + \beta_{AX} A_j X_j + \beta_{BX} B_j X_j + \beta_{XX} X_j^2 + \varepsilon_{ij} \quad (10.9)$$

Thus all main and first order interactions between two level and three level factor effects can be estimated as can the main non-linear effect for X_j. An experiment that has two factors at two levels (A and B) and one factor at four levels (X) can also be analysed using the above second order model, but this time variable X has four rather than three levels. The variable X_j^3 can be added:

$$Y_{ij} = \beta_0 + \beta_A A_j + \beta_{AB} A_j B_j + \beta_{AX} A_j X_j + \beta_{BX} B_j X_j + \beta_{XX} X_j^2 + \beta_{XXX} X_j^3 + \varepsilon_{ij} \quad (10.10)$$

10.3 ESTIMATING RESPONSE SURFACE MODELS

10.3.1 ESTIMATING A SECOND ORDER RESPONSE SURFACE MODEL

A second order response surface model is always estimated using the responses (Y_{ij}) obtained from one of the experimental designs discussed in the last chapter. These experiments contain $j = 1$ to M separate test conditions. If the 3^k design is used the number of tests conditions are $M = 3^k$, where k are the number of factors within the experiment. But if a central composite design or a face centred design is used the number of test conditions are $M = 2^{k-p} + 2k + N_c$, where N_c is the number of centre point tests and p is the design fraction. If a Box Behnken design is used the number of test conditions are

$$M = \frac{k! \times 4}{2!(k-2)!} + N_c.$$

The easiest way to obtain values for the parameters of a second order response surface model is to use the least squares method. Remember that this involves choosing values for the parameters β, λ and δ that minimise the sum of the squared prediction errors, i.e. $\Sigma \varepsilon_{ij}^2$ in equation (10.7a). It can be shown that the solution to this minimisation problem takes the form

$$B = (X'X)^{-1} X'Y \quad (10.11)$$

X is a matrix containing the coded test values at each test condition for all the variables contained in equation (10.7a). This matrix will have $P + 1$ columns (as defined by equation (10.7d)) and $j = 1$ to NM rows, where the value of M depends on the experimental design selected in the way described above. However, the rows of the X matrix must correspond to the standard order form of the experiment. All the experimental designs in the last chapter were written in standard order form. For example, the standard order form for a central composite design is the 2^{k-p} tests (in their standard form) followed by all the axial tests in alphabetical order, followed by the centre point tests. So for such a design, and with $N = 1$ and $P = 3$, the first row of X corresponds to all factors low, the 2^{k-p} row corresponds to all factors high, the two rows after this will contain the two axial point tests for factor X_1, the next two the axial point tests for factor X_2, the next two the axial point test for factor X_3 and the last rows the centre points tests. If there are N replicates made at each test condition each of these rows are replicated N times.

The first column of X will always be a column containing NM ones. The remaining columns will contain, in the same order, the test values and products thereof for all the variables shown in equation (10.7a). Thus the second column of X will contain the test values for X_{1j} corresponding to all the M test conditions, the third column of X will contain the test values for X_{2j} corresponding to all the M test conditions and the final column of X will contain the values for Z^2_{M2j} corresponding to all the M test conditions. X' is the transpose matrix of X in which the columns of X become the rows of X'.

Y is a vector containing the values for all the NM responses made during the experiment. N responses are recorded at each of the $j = 1$ to M test condition. Thus Y will contain NM rows. The ordering of each response in vector Y is the same as the order in which the tests were carried out and thus in the same order as the rows in X. Thus, the first N rows will contain the N values for Y obtained at the $j = 1$ test condition, the next N rows will contain the N values for Y obtained at the $j = 2$ test condition, and the last N rows will contain the N values for Y obtained at the $j = M$ test condition. The exact ordering therefore depends on the standard order for the experimental design used.

B is a vector containing, in the same standard order, the values for the $P + 1$ parameters (all the β, λ and δ) of the second order response surface model given by equation (10.7a). This vector will therefore be made up of $P + 1$ rows. Row one will contain the value for β_0, row two the value for β_1, row three the value for β_2 and the $(P + 1)$th row the value for $\lambda_{M_2M_2}$.

The standard deviation, S_{Par}^2, of each of the estimated parameters of the second order response surface model (which is the equivalent of the standard deviation of each location effect in the 2^{k-p} designs) can be estimated using the formula.

$$S_{Par} = \sqrt{(X'X)^{-1} S_\varepsilon^2} \qquad (10.12)$$

In equation (10.12), S_ε^2 is the estimated standard deviation of the prediction error term, ε_{ij}, in a second order model. It can be found using the familiar standard deviation formula for any variable.

$$S_\varepsilon = \sqrt{\frac{\sum_{j=1}^{M} \sum_{i=1}^{N} [\varepsilon_{ij}]^2}{N \times M - P}} \qquad (10.13)$$

N is the number of replicates which can of course be one, in which case the ith summation operators drops out. M is the number of test condition (e.g. $M = 3^k$) and P is the number of parameters estimated. Notice that the mean error has been left out of the square bracketed term because by definition it is zero in value. The standard deviations for each parameter are then located along the diagonal of the matrix S_{par}.

A t statistic for measuring the importance of each parameter in the model is constructed in the same way as that described in Chapter 6, namely divide each parameter estimate by its standard deviation. The resulting t statistic has $V = N \times M - P$ degrees of freedom.

Equation (10.11) is executed within Excel using the LINEST function first described in Section 5.3.2. As $P + 1$ parameters together with their standard deviations need

1st order model – equation (10.6c)

$$\mathbf{B} = \begin{bmatrix} \beta_0 \\ \beta_A \\ \beta_B \\ \beta_{AB} \end{bmatrix}$$

Second order model – equation (10.6a)

$$\mathbf{B} = \begin{bmatrix} \beta_0 \\ \beta_A \\ \beta_B \\ \beta_{AB} \\ \beta_{AA} \\ \beta_{BB} \end{bmatrix}$$

Most general model – equation (10.8a)

$$\mathbf{B} = \begin{bmatrix} \beta_0 \\ \beta_A \\ \beta_B \\ \beta_{AB} \\ \beta_{AA} \\ \beta_{BB} \\ \beta_{A^2B^2} \\ \beta_{A^2B} \\ \beta_{AB^2} \end{bmatrix}$$

Fig. 10.3 The B matrix for various models using a 3^2 experiment.

estimating, simply highlight a block of two rows by $P + 1$ columns and then type LINEST (R1, R2, FALSE, TRUE) into the formula box at the top of the computer screen. If the TRUE term is replaced by the word FALSE then no standard deviation is calculated for any of the parameters in equation (10.7a). R1 is the range of cells down a single column containing the vector Y and R2 is the range of cells containing the X matrix. Remember that this function returns the β parameters in equation (10.7a) in reverse order to that shown in this equation.

10.3.2 Estimating Some Response Surface Models Using Data from a 3^2 Design

Figure 10.3 shows the composition of the B matrix for equations (10.6a), (10.6c) and (10.8a) when used in combination with a 3^2 experiment. These equations correspond to the first order, second order and most general model for analysing data from a 3^2 design. Notice the ordering of each β value in the B matrix is the same as that in equations (10.6a), (10.6c) and (10.8a).

The Y vectors for these three models are identical. This vector depends only on the number of replications (N) and the number of test conditions (M). The latter depends on the experimental design chosen to generate the responses entering the Y vector. For the 3^2 design $M = 3^2 = 9$. The resulting Y vector is illustrated fully in Figure 10.4. The Y_{ij} responses entering this vector must be in the standard order form for the experiment from which

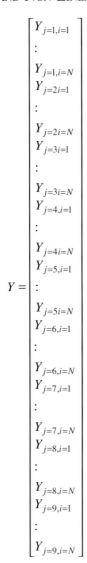

Fig. 10.4 The Y matrix for the 3^2 designs.

they were obtained. Thus $Y_{j=1}$ is the response obtained from the first test of a 3^2 design in standard order form, i.e. the response obtained when both factors A and B are set low (see Table 9.2). If N responses are recorded at this test condition, the first N rows of matrix Y must contain all the responses obtained at this test condition. Likewise $Y_{j=M=9}$ is the response obtained from the last test of a 3^2 design in standard order form, i.e. the

Most general model – equation (10.8a)

$$X = \begin{bmatrix} & A_j & B_j & A_jB_j & A^2_j & B^2_j & A^2_jB^2_j & A^2_jB_j & A_jB^2_j \\ 1 & -1 & -1 & 1 & 1 & 1 & 1 & -1 & -1 \\ 1 & 0 & -1 & 0 & 0 & 1 & 0 & 0 & 0 \\ 1 & 1 & -1 & -1 & 1 & 1 & 1 & -1 & 1 \\ 1 & -1 & 0 & 0 & 1 & 0 & 0 & 0 & 0 \\ 1 & 0 & 0 & 0 & 0 & 0 & 0 & 0 & 0 \\ 1 & 1 & 0 & 0 & 1 & 0 & 0 & 0 & 0 \\ 1 & -1 & 1 & -1 & 1 & 1 & 1 & 1 & -1 \\ 1 & 0 & 1 & 0 & 0 & 1 & 0 & 0 & 0 \\ 1 & 1 & 1 & 1 & 1 & 1 & 1 & 1 & 1 \end{bmatrix}$$

2nd order model – equation (10.6a)

$$X = \begin{bmatrix} & A_j & B_j & A_jB_j & A^2_j & B^2_j \\ 1 & -1 & -1 & 1 & 1 & 1 \\ 1 & 0 & -1 & 0 & 0 & 1 \\ 1 & 1 & -1 & -1 & 1 & 1 \\ 1 & -1 & 0 & 0 & 1 & 0 \\ 1 & 0 & 0 & 0 & 0 & 0 \\ 1 & 1 & 0 & 0 & 1 & 0 \\ 1 & -1 & 1 & -1 & 1 & 1 \\ 1 & 0 & 1 & 0 & 0 & 1 \\ 1 & 1 & 1 & 1 & 1 & 1 \end{bmatrix}$$

1st order model – equation (10.6c)

$$X = \begin{bmatrix} & A_j & B_j & A_jB_j \\ 1 & -1 & -1 & 1 \\ 1 & 0 & -1 & 0 \\ 1 & 1 & -1 & -1 \\ 1 & -1 & 0 & -1 \\ 1 & 0 & 0 & 0 \\ 1 & 1 & 0 & 0 \\ 1 & -1 & 1 & -1 \\ 1 & 0 & 1 & 0 \\ 1 & 1 & 1 & 1 \end{bmatrix}$$

Fig. 10.5 The X matrix for various models of an experiments with two factors at three levels.

response obtained when both factors A and B are set high. If N responses are recorded at this test condition, the last N rows of matrix Y must contain all the responses obtained at this test condition.

The X matrix in Figure 10.5 is a matrix of coded test conditions for the unreplicated 3^2 design where $M = 9$ for the three models shown in Figure 10.3. In a replicated design (with N replications) each row of the X matrix in Figure 10.5 is replicated N times. The ordering of the columns of X for the three models in Figure 10.5 are the same as the ordering of the variables in equations (10.6a), (10.6c) and (10.8a). The first column of each X matrix in Figure 10.5 is a column of ones to enable β_0 in these equations to be estimated. For the first order model, the ordering of the variables in equation (10.6c) is factor A, then factor B and then the AB interaction. Thus the second column of X will contain test information on factor A, the third column will contain test information on factor B, and column four will contain the product of these two variables. The X matrix for the second order model given by equation (10.6a) starts of the same as that for the first order model. However, the fifth column of X will contain the square of variable A and the sixth column the square of variable B. These are the last two variables in equation (10.6a). Notice that the column headings for the X matrix correspond to the order of the variables in the most general model for this design as given by equation (10.8a).

Each row of X will correspond to a particular test condition. The rows of X are also in the standard order form of the experiment. Thus the first row of the X matrices in Figure 10.5 gives the first test of the 2^3 design when written out in standard order form. This test has both factors low and so a –1 is inserted into columns A and B of the X matrix (see Table 9.2 for a recap of this order). The last row of the X matrices in Figure 10.5 gives the last test of the 2^3 design when written out in standard order form. This test has both factors high and so a +1 is inserted into the last row of columns A and B in the X matrix.

The numbers in the columns headed AB are calculated by multiplying together the numbers in the A and B columns of matrix X. The numbers in the columns headed A^2 are calculated by squaring the numbers in the A column of matrix X. The numbers in the columns headed B^2 are calculated by squaring the numbers in the B column of matrix The numbers in the columns headed A^2B are calculated by multiplying together the numbers in the A^2 and B columns of matrix X. The numbers in the columns headed AB^2 are calculated by multiplying together the numbers in the A and B^2 columns of matrix X. The numbers in the columns headed A^2B^2 are calculated by multiplying together the numbers in the A^2 and B^2 columns of matrix X.

10.3.3 Estimating Some Response Surface Models Using Data from a 3^3 Design

For a 3^3 design and using a second order model, the first column of X will contain a series of ones to allow the estimation of β_0. The next three columns will contain $3^3 = 27$ rows of coded test conditions in standard order form for factors A, B and C, i.e. the A, B and C columns of Table 9.3. The next column contains the product of the second and third columns (i.e. variable $A_j \times B_j$), the next column contains the product of the second and fourth columns (i.e. variable $A_j \times C_j$) and the next column contains the product of the third and fourth columns (i.e. variable $B_j \times C_j$). The final three columns contain the squares of the second to fourth columns of X respectively (i.e. variables A_j^2, B_j^2, C_j^2). This is illustrated in Figure 10.6 for an unreplicated design. Notice that the ordering of the columns in the X matrix is the same as the ordering of the variables in the second order model given by equation (10.8c). In a replicated design with N replications each row of the X matrix in Figure 10.6 is replicated N times.

$$X = \begin{bmatrix}
 & A_j & B_j & C_j & A_jB_j & A_jC_j & B_jC_j & A_j^2 & B_j^2 & C_j^2 \\
1 & -1 & -1 & -1 & 1 & 1 & 1 & 1 & 1 & 1 \\
1 & 0 & -1 & -1 & 0 & 0 & 1 & 0 & 1 & 1 \\
1 & 1 & -1 & -1 & -1 & -1 & 1 & 1 & 1 & 1 \\
1 & -1 & 0 & -1 & 0 & 1 & 0 & 1 & 0 & 1 \\
1 & 0 & 0 & -1 & 0 & 0 & 0 & 0 & 0 & 1 \\
1 & 1 & 0 & -1 & 0 & -1 & 0 & 1 & 0 & 1 \\
1 & -1 & 1 & -1 & -1 & 1 & -1 & 1 & 1 & 1 \\
1 & 0 & 1 & -1 & 0 & 0 & -1 & 0 & 1 & 1 \\
1 & 1 & 1 & -1 & 1 & -1 & -1 & 1 & 1 & 1 \\
1 & -1 & -1 & 0 & 1 & 0 & 0 & 1 & 1 & 0 \\
1 & 0 & -1 & 0 & 0 & 0 & 0 & 0 & 1 & 0 \\
1 & 1 & -1 & 0 & -1 & 0 & 0 & 1 & 1 & 0 \\
1 & -1 & 0 & 0 & 0 & 0 & 0 & 1 & 0 & 0 \\
1 & 0 & 0 & 0 & 0 & 0 & 0 & 0 & 0 & 0 \\
1 & 1 & 0 & 0 & 0 & 0 & 0 & 1 & 0 & 0 \\
1 & -1 & 1 & 0 & -1 & 0 & 0 & 1 & 1 & 0 \\
1 & 0 & 1 & 0 & 0 & 0 & 0 & 0 & 1 & 0 \\
1 & 1 & 1 & 0 & 1 & 0 & 0 & 1 & 1 & 0 \\
1 & -1 & -1 & 1 & 1 & -1 & -1 & 1 & 1 & 1 \\
1 & 0 & -1 & 1 & 0 & 0 & -1 & 0 & 1 & 1 \\
1 & 1 & -1 & 1 & -1 & 1 & -1 & 1 & 1 & 1 \\
1 & -1 & 0 & 1 & 0 & -1 & 0 & 1 & 0 & 1 \\
1 & 0 & 0 & 1 & 0 & 0 & 0 & 0 & 0 & 1 \\
1 & 1 & 0 & 1 & 0 & 1 & 0 & 1 & 0 & 1 \\
1 & -1 & 1 & 1 & -1 & -1 & 1 & 1 & 1 & 1 \\
1 & 0 & 1 & 1 & 0 & 0 & 1 & 0 & 1 & 1 \\
1 & 1 & 1 & 1 & 1 & 1 & 1 & 1 & 1 & 1 \\
\end{bmatrix}$$

Fig. 10.6 The X matrix for the unreplicated 3^3.

The Y matrix will be the column of responses corresponding to the M test conditions in the X matrix. The responses within the Y matrix are thus in standard order form. For the 3^3 design the Y vector will have a length equal to $27 \times N$. Finally, the B matrix contains all the parameters in equation (10.8c). The ordering in B will be the same as the ordering of the parameters in this equation. This is shown below.

$$B = \begin{bmatrix} \beta_0 \\ \beta_A \\ \beta_B \\ \beta_C \\ \beta_{AB} \\ \beta_{AC} \\ \beta_{BC} \\ \beta_{AA} \\ \beta_{BB} \\ \beta_{CC} \end{bmatrix}$$

10.3.4 Estimating Some Response Surface Models Using Data from a Central Composite Design

To illustrate consider a process containing three factors that has been analysed using a central composite design, where the 2^{k-p} part of the design is a full factorial. For this design and using a second order model to study the results, the first column of X will contain a series of ones to allow the estimation of β_0. The next three columns will contain the $M = 2^k + 2k + N_c$ rows of coded test conditions in standard order form for factors A, B and C, i.e. the factor A, B and C columns of Table 9.7. In this example N_c is equal to three. The first eighth rows of columns A to C thus contain the coded test conditions for the 2^3 design. The next six rows of columns A to C contain the axial point tests and the final three rows the replicated centre point test of Table 9.7. The next column contains the product of the second and third columns (i.e. variable $A_j \times B_j$), the next column contains the product of the second and fourth columns (i.e. variable $A_j \times C_j$) and the next column contains the product of the third and fourth columns (i.e. variable $B_j \times C_j$). The final three columns contain the squares of the second to fourth columns of X respectively (i.e. variables A_j^2, B_j^2, C_j^2). This is illustrated in Figure 10.7 for unreplicated designs. Notice that the ordering of the columns in the X matrix is the same as the ordering of the variables in the second order model given by equation (10.8c). In a replicated design with N replications each row of the X matrix in Figure 10.7 is replicated N times.

The Y matrix will be the column of responses corresponding to the M test conditions in the X matrix. The responses within the Y matrix are thus in standard order form. Finally, the B matrix contains all the parameters in equation (10.8c). The ordering in B will be the same as the ordering of the parameters in this equation and so will be the same as the last B matrix above.

The second order model for the unreplicated central composite design with k = 3

$$X = \begin{bmatrix} 1 & A_j & B_j & C_j & A_jB_j & A_jC_j & B_jC_j & A_j^2 & B_j^2 & C_j^2 \\ 1 & -1 & -1 & -1 & 1 & 1 & 1 & 1 & 1 & 1 \\ 1 & 1 & -1 & -1 & -1 & -1 & 1 & 1 & 1 & 1 \\ 1 & -1 & 1 & -1 & -1 & 1 & -1 & 1 & 1 & 1 \\ 1 & 1 & 1 & -1 & 1 & -1 & -1 & 1 & 1 & 1 \\ 1 & -1 & -1 & 1 & 1 & -1 & -1 & 1 & 1 & 1 \\ 1 & 1 & -1 & 1 & -1 & 1 & -1 & 1 & 1 & 1 \\ 1 & -1 & 1 & 1 & -1 & -1 & 1 & 1 & 1 & 1 \\ 1 & 1 & 1 & 1 & 1 & 1 & 1 & 1 & 1 & 1 \\ 1 & 1.682 & 0 & 0 & 0 & 0 & 0 & 2.826 & 0 & 0 \\ 1 & -1.682 & 0 & 0 & 0 & 0 & 0 & 2.826 & 0 & 0 \\ 1 & 0 & 1.682 & 0 & 0 & 0 & 0 & 0 & 2.826 & 0 \\ 1 & 0 & -1.682 & 0 & 0 & 0 & 0 & 0 & 2.826 & 0 \\ 1 & 0 & 0 & -1.682 & 0 & 0 & 0 & 0 & 0 & 2.826 \\ 1 & 0 & 0 & 1.682 & 0 & 0 & 0 & 0 & 0 & 2.826 \\ 1 & 0 & 0 & 0 & 0 & 0 & 0 & 0 & 0 & 0 \\ 1 & 0 & 0 & 0 & 0 & 0 & 0 & 0 & 0 & 0 \\ 1 & 0 & 0 & 0 & 0 & 0 & 0 & 0 & 0 & 0 \end{bmatrix}$$

Fig. 10.7 The X matrix for the unreplicated 3^3.

10.4 ANALYSIS OF THE FRICTION WELDING EXPERIMENT

10.4.1 THE FIRST TWO PROCESS VARIABLES

In Section 9.2, a 3^2 experiment was set up to study the effect of two of the process variables on the degree of upset. These factors were the axial force (factor A) and the frequency of vibration (factor B). The third process variable, the amplitude of vibration was held fixed at 2 mm in the experiment. Table 9.4 shows the test conditions and results obtained (upset in mm) from carrying out this experiment. A clear objective is therefore to try and maximise the degree of upset.

Table 10.1 shows the complete set of workings required to estimate a second order model from this design. Notice how the Y and X matrices are as illustrated in

Section 10.3.2. The X matrix is next transposed, (X'), and multiplied by the X matrix to give $X'X$ which is then inverted. Vector B is then obtained by multiplying this inverted matrix by the $X'Y$ vector. The individual parameters in this vector are in the same order as those in equation (10.6a). Substituting the values for each parameter into equation (10.6a) gives

$$Y_j = 4.326 + 2.485A_j + 2.338B_j + 0.535A_jB_j + 0.272A_j^2 - 0.108B_j^2 + \varepsilon_j \quad (10.14a)$$

Notice that this equation does not predict the recorded response exactly because three terms have been omitted from the equation – $A_j^2B_j^2$, $A_j^2B_j$, $A_jB_j^2$. If these had been included the resulting full model would exactly predict the response because ε_j would disappear. ε_j in equation (10.14a) therefore picks up these omitted terms which hopefully are small in magnitude and on average equal zero. On average then, the following equation gives a prediction of the degree of upset

$$\hat{Y}_j = 4.326 + 2.485A_j + 2.338B_j + 0.535A_jB_j + 0.272A_j^2 - 0.108B_j^2 \quad (10.14b)$$

The interpretation to be placed on each of these numbers can be best seen through use of a **three way diagram**. The three way diagram for the friction welding experiment is shown in Figures 10.8 to 10.10. All of the three segmented lines could have been plotted on one graph, but to keep things clearer they have been separated out. These diagrams simply plot out the nine data points shown in Table 9.4 together with the nine predicted values derived from equation (10.14b). To see how these nine predicted values are obtained take the first test condition. Here $A_j = -1$, $B_j = -1$, and so $A_jB_j = 1$, $A_j^2 = 1$, $B_j^2 = 1$ $A_j^2B_j^2 = 1$. Substituting these values into equation (10.14b) gives the predicted response for the first test condition.

$$Y_1 = 4.326 - 2.485 - 2.338 + 0.535 + 0.272 - 0.108 = 0.202$$

The prediction error associated with this test condition is therefore

$$\varepsilon_1 = 0.4378 - 0.202 = 0.2358$$

All the other predicted values and errors can be calculated in a similar way and are shown in Figures 10.8 to 10.10. Each of these figures corresponds to a different level for factor B, the frequency of vibration. An interpolated prediction is also shown. This is given by equation (10.14b) which traces out a smooth quadratic polynomial. Each half of each segmented line has been labelled (1) through to (6).

In Figure 10.8 the slope of the quadratic polynomial at any level for factor A and B is given by the A'_{Loc} effect or

$$A'_{Loc} = \beta_A + \beta_{AB}B_j + 2\beta_{AA}A_j$$

In Figure 10.8 the level for factor B is always at its lowest so that $B_j = -1$. Substituting in the value for these parameters with this extra constraint in mind gives

$$A'_{Loc} = 2.485 - 0.535 + 2 \times 0.272A_j$$

Table 10.1 The least squares procedure for estimating a second order model in the friction welding experiment.

$$Y = \begin{bmatrix} 0.4378 \\ 1.6496 \\ 4.0982 \\ 1.7164 \\ 4.641 \\ 7.165 \\ 3.9696 \\ 6.4726 \\ 9.7698 \end{bmatrix}$$

	A_i	B_i	A_iB_i	A_i^2	B_i^2
1	-1	-1	1	1	1
1	0	-1	0	0	1
1	1	-1	-1	1	1
1	-1	0	0	1	0
1	0	0	0	0	0
1	1	0	0	1	0
1	-1	1	-1	1	1
1	0	1	0	0	1
1	1	1	1	1	1

$X =$ (above)

$$X' = \begin{bmatrix} 1 & 1 & 1 & 1 & 1 & 1 & 1 & 1 & 1 \\ -1 & 0 & 1 & -1 & 0 & 1 & -1 & 0 & 1 \\ -1 & -1 & -1 & 0 & 0 & 0 & 1 & 1 & 1 \\ 1 & 0 & -1 & 0 & 0 & 0 & -1 & 0 & 1 \\ 1 & 0 & 1 & 1 & 0 & 1 & 1 & 0 & 1 \\ 1 & 1 & 1 & 0 & 0 & 0 & 1 & 1 & 1 \end{bmatrix}$$

$$X'X = \begin{bmatrix} 9 & 0 & 0 & 0 & 6 & 6 \\ 0 & 6 & 0 & 0 & 0 & 0 \\ 0 & 0 & 6 & 0 & 0 & 0 \\ 0 & 0 & 0 & 4 & 0 & 0 \\ 6 & 0 & 0 & 0 & 6 & 4 \\ 6 & 0 & 0 & 0 & 4 & 6 \end{bmatrix}$$

$$X'Y = \begin{bmatrix} 39.92 \\ 14.9092 \\ 14.0264 \\ 2.1398 \\ 27.1568 \\ 26.3976 \end{bmatrix}$$

$$(X'X)^{-1} = \begin{bmatrix} 0.55556 & 0 & 0 & 0 & -0.3333 & -0.3333 \\ 0 & 0.16667 & 0 & 0 & 0 & 0 \\ 0 & 0 & 0.16667 & 0 & 0 & 0 \\ 0 & 0 & 0 & 0.25 & 0 & 0 \\ -0.3333 & 0 & 0 & 0 & 0.5 & 0 \\ -0.3333 & 0 & 0 & 0 & 0 & 0.5 \end{bmatrix}$$

$$B = \begin{bmatrix} 4.32631 \\ 2.48487 \\ 2.33773 \\ 0.53495 \\ 0.27173 \\ -0.1079 \end{bmatrix} = \begin{bmatrix} \beta_0 \\ \beta_A \\ \beta_B \\ \beta_{AB} \\ \beta_{AA} \\ \beta_{BB} \end{bmatrix}$$

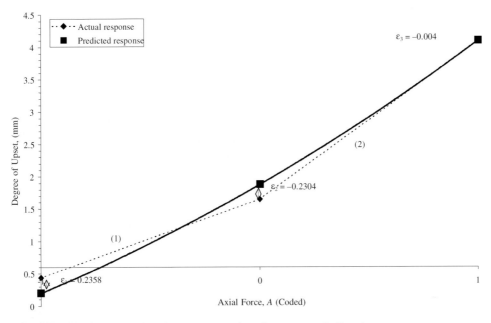

Fig. 10.8 Actual and predicted responses at low frequency of vibration.

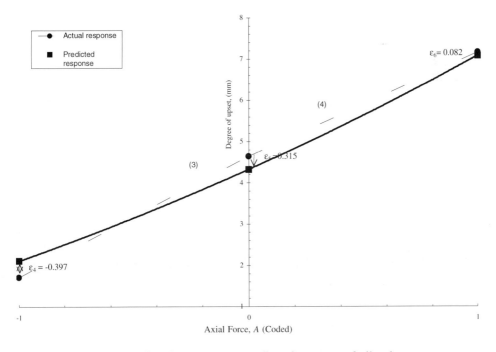

Fig. 10.9 Actual and predicted responses at medium frequency of vibration.

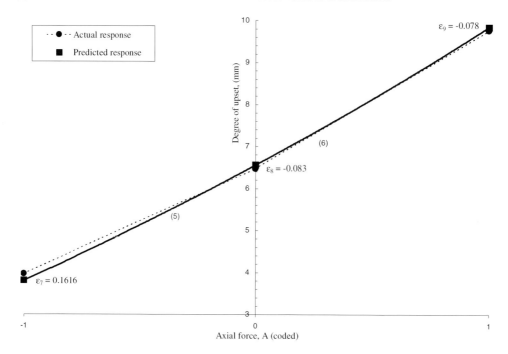

Fig. 10.10 Actual and predicted responses at high drequency of vibration.

The linear lines (1) and (2) can be though of as an average value for this effect over the relevant levels. So if there was no non-linearity present (so $\beta_{AA} = 0$) the curve in Figure 10.8 would collapse to these two segmented lines and if there was no interaction ($\beta_{AB} = 0$) these two segmented lines would collapse to a single linear line with slope $\beta_A = 2.485$. Note the value for ε_1 is as calculated above.

In Figure 10.9, the slope of the quadratic polynomial at any level for factor A and B is given by the A'_{Loc} effect or

$$A'_{Loc} = \beta_A + \beta_{AB} B_j + 2\beta_{AA} A_j$$

In Figure 10.9 the level for factor B is always at its middle level so that $B_j = 0$. Substituting in the value for these parameters with this extra constraint in mind gives

$$A'_{Loc} = 2.485 + 2 \times 0.272 A_j$$

Lines (3) and (4) can be though of as an average value for this effect over the relevant levels. So if there was no non-linearity's present (so $\beta_{AA} = 0$) the curve in Figure 10.9 would collapse to these two segmented lines and if there was no interaction

($\beta_{AB} = 0$) these two segmented lines would collapse to a single linear line with slope $\beta_A = 2.485$.

In Figure 10.10 the slope of the quadratic polynomial at any level for factor A and B is given by the A'_{Loc} effect or

$$A'_{Loc} = \beta_A + \beta_{AB} B_j + 2\beta_{AA} A_j$$

In Figure 10.10 the level for factor B is always at its high level so that $B_j = 1$. Substituting in the value for these parameters with this extra constraint in mind gives

$$A'_{Loc} = 2.485 + 0.535 + 2 \times 0.2721 A_j$$

Lines (5) and (6) can be though of as an average value for this effect over the relevant levels. So if there was no non-linearity present (so $\beta_A = 0$) the curve in Figure 10.10 would collapse to these two segmented lines and if there was no interaction ($\beta_{AB} = 0$) these two segmented lines would collapse to a single linear line with slope $\beta_A = 0.12$.

Finally, it is possible to test the significance of the linear and non-linear effects in equation (10.14a). The standard deviation associated with each estimated parameter can be found from equations (10.12) and (10.13). S_ε, the standard deviation of the error, is from these equations and the nine errors shown in Figure 10.8 to 10.10.

$$S_\varepsilon = \sqrt{\frac{(0.2358)^2 + (-0.2304)^2 + (-0.004)^2 + \ldots\ldots + (-0.083)^2 + (-0.078)^2}{9-6}} = \sqrt{\frac{0.411}{3}} = 0.37.$$

Combining this with the value for $(X'X)^{-1}$ shown in Table 10.1 gives, when substituted into equation (10.12).

$$S_{Par} = \begin{bmatrix} 0.276 & & & & & \\ & 0.151 & & & & \\ & & 0.151 & & & \\ & & & 0.185 & & \\ & & & & 0.262 & \\ & & & & & 0.262 \end{bmatrix}$$

The standard deviation for each of the six parameters is therefore

$$S_{\beta_0} = 0.276, \quad S_{\beta_A} = 0.151, \quad S_{\beta_B} = 0.151, \quad S_{\beta_{AB}} = 0.185, \quad S_{\beta_{AA}} = 0.262, \quad S_{\beta_{BB}} = 0.262$$

A student t statistic for each coefficient can be constructed in the usual way by dividing each coefficient by its standard deviation.

$$t_{\beta_0} = \frac{4.326}{0.276} = 15.67, \quad t_{\beta_A} = \frac{2.485}{0.151} = 16.46, \quad t_{\beta_B} = \frac{2.338}{0.151} = 15.48,$$

$$t_{\beta_{AB}} = \frac{0.535}{0.185} = 2.89, \quad t_{\beta_{AA}} = \frac{0.272}{0.262} = 1.04, \quad t_{\beta_{BB}} = \frac{-0.108}{0.262} = -0.41.$$

The degrees of freedom associated with each of these tests is

$$V = M - P = 3$$

The critical (5%) t value from Table 6.3 is therefore -3.182 to $+3.182$. This suggests that the degree of upset is dependant upon the axial force, the frequency of vibration and possibly the linear interaction between the two. However any non-linearity is very week and confined, if anything, to the axial force.

10.4.2 ALL THREE PROCESS VARIABLES

Now consider all three factors in the friction welding experiment of Section 2.5. Let C represent the third factor, the amplitude of vibration. Table 9.7 of Section 9.4 showed a central composite design for this three factor process, together with the results on the degree of upset obtained from this experiment.

Table 10.2 shows the results of estimating a second order model using the least squares procedure. Note carefully how the X matrix is set up. The first column of X is a vector of one so that β_o can be estimated. The next three columns of X represent the test conditions for the central composite design, i.e. are the same as factor columns A to C in Table 9.7. The next three columns of X represent the linear interaction terms and so are found by multiplying together the elements in the A to C columns of X. Finally, the last three columns of X represent the quadratic effects and are found by squaring the elements in the A to C columns of X. The rest of Table 10.2 shows the calculations required for B. The B vector in Table 10.2 gives the following estimated second order response surface model of the friction welding process.

$$Y_j = 7.811 + 2.526 A_j + 2.064 B_j + 2.471 C_j - 0.015 A_j B_j - 0.03 A_j C_j - 0.385 B_j C_j$$
$$- 0.05 A_j^2 - 0.384 B_j^2 - 0.514 C_j^2 + \varepsilon_j. \tag{10.15}$$

This model can be simplified by testing which of these effects are important using the t test discussed in Section 10.3.1. The standard deviation for the prediction error is

$$S_\varepsilon = \sqrt{\frac{(1.066)^2 + (-0.415)^2 + (-0.329)^2 + \ldots + (-0.51)^2 + (-0.004)^2 + (-0.314)^2}{17 - 10}} = \sqrt{\frac{4.9836}{7}} = 0.844$$

Combining this with the $(X'X)^{-1}$ matrix in Table 10.2 gives the following standard deviations for each of the parameters of the second order model.

$$S_{\beta_o} = 0.844\sqrt{0.332} = 0.486,$$

$$S_{\beta_A} = 0.844\sqrt{0.0732} = 0.228,$$

Table 10.2 Least squares procedure for estimating a second order model in the friction welding process.

$Y = \begin{bmatrix} 0.4378 \\ 4.0982 \\ 3.9696 \\ 9.7698 \\ 4.9792 \\ 10.721 \\ 9.1712 \\ 12.653 \\ 7.0845 \\ 7.8286 \\ 8.5727 \\ 12.212 \\ 2.8096 \\ 10.388 \\ 2.7438 \\ 10.508 \\ 1.8844 \end{bmatrix}$

$X = $

	A_j	B_j	C_j	AB_j	AC_j	BC_j	A_j^2	B_j^2	C_j^2
1	-1	-1	-1	1	1	1	1	1	1
1	1	-1	-1	-1	-1	1	1	1	1
1	-1	1	-1	-1	1	-1	1	1	1
1	1	1	-1	1	-1	-1	1	1	1
1	-1	-1	1	1	-1	-1	1	1	1
1	1	-1	1	-1	1	-1	1	1	1
1	-1	1	1	-1	-1	1	1	1	1
1	1	1	1	1	1	1	1	1	1
1	0	0	0	0	0	0	0	0	0
1	0	0	0	0	0	0	0	0	0
1	0	0	0	0	0	0	0	0	0
1	1.682	0	0	0	0	0	2.8291	0	0
1	-1.682	0	0	0	0	0	2.8291	0	0
1	0	1.682	0	0	0	0	0	2.8291	0
1	0	-1.682	0	0	0	0	0	2.8291	0
1	0	0	1.682	0	0	0	0	0	2.8291
1	0	0	-1.682	0	0	0	0	0	2.8291

$X'Y = \begin{bmatrix} 119.83 \\ 34.5 \\ 28.185 \\ 33.755 \\ -0.12 \\ -0.237 \\ -3.079 \\ 98.299 \\ 92.952 \\ 90.86 \end{bmatrix}$

$B = \begin{bmatrix} 7.8105 \\ 2.5259 \\ 2.0636 \\ 2.4714 \\ -0.015 \\ -0.03 \\ -0.385 \\ -0.05 \\ -0.384 \\ -0.514 \end{bmatrix} \begin{matrix} = \beta_0 \\ = \beta_A \\ = \beta_B \\ = \beta_C \\ = \beta_{AB} \\ = \beta_{AC} \\ = \beta_{BC} \\ = \beta_{AA} \\ = \beta_{BB} \\ = \beta_{CC} \end{matrix}$

$(X'X) = \begin{bmatrix} 17 & 0 & 0 & 0 & 0 & 0 & 0 & 13.658 & 13.658 & 13.658 \\ 0 & 13.658 & 0 & 0 & 0 & 0 & 0 & 0 & 0 & 0 \\ 0 & 0 & 13.658 & 0 & 0 & 0 & 0 & 0 & 0 & 0 \\ 0 & 0 & 0 & 13.658 & 0 & 0 & 0 & 0 & 0 & 0 \\ 0 & 0 & 0 & 0 & 8 & 0 & 0 & 0 & 0 & 0 \\ 0 & 0 & 0 & 0 & 0 & 8 & 0 & 0 & 0 & 0 \\ 0 & 0 & 0 & 0 & 0 & 0 & 8 & 0 & 0 & 0 \\ 13.658 & 0 & 0 & 0 & 0 & 0 & 0 & 24.008 & 8 & 8 \\ 13.658 & 0 & 0 & 0 & 0 & 0 & 0 & 8 & 24.008 & 8 \\ 13.658 & 0 & 0 & 0 & 0 & 0 & 0 & 8 & 8 & 24.008 \end{bmatrix}$

$(X'X)^{-1} = \begin{bmatrix} 0.332 & 0 & 0 & 0 & 0 & 0 & 0 & -0.113 & -0.113 & -0.113 \\ 0 & 0.0732 & 0 & 0 & 0 & 0 & 0 & 0 & 0 & 0 \\ 0 & 0 & 0.0732 & 0 & 0 & 0 & 0 & 0 & 0 & 0 \\ 0 & 0 & 0 & 0.0732 & 0 & 0 & 0 & 0 & 0 & 0 \\ 0 & 0 & 0 & 0 & 0.125 & 0 & 0 & 0 & 0 & 0 \\ 0 & 0 & 0 & 0 & 0 & 0.125 & 0 & 0 & 0 & 0 \\ 0 & 0 & 0 & 0 & 0 & 0 & 0.125 & 0 & 0 & 0 \\ -0.113 & 0 & 0 & 0 & 0 & 0 & 0 & 0.0887 & 0.0262 & 0.0262 \\ -0.113 & 0 & 0 & 0 & 0 & 0 & 0 & 0.0262 & 0.0887 & 0.0262 \\ -0.113 & 0 & 0 & 0 & 0 & 0 & 0 & 0.0262 & 0.0262 & 0.0887 \end{bmatrix}$

$$S_{\beta_B} = 0.844\sqrt{0.0732} = 0.228,$$

$$S_{\beta_C} = 0.844\sqrt{0.0732} = 0.228,$$

$$S_{\beta_{AB}} = 0.844\sqrt{0.125} = 0.298,$$

$$S_{\beta_{AC}} = 0.844\sqrt{0.125} = 0.298,$$

$$S_{\beta_{BC}} = 0.844\sqrt{0.125} = 0.298,$$

$$S_{\beta_{AA}} = 0.844\sqrt{0.0887} = 0.251,$$

$$S_{\beta_{BB}} = 0.844\sqrt{0.0887} = 0.251,$$

$$S_{\beta_{CC}} = 0.844\sqrt{0.0887} = 0.251,$$

A student t statistic for each parameter can be constructed in the usual way by dividing each coefficient by its standard deviation

$$t_{\beta_0} = \frac{7.811}{0.486} = 16.07, \quad t_{\beta_A} = \frac{2.526}{0.228} = 11.08, \quad t_{\beta_B} = \frac{2.064}{0.228} = 9.05,$$

$$t_{\beta_C} = \frac{2.471}{0.228} = 10.84, \quad t_{\beta_{AB}} = \frac{-0.015}{0.298} = -0.05, \quad t_{\beta_{AC}} = \frac{-0.03}{0.298} = -0.10,$$

$$t_{\beta_{BC}} = \frac{-0.385}{0.298} = -1.29, \quad t_{\beta_{AA}} = \frac{-0.05}{0.251} = -0.20, \quad t_{\beta_{BB}} = \frac{-0.384}{0.251} = -1.53,$$

$$t_{\beta_{CC}} = \frac{-0.514}{0.251} = -2.05.$$

The degrees of freedom associated with each of these tests is

$V = M - P = 17 - 10 = 7$

The critical (10%) t value from Table 6.3 is therefore -1.9 to $+1.9$. This suggests that the degree of upset is linearly related to axial force and the frequency of vibration, but non-linearly dependant upon the amplitude of vibration. Possibly of importance is the interaction between the amplitude and frequency of vibration and the square of the frequency of vibration.

The above results could have been found using the LINEST function in Excel. Figure 10.11 shows how this is done.

There are ten parameters to estimate in the second order response surface model of equation (10.15). The standard deviations of these parameters are also required so that

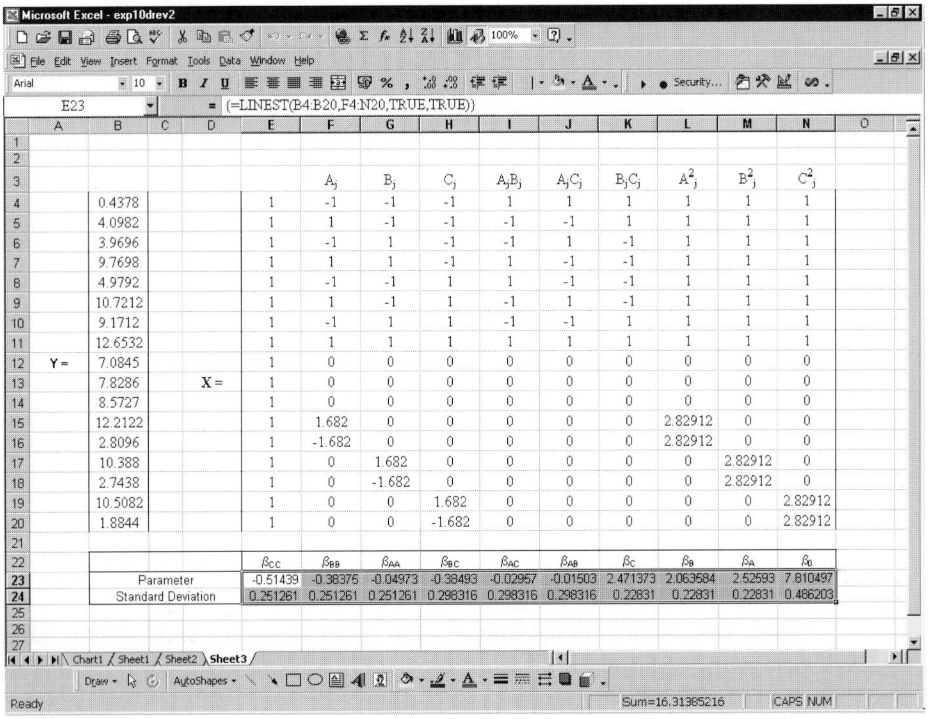

Row	B		E	F A_j	G B_j	H C_j	I A_jB_j	J A_jC_j	K B_jC_j	L A^2_j	M B^2_j	N C^2_j
4	0.4378		1	-1	-1	-1	1	1	1	1	1	1
5	4.0982		1	1	-1	-1	-1	-1	1	1	1	1
6	3.9696		1	-1	1	-1	-1	1	-1	1	1	1
7	9.7698		1	1	1	-1	1	-1	-1	1	1	1
8	4.9792		1	-1	-1	1	1	-1	-1	1	1	1
9	10.7212		1	1	-1	1	-1	1	-1	1	1	1
10	9.1712		1	-1	1	1	-1	-1	1	1	1	1
11	12.6532		1	1	1	1	1	1	1	1	1	1
12 Y =	7.0845		1	0	0	0	0	0	0	0	0	0
13	7.8286	X =	1	0	0	0	0	0	0	0	0	0
14	8.5727		1	0	0	0	0	0	0	0	0	0
15	12.2122		1	1.682	0	0	0	0	0	2.82912	0	0
16	2.8096		1	-1.682	0	0	0	0	0	2.82912	0	0
17	10.388		1	0	1.682	0	0	0	0	0	2.82912	0
18	2.7438		1	0	-1.682	0	0	0	0	0	2.82912	0
19	10.5082		1	0	0	1.682	0	0	0	0	0	2.82912
20	1.8844		1	0	0	-1.682	0	0	0	0	0	2.82912

	β_{CC}	β_{BB}	β_{AA}	β_{BC}	β_{AC}	β_{AB}	β_C	β_B	β_A	β_0
Parameter	-0.51439	-0.38375	-0.04973	-0.38493	-0.02957	-0.01503	2.471373	2.063584	2.52593	7.810497
Standard Deviation	0.251261	0.251261	0.251261	0.298316	0.298316	0.298316	0.22831	0.22831	0.22831	0.486203

Fig. 10.11 Results of student t statistics using LINEST function in Excel.

an area of ten columns by two rows is highlighted in Excel and the LINEST function shown towards the top of the screen is typed in. Notice the range R1 in the LINEST command contains the Y matrix and the R2 range the X matrix excluding the initial column of ones. Hence the inclusion of the first TRUE statement in the function (this would be set to FALSE if the column of ones had been included in the R2 range). Once this function is typed in, the ctrl shift and return keys are hit to show the highlighted results. They are the same numbers as those shown in Table 10.2 (but to more decimal places).

A simplified version of the second order model for this process is obtained by deleting all those variables in equation (10.15) that have a t value less than 1.9 in absolute value and by dropping the prediction error

$$Y_j = 7.811 + 2.526A_j + 2.064B_j + 2.471C_j - 0.514C_j^2 \tag{10.16}$$

This model shows how upset varies with test conditions on the average. The ultimate objective of this experiment was to maximise the average degree of upset and equation (10.16) suggests that this can be achieved by setting the axial force at its high level of 67 kN (+1.682 in coded units) and the frequency of vibration at 38 Hz (+1.682 in coded units). To find the amplitude of vibration that maximises the average degree of upset insert these conditions into equation (10.16) to give

$$Y_j = 7.811 + 2.526(+1.682) + 2.064(+1.682) + 2.471C_j - 0.514C_j^2,$$

or

$$Y_j = 15.531 + 2.471C_j - 0.514C_j^2 \qquad (10.17)$$

Differentiating equation (10.17) with respect to factor C (the amplitude of vibration) and setting the result equal to zero to find a maximum gives

$$\frac{\partial Y_j}{\partial C_j} = 2.417 - 1.028C_j = 0, \text{ or } C_j = 2.35$$

That is, the amplitude of vibration should be set at the coded level of 2.35 to maximise the degree of upset. This is outside the range of the experiment and so C should be set at the highest level for the central composite design, i.e. +1.682. This corresponds to an amplitude of vibration of 2.5 + 1.682(0.5) = 3.34 mm.

PART V
OPTIMISATION OF NON–LINEAR PROCESSES

11. Sequential Testing

The second order response surface model fits the data obtained from non-linear designs best when that design has been set up near the conditions for the process variables that are required to optimise the manufacturing process. Thus achieving process optimisation, using a second order response surface model for the mean and variability in response, will usually be attempted only at the end of a series of sequential tests that have located the approximate point of optimality.

So in Section 11.1 below the idea of sequential testing will be introduced. In this approach the path of steepest ascent is used to plan a series of tests along the quickest route to the optimum. This will be illustrated in Section 11.2 using the results obtained in previous chapters on the ausforming process.

11.1 SEQUENTIAL TESTING AND THE PATH OF STEEPEST ASCENT

Consider Figure 11.1 that describes the relationship existing between two design factors (A and B) and a quality characteristic (Y) of a product manufactured from a hypothetical process. This response surface is actually generated from the following non-linear equation

$$Y_j = \rho_1 \exp\left[-0.5\left(\frac{A_j - \rho_2}{\rho_3}\right) - 0.5\left(\frac{B_j - \rho_4}{\rho_5}\right)\right] \tag{11.1}$$

ρ_1 to ρ_5 are parameter constants and A_j and B_j are the levels for the design factors. If the objective is to maximise Y and if an engineer knew this equation, then choosing values for A_j and B_j so as to maximise Y would be a very straightforward exercise. The highest point in Figure 11.1 shows what is referred to as a **global maximum**. No other point around it is higher. In reality engineers will know very little about the precise nature of the response surface. Experimentation is needed to find this out. There are two key characteristics of most manufacturing processes that can provided guidance on how best to carry out those experiments that will enable the global maximum to be found.

i. Processes that are operating a long way from the global optimum are likely to be located on a flat part of the response surface. If, for example, the manufacturing process is operating towards the bottom right hand corner in Figure 11.1, then around that region the surface is quite flat. Hence in such regions a change in the level of a design factor is likely to have a predominately linear impact upon the response.

ii. Processes that are operating close to the global optimum are likely to be located on a very curved part of the response surface. In Figure 11.1 the slope of the surface

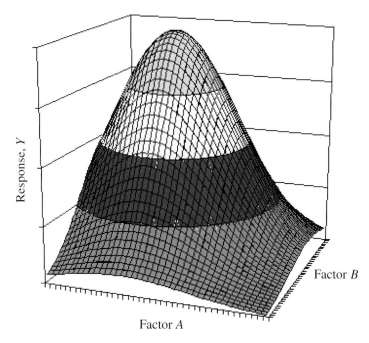

Fig. 11.1 Complete response surface for a hypothetical manufacturing process.

around the global maximum is quite steep. Consequently, a change in the level of a design factor is likely to have a predominately non-linear impact on the response. These two characteristics imply that the following **sequential** approach to experimentation should be extremely successful in locating the global maximum.

A. Set up a simple linear design with the current operating conditions of the manufacturing process being used as the centre point for the design. Depending on the number of design factors being looked at, this design should either be a 2^k factorial or 2^{k-p} fractional factorial.

B. Carry out a number of centre point tests and conduct the test for curvature discussed in Section 9.4 (equation 9.4). If, as is likely at this stage, there are no non-linear effects then the current operating conditions are a long way from the optimum and so proceed to step C below. If curvature appears to be present, the process is operating close to the optimum point so jump to step E below.

C. From the 2^{k-p} or 2^k design eliminate all unimportant effects using a probability plot and identify the **path of steepest ascent** from the factors remaining. Then carry out a set of single tests at equally spaced points along this path. As this linear design only approximates the shape of the response surface at this stage (the interaction terms crudely model any curvature on the surface at this point) the path of steepest ascent will not go straight to the global maximum. Rather it will go in its general direction. For example, in Figure 11.1 the path derived from a 2^2 design around the bottom right hand corner of the surface is likely to move up and around the hill. Because of this, there will come a time when the recorded responses obtained from individual test carried out along the path will start to decline.

D. Set up a new linear 2^k or 2^{k-p} design around the point of decline on the path of steepest ascent. Then repeat steps B and C until conditions are such that step E should be carried out.

E. At this point on the surface, non-linear effects have been identified from a 2^{k-p} design with centre points. Consequently, the global optimum is in the vicinity of this centre point. To locate it precisely, augment the last 2^{k-p} design with axial points. Alternatively, if k is small carry out a fresh replicated 3^k design. From such designs estimate a second order response surface model for the mean and variance. These will be used to optimise the process using the techniques discussed in Chapter 8 and a new technique to be discussed in the next chapter. If the technique chosen requires replication then repeat the last design.

In this sequential process there are just two tasks that have not yet been dealt with. One is the use of a dual response surface model in step E and this will be discussed again in Chapter 12. The other is identifying the path of steepest ascent from a 2^{k-p} design. Before looking at an application of this to the ausforming process consider the following illustration of the path of steepest ascent.

The line of steepest ascent will always be perpendicular to the gradient of the contours on the response surface shown in Figure 11.1. To illustrate consider a simple 2^2 design with the centre of this design taken as the origin for the line of steepest ascent. Its slope will therefore be given by the negative reciprocal of the slope of the contour line at this centre point. As the response surface is likely to be linear a long way from the optimum, a first order response surface model can be used to analyse the data from this 2^2 design. A_j and B_j in this model are the coded test conditions for factor A and B.

$$Y_j = \beta_0 + \beta_A A_j + \beta_B B_j + \beta_{AB} A_j B_j \qquad (11.2a)$$

Now at the design centre $A_j = B_j = 0$ so that at the design centre $\beta_0 = Y_j$

$$Y_j = \beta_0 + \beta_A \times 0 + \beta_B \times 0 + \beta_{AB} \times 0 \times 0 = \beta_0 \qquad (11.2b)$$

Substituting this value for Y_j into equation (11.2a) and rearranging gives

$$\beta_A A_j = -\beta_B B_j - \beta_{AB} A_j B_j$$

and upon further rearrangement

$$A_j = \frac{-\beta_B B_j}{\beta_A + \beta_{AB} B_j} \qquad (11.2c)$$

This is the centre point line contour in that is shows all those values for the coded levels of factors A and B that give a response value of β_0. The contour line associated with any value for Y is given by

$$A_j = \frac{\tilde{Y} - \beta_0 - \beta_B B_j}{\beta_A + \beta_{AB} B_j} \quad (11.2d)$$

where $\tilde{y}$ is the value for Y associated with a particular contour.

The slope of the contour line at the centre of the design is found by applying the quotient rule of differentiation to equation (11.2c).

$$\frac{dA_j}{dB_j} = \frac{-\beta_B}{\beta_A + \beta_{AB} B_j} + \frac{\beta_B \beta_{AB} B_j}{[\beta_A + \beta_{AB} B_j]^2} \quad (11.2e)$$

The path of steepest ascent out from the centre of the design is perpendicular to this slope and so is given by

$$A_j = -\left[\frac{1}{\frac{-\beta_B}{\beta_A + \beta_{AB} B_j} + \frac{\beta_B \beta_{AB} B_j}{[\beta_A + \beta_{AB} B_j]^2}}\right] B_j \quad (11.3a)$$

This gives the test levels, in coded units, for factors A and B along the path of steepest ascent. These coded levels are easily converted into actual levels using equation (3.2). Equation (11.3a) is used to find the levels for factor A on the path of steepest ascent given a chosen increment for factor B from the design centre. For simplicity the interaction effects are often ignored ($\beta_{AB} = 0$) and the approximate path of steepest ascent used.

$$A_j = -\left[\frac{1}{\frac{-\beta_B}{\beta_A}}\right] B_j \quad (11.3b)$$

11.2 SEQUENTIAL EXPERIMENTATION FOR THE AUSFORMING PROCESS

11.2.1 THE FIRST TWO FACTORS ONLY

In most manufacturing experiments there are natural limits to the factors under consideration. This is certainly true for the ausforming process. In Section 2.1 it was

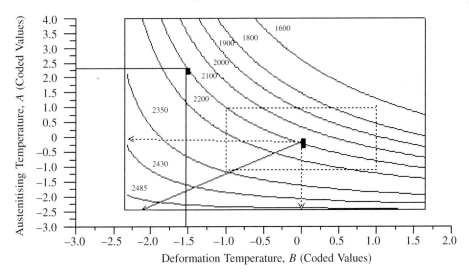

Fig. 11.2 Contour map for part of the ausforming process.

made clear that steels only have the allotropic form in the temperature range 850 to 1200°C. Further, the austenite is only unstable in the range 300 to 600°C. These two ranges then form the natural limits for the austenitising and deformation temperature respectively. The objective is then to try and maximise strength within this feasible range of test conditions. In particular the target quoted in Section 8.2.1 was to achieve a tensile strength of at least 2600 MPa (and to minimise the variability around this target). In Figure 11.2 this feasible range of test conditions has been boxed in with a solid line. Notice that the test conditions are in coded values as derived from equation (3.1). Thus the upper limit of 1200°C for austenitising temperature corresponds to a coded value of

$$\frac{1200-985}{55} = 3.91$$

where 985 is the average austenitising temperature obtained from the upper and lower limits of the 2^2 design shown in Table 3.2, i.e. 1040 and 930°C, and 55 is half the distance between these upper and lower austenitising temperatures. Similarly, the lower limit of 300°C for deformation temperature corresponds to a coded value of

$$\frac{300-475}{75} = -2.33$$

where 475 is the average temperature obtained from the upper and lower limits of the 2^2 design shown in Table 3.2, i.e. 550 and 400°C, and 75 is half the distance between these upper and lower deformation temperatures.

Shown within this feasible range is a further box (the dashed lines) the corners of which represent the actual test conditions of the 2^2 design discussed in Section 3.1. That experiment held all the other factors in the process fixed at their low levels. Nevertheless, this experiment clearly covered only part of the feasible range defining the complete ausforming process. In Chapter 8 we optimised this process within this smaller region of the complete response surface. Thus whilst we were able to meet the target requirement within this range it is not likely to represent a maximum possible average strength. To find this we must search within the bigger feasible region (i.e. the complete response surface) of Figure 11.2 using the path of steepest ascent. The results of the experiment can now be used to predict the steels strength elsewhere in this feasible range. To do this, remember that in Section 5.5.1 the following first order response surface model was identified and it can be used to describe the response surface within the inner box of Figure 11.2.

$$Y_j = \beta_0 + \beta_A A_j + \beta_B B_j + \beta_{AB} A_j B_j = 2061.25 - \frac{354.5}{2} A_j - \frac{308.5}{2} B_j - \frac{123.5}{2} A_j B_j$$

Substituting the β values shown in this equation into equation (11.2d) gives

$$A_j = \frac{\tilde{Y} - 2061.25 - (-154.25 B_j)}{-177.25 + (-61.75 B_j)} \tag{11.4}$$

This equation was used to draw all the contour lines in Figure 11.2. For example, consider a tensile strength of 2100 MPa. It is already known (from Table 3.2) that setting factor A high and B low will yield this particular strength so this contour line must pass through the top left hand corner of the inner box in Figure 11.2. Other points along this curve can be found using equation (11.4). For example consider −1.5 for the coded level of factor B. The level for factor A which in combination with this level for factor B that will give a strength of 2100 MPa can be found by substituting $B_j = -1.5$ and $\tilde{Y} = 2100$ into equation (11.4).

$$A_j = \frac{2100 - 2061.25 - (-154.25) \times (-1.5)}{-177.25 + (-61.75) \times (-1.5)} = 2.28$$

Thus factor A = 2.28 and factor B = −1.5 should result in a steel with an average tensile strength of 2100 MPa. This point is highlighted in Figure 11.2. Carrying on in this way all other points on this contour can be found together with all the points defining the other contours shown in Figure 11.2. Curves of higher strength are located towards the bottom left hand corner of the feasible region.

To identify the path of steepest ascent from the centre of the 2^2 design of Section 3.1.2, first choose a step size for one of the factors. This will either be a factor whose minimum step size is technically constrained, or more usually the factor with the largest estimated effect. For example, suppose process conditions are such that the deformation temperature can only be increased in increments of 20.5°C or more. In coded units this corresponds to an increment of 0.273. That is, 20.5°C represents 13.66%

of the full deformation temperature differential, (400 to 550°C). This differential in coded terms is –1 to +1 or 2. Now 13.66% of 2 is in turn 0.273. Let this be the chosen step size for factor B. Because the parameter β_B is negative in the above equation, B_j must be reduced in increments of 0.273 coded units to increase strength.

Next work out the coded value step size for the factor A that will trace out the path of steepest ascent using equations (11.3a). Starting at the centre of the design, where $B_j = 0$, the contour has a slope.

$$\frac{dA_j}{dB_j} = \frac{-\beta_B}{\beta_A + \beta_{AB}B_j} + \frac{\beta_B\beta_{AB}B_j}{[\beta_A + \beta_{AB}B_j]^2} = -\frac{-[-154.25]}{-177.25 - 61.75(0)} + \frac{-154.25[-61.75(0)]}{[-177.25 - 61.75(0)]^2} = -0.8702$$

Factor A must change by 0.8702 coded units to leave strength unchanged following a one unit coded change in factor B. With the new level for, $B_j = -0.273$, the level for A_j required to stay on the path of steepest ascent is from equation (11.3a)

$$A_j = \frac{1}{0.8702}(-0.273) = -0.314$$

Consequently, the new coded levels for factors A and B are respectively, $A_j = -0.314$ and $B_j = -0.273$. These levels will put the ausforming process on a new higher strength contour line with slope.

$$\frac{dA_j}{dB_j} = \frac{-\beta_B}{\beta_A + \beta_{AB}B_j} + \frac{\beta_B\beta_{AB}B_j}{[\beta_A + \beta_{AB}B_j]^2} = -\frac{-[-154.25]}{-177.25 - 61.75(-0.273)} + \frac{-154.25[-61.75(-0.273)]}{[-177.25 - 61.72(-0.273)]^2} = -1.063$$

Factor A must change by 1.063 coded units to leave strength unchanged following a one unit coded change in factor B. With the level of B_j decreasing by a further 0.273 coded units to $B_j = -0.546$, the level for A_j required to stay on the path of steepest ascent is

$$A_j = \frac{1}{1.063}(-0.546) = -0.514$$

Consequently, the new coded levels for factors A and B are respectively, $A_j = -0.514$ and $B_j = -0.546$. The complete path of steepest ascent is then found by continually repeating this process and is shown in Table 11.1.

The coded path of steepest ascent can be converted into the uncoded path of steepest ascent using equation (3.2). That is, by substituting in the coded test conditions along the path of steepest ascent into

Actual test condition = (Coded test condition × {range/2}) + mean test condition.

This uncoded path of steepest ascent is shown in columns four and five of Table 11.1.

Having identified the path of steepest ascent an experiment should be carried out at various points along it, i.e. at the test conditions shown in Table 11.1. These experiments should continue along the path of steepest ascent until either the recorded response starts to diminish from experiment to experiment or until one of the boundaries defining the

Table 11.1 Path of steepest ascent for part of the ausforming process.

Test No.	Coded Test Condition		Actual Test Condition		Predicted Results
	Factor A	Factor B	Factor A, °C	Factor B, °C	MPa
1	0	0	985	475	2061.25
2	−0.314	−0.273	967.75	454.52	2153.676
3	−0.514	−0.546	956.74	434.05	2219.211
4	−0.617	−0.819	951.05	413.57	2265.758
5	−0.64	−1.092	949.75	393.1	2300.076
6	−0.602	−1.365	951.88	372.63	2327.774
7	−0.517	−1.638	956.52	352.15	2353.313
8	−0.404	−1.911	962.74	331.68	2380.005
9	−0.28	−2.184	969.58	311.2	2410.017
10	−0.161	−2.457	976.12	290.73	2444.365

feasible range of test conditions is reached. The last column of Table 11.1 shows the tensile strengths that are predicted to occur along the path of steepest ascent. Note that the experiments start at the centre point of the two level design. The predicted responses are found by substituting the coded test conditions shown in Table 11.1 into

$$Y_j = \beta_0 + \beta_A A_j + \beta_B B_j + \beta_{AB} A_j B_j = 2061.25 - \frac{354.5}{2} A_j - \frac{308.5}{2} B_j - \frac{123.5}{2} A_j B_j \quad (11.5)$$

Take the first experiment, which is at the centre point of the design. The coded test conditions are $A_j = 0$ and $B_j = 0$. Substituting these values into equation (11.5) gives

$$Y_j = 2061.25 - 177.25 \times 0 - 154.25 \times 0 - 61.75 \times 0 = 2061.25$$

After carrying out this first experiment, a strength of around 2061.25 MPa should be observed. The actual test conditions for this first experiment can be found from equation (3.2). Thus $A_j = 0$ corresponds to a true austenitising temperature of

$$985 + \left(0 \times \left(\frac{110}{2}\right)\right) = 985°C,$$

with a deformation temperature of

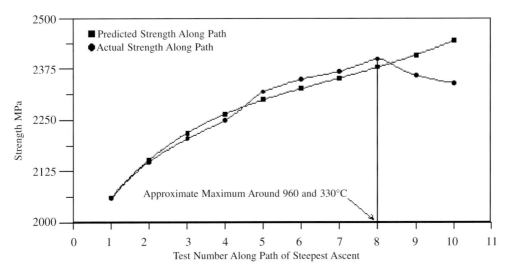

Fig. 11.3 Actual and predicted strengths along path of steepest ascent.

$$475 + \left(0 \times \left(\frac{150}{2}\right)\right) = 475°C$$

Note that the experiments should stop either before test nine, if a maximum has already been observed by then, or at the ninth test because any further experimentation would involve going past the lower limit of 300°C for the deformation temperature. It is therefore predicted that a strength of around 2410 MPa can be achieved through the manipulation of these two factors.

The actual results obtained under the test conditions described by the path of steepest ascent could of course differ from these predicted values. They should not for those tests within the limits of the original experiment, i.e. within the inner box of Figure 11.2. Outside of this range it is quite likely that the responses recorded during the experiments along the path of steepest ascent will differ from those predicted. This is because the above results assume the response surface for strength is the same at all points in the feasible test condition range and this is not likely to be the case. The surface is likely to become more curved as the maximum is approached.

Figure 11.3 plots the predicted path of steepest accent shown in Table 11.1 together with the strength measurements actually made at the test conditions along the path of steepest ascent. There is only a real difference between these two after test eight (which is beyond the limits of the 2^2 design) and a maximum strength seems to be located somewhere around 960°C and 330°C for the austenitising and deformation temperatures respectively. A new two level design should be set up next using these temperatures as

Table 11.2 Path of steepest ascent (coded values) for the ausforming process.

	Factor A	Factor B	Factor C	Factor D	Factor E
1 Unit Change in E	$\dfrac{\beta_A}{\beta_E} = -3.02 =$	$\dfrac{\beta_B}{\beta_E} = -1.03 =$	$\dfrac{\beta_C}{\beta_E} = 1.31 =$	$\dfrac{\beta_D}{\beta_E} = 0.13 =$	-
0.2 Unit Change in E	$0.2 \times -3.02 =$ -0.604	$0.2 \times -1.03 =$ -0.206	$0.2 \times 1.31 =$ 0.262	$0.2 \times 0.13 =$ 0.026	0.2

the central point of the design and a new path of steepest ascent identified. This can be repeated until the approximate optimal conditions are identified.

Notice however that the maximum on the path of steepest ascent is well below the required target of 2600 and so it is likely that the other factors will need to be varied from their low levels to meet such a target.

11.2.2 THE FIRST FIVE FACTORS

The above analysis only provides a partial insight into the ausforming process because all the other potentially important factors were fixed at their low levels. To see whether a strength of 2600 MPa can be obtained by this process, consider again five of the relevant factors. The results of Section 5.5.3 can be used to do this. Here a 2^5 experiment was conducted with the remaining two process variables held fixed at their low levels. The results of the 2^5 design have already been presented in Table 5.15. In Section 6.5.3 a probability plot was used to identify the following simplified model for the process.

$$Y_j = \beta_0 + \beta_A A_j + \beta_B B_j + \beta_C C_j + \beta_E E_j + \beta_{CE} C_j E_j + \beta_{DE} D_j E_j + \varepsilon_j$$

or

$$Y_j = 2261.47 - 246.47 A_j - 84.34 B_j + 106.59 C_j + 81.53 E_j - 168.34 C_j E_j + 175.16 D_j E_j + \varepsilon_j$$

The most technically binding of the five factors present in the above model is the rate of deformation (factor E). Process conditions are such that 0.5 mm per minute is the smallest possible change that can be implemented for factor E. This represents approximately 10% of the actual test condition range considered in the 2^5 design, (0.3 mm per minute to 5 mm per minute). In coded terms this represent an incremental change of 0.2 units, (i.e. 10% of 2). Because the sign in front of E_j in the equation above is positive, strength will be increased by increasing factor E by 0.2 units.

Table 11.3 Test conditions for the path of steepest ascent.

Factor A		Factor B		Factor C		Factor D		Factor E		Predicted Results
Coded	Actual	Coded	Actual	Coded	Actual	Coded	Actual	Coded	Actual	
0	985	0	475	0	65	0	4	0	2.65	2261.47
−0.604	951.8	−0.206	460	0.262	68.92	0.026	4.07	0.2	3.15	2464.246
−1.208	918.6	−0.412	444	0.524	72.85	0.05	4.15	0.40	3.65	2651.38
−1.812	885.3	−0.618	429	0.786	76.78	0.075	4.22	0.60	4.15	2822.329
−2.416	852.1	−0.824	413	1.048	80.72	0.1	4.3	0.80	4.65	2977.674

To move along the coded path of steepest ascent each of the main and first order interaction effects must be considered. The calculations soon become very lengthy. To simplify matters it is common to ignore all interaction effects, i.e. assume $\beta_{CE} = \beta_{DE} = 0$. Then equation (11.3b) can be used. Using this simplifying assumption, each of the other factors must be changed by the amounts shown in Table 11.2 for every 0.2 coded unit increase made to factor E.

From Table 11.2, a change of −0.604 for factor A represents 30.2% of the coded test range for factor A, so that factor A should be decreased in increments of approximately 33°C if the objective is to increase strength. (33°C represents 30.2% of the test condition range 1040 to 930°C). The coded change of −0.206 for factor B represents 10.3% of the coded test range for factor B, so that deformation temperature should be decreased in increments of approximately 15°C. The coded change of 0.262 for factor C represents 13.1% of the coded test range and so the amount of deformation should be increased in increments of approximately 3.9% points. The coded change of 0.026 for factor D represents 1.3% of the coded test range and so the quench rate should be increased in increments of 0.08°C per second.

The path of steepest ascent along which additional experiments should be carried out, is shown in Table 11.3, together with the responses predicted from these experiments. These predicted responses are found by substituting in the coded test values shown in Table 11.3 into the previous equation.

These additional experiments are likely to stop after the fifth test because at this stage the lower boundary for the austenitising temperature is reached. The important point however is that a strength in excess of 2600 MPa is likely to be achieved by the ausforming process. To complete the sequential testing programme, a further 2^5 design should be set up using the fifth test condition (in Table 11.3) as the centre point. If tests for non-linearity suggest considerable curvature, a three level design should be carried out and the maximum identified.

12. Dual Response Surface Methodologies

This chapter introduces a technique that can be used when replicated results are available. It should therefore be seen as an extension to the PerMIA summary statistics approach of Chapters 7 and 8. The extension takes the form of using second, rather than first order response surface models to predict the mean and variance in response. As such this comes at the expense of increased computational burden and complexity in analysis. This dual response surface methodology makes use of replicated tests for calculating process variability, noise factors should only be used as a means of obtaining the replicates. This technique is not suited to the direct inclusion of noise variables in the response surface models for the mean and variance of the process. Further, the technique is only suitable near the global optimum of the process and so will usually be applied at the end of a series of sequential tests that have located the approximate point of optimality. Section 12.1 discusses the dual surface response methodology and Section 12.2 applies this technique to the printing process case study of Section 2.8.

12.1 THE DUAL RESPONSE SURFACE METHODOLOGY

This technique is an extension of the generalised linear model approach to process optimisation discussed in Section 7.6 and Chapter 8. It is extended to take into account any non-linearities that may be present in the manufacturing process being studied. As such, the technique works with second rather than first order response surface models for the mean and variance stabilised transformation. The dual response surface methodology is designed to overcome some of the main shortcomings of the generalised linear model. One of the main reasons for using a PerMIA statistic is that such a transformation enables a distinction to be made between a control and a signal factor. The underlying philosophy of the approach is that the PerMIA, and thus variability independent of the mean, is not affected by a subset of variables that determines the mean. It is then possible to use the control factors to minimise variability and then use the signal factors to attain the target condition for the mean.

However, there is no guarantee that a PerMIA statistic that leads to separate control and signal factors can be found. It will often be the case that those factors that influence the mean will also influence the variability that is independent of the mean. Put another way, any apparent dependency of variability on the mean is just as likely to reflect a common factor influencing both the mean and variability of the process, as it is to reflect the presence of a non normal distribution for the response variable.

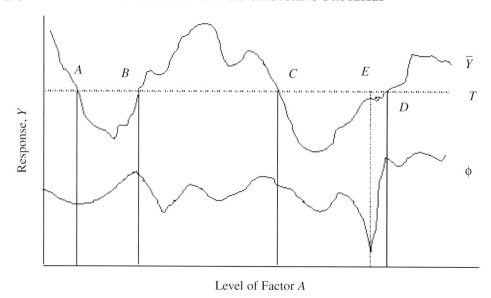

Fig. 12.1 Illustration of constrained optimisation problem.

12.1.1 MINIMISE VARIABILITY SUBJECT TO A MEAN CONSTRAINT

In this more general setting it is impossible to use a two step procedure, as an adjustment of the mean to target is also likely to alter the variability minimised in step one. Vining and Myers[21] have suggested a solution to this problem in the form of a **non-linear constrained optimisation** problem. They suggest that the levels for the design factors should be set so as to minimise variability subject to the constraint that the mean exactly equals some specified target value. This is illustrated in Figure 12.1 for the simple case of a process that has just one design factor, factor A, and no noise factor. Such a constrained optimisation approach requires that each of the factors has at least three levels. In Figure 12.1 factor A has many levels. It should therefore only be used when the manufacturing process is operating at conditions close to a global optimum where curvature exists to justify the use of an extra level.

Figure 12.1 shows that both the mean ($\bar{Y}$) and the variance stabilised transformation, (ϕ) in response are dependant upon the same design factor A. It then becomes impossible to use the two step approaches of Chapter 8. In that chapter, the generalised linear model for the variance stabilised transformation was assumed to contain a subset of factors that did not influence the mean. Then this subset could be used to minimize the variability that was independent of the mean (via the generalised linear model), and the remaining factors could be used to put the mean on target without changing this minimised variance. For the process shown in Figure 12.1 this two step procedure is impossible. Factor A

influences both the variance independent of the mean (ϕ) and the mean itself. So using factor A to put the mean on target will also alter this measure of variability.

Thus in Figure 12.1 there is no signal factor and a two step procedure can't be used. A solution is obtained however using the constrained optimisation rule of Vining and Myers. In Figure 12.1 the constraint that the mean is on target ($T = \overline{Y}$) can be satisfied by setting the level for factor A at one of four possible values. Among them, point A has the minimum variance so that factor A would be set at the level associated with this point.

To formalise this constrained optimisation problem let X_v stand for all the design factors of the experiment. If a process has noise factors these should be used in the laboratory to obtain replicates for the response at each design factor test condition using the crossed array designs shown in subsection 7.6.1. Let Y_{ij} be the $i = 1$ to N responses recorded at the $j = 1$ to M test conditions of the non-linear experimental design. (Either a 3^k, central composite or Box-Behnken design). The scatter in the N responses at any one test condition should reflect the variability in the noise factors and common cause variation. Vining and Myers suggest the use of a second order model to represent the mean and variance stabilised transformation of the process. The mean $\overline{Y}_j$ is calculated from the N replicates of the response at each test condition.

$$\overline{Y}_j = \frac{\sum_{i=1}^{N} Y_{ij}}{N} \tag{12.1}$$

The variance independent of the mean is also calculated from the N replicates of the response at each test condition through the formula (see Section 7.6.4).

$$\phi_j = \frac{S_{Y_j}^2}{\overline{Y}_j^\theta} \tag{12..2}$$

$S_{Y_j}^2$ is the variance calculated from the N responses at test condition j.

$$S_{Y_j} = \sqrt{\frac{\sum_{i=1}^{N}\left[Y_{ij} - \overline{Y}_j\right]^2}{N-1}} \tag{12.3}$$

The value for θ is estimated from the regression

$$\ln(S_{Y_j}^2) = \ln(\phi) + \theta \ln(\overline{Y}_j) + \varepsilon_j \tag{12.4}$$

Then for a process with M_1 design factors, the second order response surface models for the mean and variance stabilised transformations as calculated above are

$$\overline{Y}_j = \beta_0 + \sum_{v=1}^{M_1} \beta_v X_{vj} + \sum_{v=1}^{M_1}\sum_{w=v+1}^{M_1} \beta_{vw} X_{vj} X_{wj} + \sum_{v=1}^{M_1} \beta_{vv} X_{vj}^2 + \varepsilon_{mj} \tag{12.5a}$$

$$\phi_j = \gamma_0 + \sum_{v=1}^{M_1} \gamma_v X_{vj} + \sum_{v=1}^{M_1} \sum_{w=v+1}^{M_1} \gamma_{vw} X_{vj} X_{wj} + \sum_{v=1}^{M_1} \gamma_{vv} X_{vj}^2 + \varepsilon_{sj} \qquad (12.5b)$$

These models should be familiar from Chapters 7 and 10 where A_j, B_j etc. has been replaced by X_{ij}. Both the mean and variance are made dependant upon the level of each factor, the square of the level of each factor and all possible first order interactions between the factors. Notice that in each equation a prediction error exists because the second order model always excludes variables. It is a simplified model in which all higher linear interactions (such as that between factors X_1, X_2 and X_3) and non-linear interactions (such as that between factors X_1, X_2^2) are left out of the model. The constrained optimisation procedure discussed above can be implemented in the following way.

i. Carry out a non-linear design. This should either be a 3^k design if k is small (less than 4), or a central composite design or Box–Behnken design, if k is large.

ii. Estimate equations (12.5a) and (12.5b) using least squares. This involves choosing values for all the β parameters so as to minimise $\Sigma \varepsilon_{mj}^2$ and all the γ parameters so as to minimise $\Sigma \varepsilon_{sj}^2$. Let a hat designate a least squares estimate so that $\hat{\beta}_1$ is the least squares estimate of β_1 for example. Notice that prior to this a variance stabilising transformation must have been selected.

iii. Simplify the estimated second order model by eliminating all those parameters that have t values less than the critical 5% value. Let $\hat{\bar{Y}}$ and $\hat{\phi}_j$ represent the resulting fitted response surfaces. For example, an analyses of t statistics may reveal that only β_0, β_1, β_{11}, γ_0, γ_2 and γ_{22} are significant. The fitted response surfaces are then given by

$$\hat{\bar{Y}}_j = \hat{\beta}_0 + \hat{\beta}_1 X_{1j} + \hat{\beta}_{11} X_{1j}^2$$

$$\hat{\phi}_j = \hat{\gamma}_0 + \hat{\gamma}_2 X_{2j} + \hat{\gamma}_{22} X_{2j}^2$$

iv. Next minimise the fitted response surface for ϕ subject to the fitted response surface for $\bar{Y}$ equalling the target T. That is

Minimise $\hat{\phi}_j$ \hfill (12.6a)

Subject to $\hat{\bar{Y}}_j = T$ \hfill (12.6b)

This can be achieved by using the Lagrange multiplier

$$L_j = \hat{\phi}_j + \lambda(\hat{\bar{Y}}_j - T) \qquad (12.6c)$$

where λ is the Lagrange multiplier. Such an optimum can be identified by finding values for $\hat{\phi}_j, \hat{\bar{Y}}_j$ that satisfy

$$\frac{\partial L}{\partial \hat{\phi}} = 0 \quad \text{and} \quad \frac{\partial L}{\partial \hat{\bar{Y}}} = 0$$

These derivatives are often difficult to solve analytically, but a solution can easily be found using the Solver option in Excel. Optimisation can be carried out at two levels. The first places no constraints on the levels for X_v. This allows for the possibility of the optimum being outside of the test conditions of the experiment. The second places constraints on the levels for X_v so that the optimum must be within the levels set by the experimental design. Both possibilities can be dealt with within Excel. This will be demonstrated in Section 12.2.

12.1.2 MINIMISE THE MEAN SQUARE ERROR

In recent years Lin and Tu[22] have proposed a slight modification of this optimisation procedure. Refer again to Figure 12.1. Point A would be obtained from carrying out the above optimisation procedure proposed by Vining and Myers. Yet it is easy to see that point E would be a better setting for design factor A. By allowing the mean to be a small way off target, the variability can be reduced a great deal below that given by point A. Whenever the mean is off target in this way a **bias** is said to exist.

Point E, if such an advantageous solution actually exists, can be found by trying to minimise the variability plus this bias. This sum known as the **mean square error** (MSE).

$$MSE = \left(\hat{\bar{Y}} - T\right)^2 + \hat{\phi} \qquad (12.6d)$$

A better solution can sometimes be found by trying to minimise the mean square error as defined above. This can be achieved by carrying out steps i to iv above, but in step iv the objective function becomes minimize MSE. Again the Solver tool in Excel can accomplish this optimisation problem.

12.2 THE PRINTING PROCESS CASE STUDY

12.2.1 THE EXPERIMENT

In Section 2.8, the process of printing coloured inks onto packaging labels was discussed in detail. The quality of the finished label was considered to be dependant upon the speed at which the label passes through the spray equipment (X_1), the pressure of the spray from the gun (X_2) and the distance from the spray gun (X_3). All of these are design factors. The measured response, Y_{ij}, was the number of printing defects over the surface of the label. Three replicates were made at each test condition ($N = 3$). After carrying out a sequential testing program the engineers at the printing shop were able to locate the approximate point on the response surface containing the minimum number of defects. They then used this point as the centre point for a 3^3 design. The aim of this non-linear design was to find those operating conditions that would result in no more than 500 surface defects, which is the target specification, T. This will only be achieved

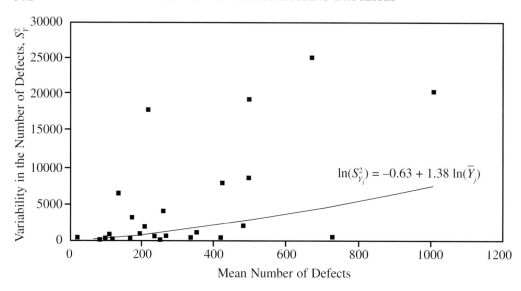

Fig. 12.2 Variability in number of defects as a function of the mean.

on a consistent basis if the mean number of surface defects meets this target with the minimum amount of variation around that mean. Hence this experiment aimed to find those process conditions that ensure $\bar{Y} \leq 500 = T$ with minimum variation. The results from this 3^3 design are shown in Table 12.1.

12.2.2 THE PERMIA

The choice of variance stabilising transformation was more or less dictated by the data. The value for θ could not be determined in the usual way from equation (12.4) because $S^2_{Y_j}$ equalled zero for two of the tests. As a result $\ln(S^2_Y)$ could not be calculated. Faced with this problem two solutions presented themselves. First, a quality control engineer can assume that a certain transformation stabilises the variance. When this approach is taken, a common transformation selected is the square root of the variance, i.e. the standard deviation. This is often used because such a transformation can be calculated when there are zero variances in the data set. The transformation $\sqrt{\phi_j} = S_{Y_j}$ is obtained when the true value of θ is zero.

Alternatively, the data points with zero variances can be eliminated and equation (12.4) estimated using the least squares method. Such an approach yields an estimate for θ of 1.38. Whilst this is significantly different from zero, it is still quite small. In fact, Figure 12.2 shows the relationship between the variance and the mean is

Table 12.1 The printing process study data.

Test	Factor Levels			Response, Y			Mean	Variance	Standard Deviation
	X_1	X_2	X_3	Y_{j1}	Y_{j2}	Y_{j3}	$\bar{Y}_j$	$S^2_{Y_j}$	$\sqrt{\phi_j}$
0 0 0	−1	−1	−1	34	10	28	24	156	12.49
1 0 0	0	−1	−1	115	116	130	120.3333	70.3333	8.3865
2 0 0	1	−1	−1	192	186	263	213.6667	1834.333	42.8291
0 1 0	−1	0	−1	82	88	88	86	12	3.4641
1 1 0	0	0	−1	44	178	188	136.6667	6465.333	80.4073
2 1 0	1	0	−1	322	350	350	340.6667	261.3333	16.1658
0 2 0	−1	1	−1	141	110	86	112.3333	760.3333	27.5741
1 2 0	0	1	−1	259	251	259	256.3333	21.3333	4.6188
2 2 0	1	1	−1	290	280	245	271.6667	558.3333	23.6291
0 0 1	−1	−1	0	81	81	81	81	0	0
1 0 1	0	−1	0	90	122	93	101.6667	312.3333	17.673
2 0 1	1	−1	0	319	376	376	357	1083	32.909
0 1 1	−1	0	0	180	180	154	171.3333	225.3333	15.0111
1 1 1	0	0	0	372	372	372	372	0	0
2 1 1	1	0	0	541	568	396	501.6667	8556.333	92.5005
0 2 1	−1	1	0	288	192	312	264	4032	63.498
1 2 1	0	1	0	432	336	513	427	7851	88.6059
2 2 1	1	1	0	713	725	754	730.6667	444.3333	21.0792
0 0 2	−1	−1	1	364	99	199	220.6667	17908.33	133.822
1 0 2	0	−1	1	232	221	266	239.6667	550.3333	23.4592
2 0 2	1	−1	1	408	415	443	422	343	18.5203
0 1 2	−1	0	1	182	233	182	199	867	29.4449
1 1 2	0	0	1	507	515	434	485.3333	1992.333	44.6356
2 1 2	1	0	1	846	535	640	673.6667	25030.33	158.2098
0 2 2	−1	1	1	236	126	168	176.6667	3081.333	55.5098
1 2 2	0	1	1	660	440	403	501	19303	138.9352
2 2 2	1	1	1	878	991	1161	1010	20293	142.4535

quite weak, so that the variance or the standard deviation itself is almost independent of the mean. Hence the justification for assuming that the standard deviation is a transformation independent of the mean. This variance stabilising transformation is shown in the last column of Table 12.1.

12.2.3 THE MODELLED RESPONSE SURFACE

Tables 12.2 and 12.3 show the least squares procedures for obtaining a second order model for the mean and standard deviation in responses respectively. They are very similar in that the only matrix to differ is the Y vector. In Table 12.2 this vector contains the mean response, whilst in Table 12.3 the Y vector contains the standard deviation in response. The X matrix is the same for both tables and contains the values for all the variables that enter the second order model. The important vectors for optimisation are those obtained for B and Γ. The values in these vectors suggest the following response surfaces.

$$\hat{\bar{Y}}_j = 327.6 + 177.0X_1 + 109.43X_2 + 131.46X_3 + 32.0X_1^2 - 22.39X_2^2 - 29.06X_3^2$$
$$+ 66.03X_1X_2 + 75.47X_1X_3 + 43.58X_2X_3, \tag{12.7a}$$

and

$$\hat{S}_{Y_j} = 34.88 + 11.53X_1 + 15.32X_2 + 29.19X_3 + 4.20X_1^2 - 1.32X_2^2 + 16.78X_3^2$$
$$+ 7.72X_1X_2 + 5.11X_1X_3 + 14.08X_2X_3. \tag{12.7b}$$

12.2.4 MINIMISE VARIABILITY SUBJECT TO A MEAN CONSTRAINT

Now consider the Vining-Myers optimisation procedure. Here the objective is to minimise the standard deviation given by equation (12.7b), subject to the mean given by equation (12.7a) equalling 500 defects. The following three additional constraints ensures that a solution is obtained within the boundaries of the experiment.

$$-1 \leq X_1 \leq 1$$

$$-1 \leq X_2 \leq 1$$

$$-1 \leq X_3 \leq 1$$

Excel can solve this problem using the set up similar to that shown in Table 12.4. This involves the creation of a row containing the values for each of the variables entering the second order models above. It does not matter what initial levels you set for factors X_1 to X_3. In Table 12.4, X_1, X_2 and X_3 are set at the levels associated with the first test shown in Table 12.1, (i.e., they are all set at −1). The other variables are then worked out from these values for X_v using Excel formulas. For example in row three of column I, a formula is inserted that computes the product of the variables X_1 and X_2. Then a formula describing the fitted response surface for the mean, standard deviation and mean square error must be typed in. This is done in cells M3:O3. Table 12.4 shows how equations (12.7) are written within these Excel cells. Once this is done, Solver can be called up by clicking on **Tools** in the menu bar and selecting **Solver** from the drop down menu. This selection is shown in Figure 12.3.

Table 12.2 Least squares estimate of the response surface for the mean.

X_1	X_2	X_3	X_1^2	X_2^2	X_3^2	X_1X_2	X_1X_3	X_2X_3		Mean			
1	-1	-1	-1	1	1	1	1	1	1		24		
1	0	-1	-1	0	1	1	0	0	1		120.333		
1	1	-1	-1	1	1	1	-1	0	-1		213.667		
1	-1	0	-1	1	0	1	0	-1	0		86		
1	0	0	-1	0	0	1	0	0	0		136.667		
1	1	0	-1	1	0	1	0	0	0		340.667		
1	-1	1	-1	1	1	1	-1	-1	-1		112.333		
1	0	1	-1	0	1	1	0	-1	0		256.333		
1	1	1	-1	1	1	1	1	-1	-1		271.667		
1	-1	-1	0	1	1	0	0	0	1		81		
1	0	-1	0	0	1	0	0	0	0		101.667		
1	1	-1	0	1	1	0	0	0	-1		357		
1	-1	0	0	1	0	0	0	0	0		171.333		
1	0	0	0	0	0	0	0	0	0		372		
1	1	0	0	1	0	0	0	0	0		501.667		
1	-1	1	0	1	1	0	0	0	-1		264		
1	0	1	0	0	1	0	0	0	0		427		
1	1	1	0	1	1	0	0	0	1		730.667		
1	-1	-1	1	1	1	1	-1	-1	1		220.667		
1	0	-1	1	0	1	1	0	0	0		239.667		
1	1	-1	1	1	1	1	1	-1	-1		422		
1	-1	0	1	1	0	1	-1	0	0		199		
1	0	0	1	0	0	1	0	0	0		485.333		
1	1	0	1	1	0	1	1	0	0		673.667		
1	-1	1	1	1	1	1	-1	1	-1		176.667		
1	0	1	1	0	1	1	0	1	0		501		
1	1	1	1	1	1	1	1	1	1		1010		

$X =$ (matrix above left), $Y =$ (column above right)

$$X'Y = \begin{bmatrix} 8496 \\ 3186 \\ 1969.67 \\ 2366.33 \\ 5856 \\ 5529.67 \\ 5489.67 \\ 792.333 \\ 905.667 \\ 523 \end{bmatrix}$$

$$B = \begin{bmatrix} 327.63 \\ 177 \\ 109.426 \\ 131.463 \\ 32 \\ -22.389 \\ -29.056 \\ 66.0278 \\ 75.4722 \\ 43.5833 \end{bmatrix} \begin{matrix} = \beta_0 \\ = \beta_1 \\ = \beta_2 \\ = \beta_3 \\ = \beta_{11} \\ = \beta_{22} \\ = \beta_{33} \\ = \beta_{12} \\ = \beta_{13} \\ = \beta_{23} \end{matrix}$$

Table 12.2 Continued.

Table 12.3 Least squares estimate of the response surface for the standard deviation.

	X_1	X_2	X_3	X_1^2	X_2^2	X_3^2	X_1X_2	X_1X_3	X_2X_3		S_y
1	-1	-1	-1	1	1	1	1	1	1		12.49
1	0	-1	-1	0	1	1	0	0	1		8.3865
1	1	-1	-1	1	1	1	-1	-1	1		42.8291
1	-1	0	-1	1	0	1	0	1	0		3.4641
1	0	0	-1	0	0	1	0	0	0		80.4073
1	1	0	-1	1	0	1	0	-1	0		16.1658
1	-1	1	-1	1	1	1	-1	1	-1		27.5741
1	0	1	-1	0	1	1	0	0	-1		4.6188
1	1	1	-1	1	1	1	1	-1	-1		23.6291
1	-1	-1	0	1	1	0	1	0	0		0
1	0	-1	0	0	1	0	0	0	0		17.673
1	1	-1	0	1	1	0	-1	0	0		32.909
1	-1	0	0	1	0	0	0	0	0		15.0111
1	0	0	0	0	0	0	0	0	0		0
1	1	0	0	1	0	0	0	0	0		92.5005
1	-1	1	0	1	1	0	-1	0	0		63.498
1	0	1	0	0	1	0	0	0	0		88.6059
1	1	1	0	1	1	0	1	0	0		21.0792
1	-1	-1	1	1	1	1	1	-1	-1		133.822
1	0	-1	1	0	1	1	0	0	-1		23.4592
1	1	-1	1	1	1	1	-1	1	-1		18.5203
1	-1	0	1	1	0	1	0	-1	0		29.4449
1	0	0	1	0	0	1	0	0	0		44.6356
1	1	0	1	1	0	1	0	1	0		158.21
1	-1	1	1	1	1	1	-1	-1	1		55.5098
1	0	1	1	0	1	1	0	0	1		138.935
1	1	1	1	1	1	1	1	1	1		142.454

$X = $ (leftmost block) $Y = $ (S_y column)

$$X'Y = \begin{pmatrix} 1295.83 \\ 207.482 \\ 275.815 \\ 525.426 \\ 889.11 \\ 855.993 \\ 964.555 \\ 92.6335 \\ 61.3111 \\ 168.981 \end{pmatrix}$$

$$\Gamma = \begin{pmatrix} 34.8833 \\ 11.5268 \\ 15.323 \\ 29.1903 \\ 4.20374 \\ -1.3159 \\ 16.7779 \\ 7.71946 \\ 5.10926 \\ 14.0817 \end{pmatrix} \begin{matrix} = \gamma_0 \\ = \gamma_1 \\ = \gamma_2 \\ = \gamma_3 \\ = \gamma_{11} \\ = \gamma_{22} \\ = \gamma_{33} \\ = \gamma_{12} \\ = \gamma_{13} \\ = \gamma_{23} \end{matrix}$$

Table 12.3 Continued.

$$X^t = \begin{bmatrix} \text{(27-column design matrix with entries } -1, 0, 1\text{)} \end{bmatrix}$$

$$X^tX = \begin{bmatrix}
27 & 0 & 0 & 0 & 18 & 18 & 18 & 0 & 0 & 0 & 0 & 0 & 0 \\
0 & 18 & 0 & 0 & 0 & 0 & 0 & 0 & 0 & 0 & 0 & 0 & 0 \\
0 & 0 & 18 & 0 & 0 & 0 & 0 & 0 & 0 & 0 & 0 & 0 & 0 \\
0 & 0 & 0 & 18 & 0 & 0 & 0 & 0 & 0 & 0 & 0 & 0 & 0 \\
18 & 0 & 0 & 0 & 18 & 12 & 12 & 0 & 0 & 0 & 0 & 0 & 0 \\
18 & 0 & 0 & 0 & 12 & 18 & 12 & 0 & 0 & 0 & 0 & 0 & 0 \\
18 & 0 & 0 & 0 & 12 & 12 & 18 & 0 & 0 & 0 & 0 & 0 & 0 \\
0 & 0 & 0 & 0 & 0 & 0 & 0 & 12 & 0 & 0 & 0 & 0 & 0 \\
0 & 0 & 0 & 0 & 0 & 0 & 0 & 0 & 12 & 0 & 0 & 0 & 0 \\
0 & 0 & 0 & 0 & 0 & 0 & 0 & 0 & 0 & 12 & 0 & 0 & 0 \\
0 & 0 & 0 & 0 & 0 & 0 & 0 & 0 & 0 & 0 & 12 & 0 & 0 \\
0 & 0 & 0 & 0 & 0 & 0 & 0 & 0 & 0 & 0 & 0 & 12 & 0 \\
0 & 0 & 0 & 0 & 0 & 0 & 0 & 0 & 0 & 0 & 0 & 0 & 12
\end{bmatrix}$$

$$(X^tX)^{-1} = \begin{bmatrix}
0.25926 & 0 & 0 & 0 & -0.1111 & -0.1111 & -0.1111 & 0 & 0 & 0 & 0 & 0 & 0 \\
0 & 0.05556 & 0 & 0 & 0 & 0 & 0 & 0 & 0 & 0 & 0 & 0 & 0 \\
0 & 0 & 0.05556 & 0 & 0 & 0 & 0 & 0 & 0 & 0 & 0 & 0 & 0 \\
0 & 0 & 0 & 0.05556 & 0 & 0 & 0 & 0 & 0 & 0 & 0 & 0 & 0 \\
-0.1111 & 0 & 0 & 0 & 0.16667 & 0 & 0 & 0 & 0 & 0 & 0 & 0 & 0 \\
-0.1111 & 0 & 0 & 0 & 0 & 0.16667 & 0 & 0 & 0 & 0 & 0 & 0 & 0 \\
-0.1111 & 0 & 0 & 0 & 0 & 0 & 0.16667 & 0 & 0 & 0 & 0 & 0 & 0 \\
0 & 0 & 0 & 0 & 0 & 0 & 0 & 0.08333 & 0 & 0 & 0 & 0 & 0 \\
0 & 0 & 0 & 0 & 0 & 0 & 0 & 0 & 0.08333 & 0 & 0 & 0 & 0 \\
0 & 0 & 0 & 0 & 0 & 0 & 0 & 0 & 0 & 0.08333 & 0 & 0 & 0 \\
0 & 0 & 0 & 0 & 0 & 0 & 0 & 0 & 0 & 0 & 0.08333 & 0 & 0 \\
0 & 0 & 0 & 0 & 0 & 0 & 0 & 0 & 0 & 0 & 0 & 0.08333 & 0 \\
0 & 0 & 0 & 0 & 0 & 0 & 0 & 0 & 0 & 0 & 0 & 0 & 0.08333
\end{bmatrix}$$

Table 12.4 A typical spreadsheet for use with solver in Excel.

	X_1	X_2	X_3	X_1^2	X_2^2	X_3^2	X_1X_2	X_1X_3	X_2X_3	$\hat{Y}_j$	S_Y	MSE
	−1	−1	−1	1	1	1	1	1	1	75.4	25.47	180934

$X_1^2 \leftarrow (C3^{\wedge}2)$
$X_2^2 \leftarrow (D3^{\wedge}2)$
$X_3^2 \leftarrow (E3^{\wedge}2)$
$X_1X_2 \leftarrow (C3*D3)$
$X_1X_3 \leftarrow (C3*E3)$
$X_2X_3 \leftarrow (D3*E3)$

=327.63+177*C3+109.43*D3+131.43*E3+32*F3-22.39*G3-29.06*H3+66.03*I3+75.47*J3+43.58*K3

=−34.88+11.53*C3+15.32*D3+29.19*E3+4.20*F3-1.32*G3+16.78*H3+7.72*I3+5.11*J3+14.08*K3

(M3-500)^2+N3^2

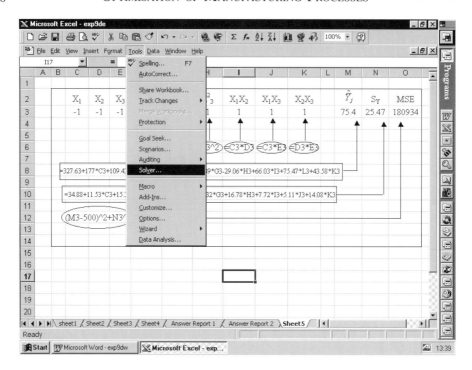

Fig. 12.3 Excel spreadsheet showing how variables in Table 12.4 are calculated.

The user friendly menu shown in the Figure 12.4 will then appear.

In the **Set Target Cell** box type in $N$3 and then click on the **Min** radio button. This tells Excel that the objective is to minimise the standard deviation. In the **By Changing Cells** box type in $C$3:$E$3. This tells Excel that the minimisation is to be done by looking for different levels for the variables X_1 to X_3. Note that because formulas for calculating X_1^2 etc. have been typed in there is no need to include these variables in this box. Next use the **Add** button to type in the various constraints shown in the **Subject to the Constraints box** in Figure 12.4. The main constraint shown above is that the mean must equal the target. Excel can be told this by using the Add button and typing in $M$3 = 500 in the Add Constraint box which then appears. The other constraints shown state that the levels for X_1 to X_3 must be within the design limits, i.e. −1 to +1. To do this use the Add button again and type in $C$3> = −1 and $C$3< = 1. This is the constraint $-1 \leq X_1 \leq 1$. This then needs to be repeated for variable X_2 and X_3. See the Subject to Constraints box in the Figure 12.4.

Once all the constraints are typed in press the Solve button. A further menu, shown in Figure 12.5 appears.

Dual Response Surface Methodologies

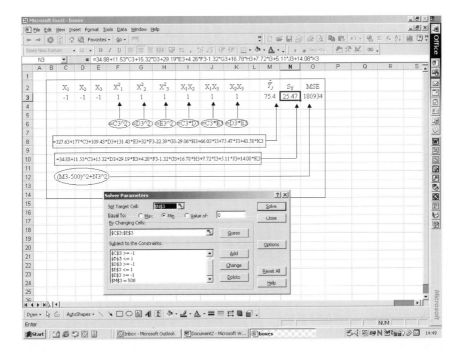

Fig. 12.4 Solve menu in Excel.

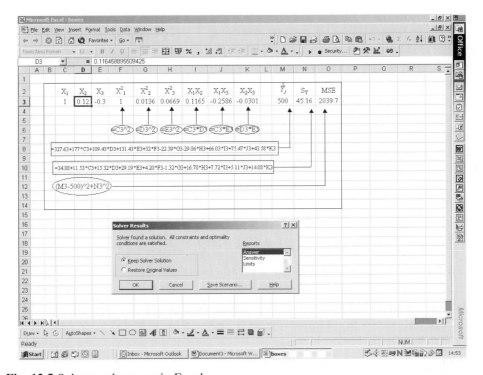

Fig. 12.5 Solve results menu in Excel.

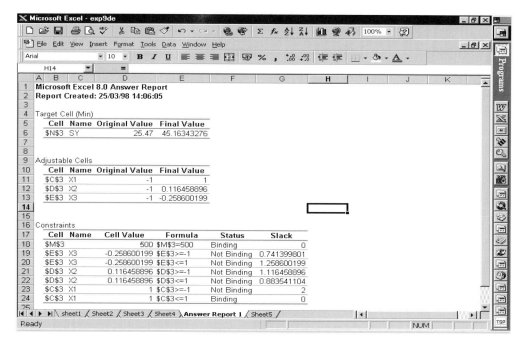

Fig. 12.6 Answer report 1.

Select restore original values and select the answer option from the reports box. Once this is done a new sheet is created in Excel with the name Answer Report 1. Figure 12.6 shows the content of this new sheet.

The first block of results states that when all factors were set low the standard deviation S_Y equalled 25.47. However, the minimum value for S_Y, consistent with being on target, is 45.16. The second block of results states that this minimum is obtained by setting the coded level for $X_1 = 1.0$, the coded level for $X_2 = 0.12$ and the coded level for $X_3 = -0.26$. The final block simply states which of the constraints were actually binding at the optimal point. For example, having to be on target influenced the optimal solution for S_Y as did having to have the value for X_1 less than one.

In summary, setting the printing speed level at 1.0, the level for spray pressure at 0.12 and the distance from spray gun at a level of –0.26 will result in the mean number of printing defects being on target, i.e. $\bar{Y} = 500$. There is no bias. The resulting variance is $45.16^2 = 2039.43$. The mean square error is also 2039.43 because the mean is on target, i.e. there is no bias.

12.2.5 THE MEAN SQUARE ERROR

The interesting question that arises from this set of results concerns the issue of whether it is possible to vastly reduce the variance at the expense of a small deviation of the mean

Dual Response Surface Methodologies

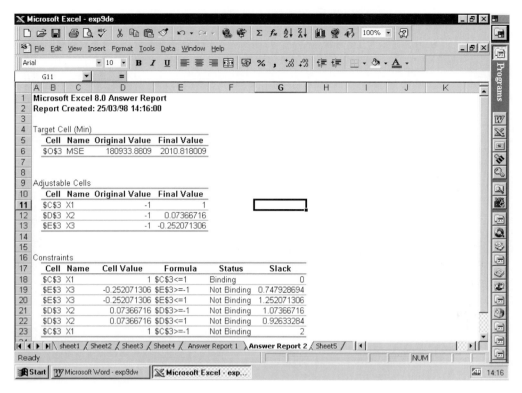

Fig. 12.7 Excel spreadsheet showing results of minimising the MSE subject only to the constraint that the levels for each field must be in the range −1 to +1.

from the target so that the sum of the two, i.e. the MSE, is actually lower than 2039.43? This is of course the Lin and Tu optimisation criteria of minimising the MSE. This is easily done in Excel by changing two options on the Solve Parameters menu. First, type in $O$3 into the **Set Target Cell** box and secondly remove $M$3 = 500 from the **Subject to Constraint** box. (Use the delete button to make this last change). Now the objective is to minimise the MSE subject only to the constraint that the levels for each factor must be in the range −1 to +1. The result is shown Figure 12.7. A lower MSE of 2010.82 is obtained. The mean number of defects is 494.625 (bias = $(500 - 494.625)^2 = 28.89$) and the variance is 1981.14, which is lower than the original above. Clearly this is a much better solution.

Finally, the objective of this experiment was to find those process conditions that resulted in a defect rate of no more than 500 (rather than exactly 500). To search for this solution add $M$3 ≤ 500 to the Subject to Constraint box and run Solver again. The

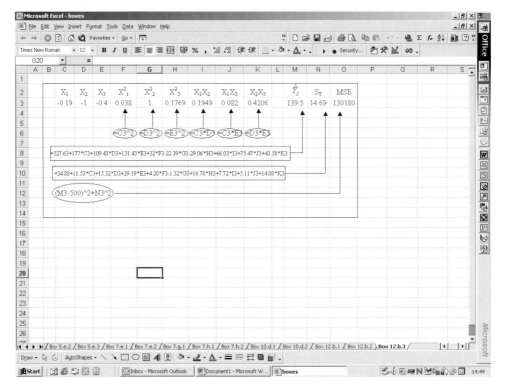

Fig. 12.8 Excel spreadsheet showing process conditions that resulted in a defect rate of no more than 500.

result is shown in Figure 12.8. The mean number of defects is 139.5 with a variance of $14.69^2 = 215.79$. The MSE is meaningless in this solution because there is no specific target. This is the best solution of the three shown and is obtained by setting the printing speed level at –0.19, the level for pressure at –1 and the distance at a level of –0.4.

References

1. G. Taguchi, *System of Experimental Design*, ASI and Quality Resources, **1** and **2**, 1987.
2. G.E.P. Box and S. Jones, 'Split-Plot for Robust Product Experimentation', *Journal of Applied Statistics*, 1992, **19**, 3–26.
3. D.E. Coleman and D.C. Montgomery, 'A Systematic Approach to Planning for a Designed Experiment (with discussion)', *Technometrics*, 1993, **35**, 1–28.
4. W.E. Duckworth and P.R. Taylor, 'Ausforming High Alloy Steels', *Iron and Steel Institute Special Report*, 1964, **86**, 61–70.
5. W.E. Duckworth, P.R. Taylor and D.A. Leak, 'Ausforming Behaviour of En24, En30B and Experimental 3Cr-Ni-Si Steel', *Journal Iron and Steel Institute*, 1964, **202**, 135–142.
6. K. Lowe, ed., *Competition between Steel and Aluminium for the Passenger Car*, International Iron and Steel Institute, 1994.
7. R.A. DePaul and A.L. Kitchen, 'The Role of Nickel, Copper, and Columbium (Niobium) in Strengthening a Low-Carbon Ferritic Steel', *Metallurgical Transactions*, 1970, **1**, 389–393.
8. R.W. Evans and G.M. McColvin, 'Hot-Forged Copper Powder Compacts', *Powder Metallurgy*, 1976, **4**, 202–209.
9. R.W. Evans, 'Sensitivity of Forging Models to Boundary Conditions', *Rolling and Forging: Measurement of Friction and Heat Transfer Coefficients; Feasibility Study Report*, M.S. Loveday ed., National Physical Laboratory, 1994.
10. R.W. Evans, 'The Influence of Operating Conditions Friction Welding', Unpublished Undergraduate Project, Department of Materials Engineering, UWS, 1996.
11. J. Engel, 'Modelling Variation in Industrial Experiments', *Applied Statistics*, 1992, **41**, 579–593.
12. J.J. Pignatiello and J.S. Ramberg, Discussion, *Journal of Quality Technology*, 1985, **17**, 198–206.
13. G.E.P. Box and N.R. Draper, *Empirical Model-Building and Response Surfaces*, John Wiley & Sons, 1987.
14. D.C. Montgomery, *Design and Analysis of Experiments*, 4th Edition. John Wiley & Sons, 1997.
15. M.S. Phadke, *Quality Engineering using Robust Design*, Prentice-Hall, NJ, 1989.
16. R.A. Fisher and F. Yates, *Statistical Tables for Biological, Agricultural and Medical Research*, Oliver & Boyd Ltd, 1946.
17. D.M. Steinberg and D. Bursztyn, 'Dispersion Effects in Robust Design Experiments with Noise Factors', *Journal of Quality Technology*, 1994, **26**, 12–20.

18. J.A. Nelder and D. Pregibon, 'An Extended Quasi-Likelihood Function', *Biometrika*, 1987, **74**, 221–232.
19. N. Logothetis, 'Box-Cox Transformations and the Taguchi Method', *Applied Statistics*, 1990, **39**, 31–48.
20. Microsoft® Excel 8 (Office 97), Microsoft Corporation.
21. G.G. Vining and R.H. Myers, 'Combining Taguchi and Response Surface Philosophies: A Dual Response Approach', *Journal of Quality Technology*, 1990, **22**, 38–45.
22. D.K.J. Lin and W. Tu, 'Dual Response Surface Optimisation', *Journal of Quality Technology*, 1995, **27**, 34–39.

Index

A

Aero engine disks, 21
Alias algebra, 70, 74, 76, 102, 175, 188, 189
Aliased effect, 69, 71, 74, 76, 97, 102, 189
Alternate fraction, 39, 40, 43, 45, 69
Ausforming process, 4, 17, 29, 30–37, 39, 40, 45–48, 76, 78–81, 87, 96, 97, 101, 103, 109 120, 132, 137, 214, 285, 288, 291, 294, 295
Austenitising temperature, 17, 29, 30, 40, 76–79, 83, 85, 95, 289, 292, 295
Austenitising time, 17, 46, 101
Axial force, 24, 235, 236, 272, 275, 278, 282
Axial points, 238, 239, 240, 241, 242, 244, 287

B

Base design, 40, 43, 46, 71, 185
Bias, 301, 312, 313
Box-Behnken design, 231, 245, 246, 247, 249, 250, 263, 299, 300
Box-Behnken design for k = 3 factors, 248, 249
Box-Behnken design for k = 4 factors, 249, 250

C

Central composite design, 231, 238, 241, 242, 244, 249, 251, 263, 300
Central limit theorem, 110
Centre points, 234, 238, 241, 244, 264, 287
Coded test condition, 31, 36, 85, 232, 233, 236
Coefficient of friction, 22, 226
Common cause, 35, 109, 137, 171, 299
Confounded interaction, 70, 72, 76
Constant contour plot, 209
Constrained optimisation, 12, 209, 212, 298
Contours of constant mean, 212, 214
Contours of constant variance, 213
Contours on the response surface, 287
Contrast, 65, 66, 74, 101, 120, 128
Control factor, 191, 193, 211
Control matrix, 60, 78
Conveyor belt speed, 25, 26

Copper, 18, 19, 128, 138, 171, 172, 209, 218
Copper Compact Experiment, 138, 171, 181
Cube, 33, 235, 240, 249, 250
Cuboidal designs, 241
Cumulative probability, 129
Cycle time, 24, 204, 222

D

Deformation, 13, 17, 24, 29, 30, 33, 40, 77, 143, 289
Deformation rate, 36, 40, 143, 214
Deformation temperature, 40, 77, 83, 135, 214
Degree of upset, 24, 235, 243, 272, 273, 280, 282
Degrees of freedom, 118, 144, 265, 278
Density, 19, 21, 171
Design factors, 141–143, 149, 160, 169, 188, 195, 204
Design fractional factorial design - 2^{k-3}, 39, 43
Design generator, 40, 42, 53, 70, 74, 185
Designed experiment, 13, 125
Disk forging process, 137, 148, 153, 156, 209, 226
Dispersion design, 9, 11
Dispersion effect, 138, 141, 147, 155, 159, 163 167, 170, 191, 207, 211, 226, 258
Distance, 24, 256, 289, 301, 314
Distribution of effect estimates, 109
Dual response surface methodology, 297

E

Emissive power, 22, 148
Energy, 4, 22, 148, 149, 226, 227
Excel, 67, 85, 169, 173–175, 182, 192
Experiment, 3, 9, 11, 14, 19, 24, 25, 29, 32
Experimental unit, 13, 14

F

Face centred central composite design, 241, 249
Factor, 5, 8, 10, 17, 29, 109, 120
Factor contrast, 65, 66, 82, 101, 106, 128
Factorial design, 14, 29, 57, 58, 67, 68, 80, 109, 137, 141, 142, 231, 250, 263

Factorial design - 2^5, 36, 85, 109
Factorial design - 2^k, 14, 36, 38, 231
Factorial design - 3^1, 232, 236
Factorial design - 3^2, 233
Factorial design - 3^3, 235
Factorial design - 3^k, 231
Factorial design - 2^2, 29, 30, 32, 33, 76, 160,
Factorial design - 2^3, 33, 34, 80, 243
Factorials with mixed levels, 15
Factors at two and four levels, 251, 253
Factors at two and three levels, 15, 251
Feasible range of test conditions, 289, 292
Finite element, 22, 24
First order interaction, 58, 59, 63, 64, 66, 71
    95, 223, 260, 295
First order response surface model, 60, 62, 63, 79,
    85, 133, 137, 141, 142
Forging, 137, 148, 149, 156, 171
Forging die, 19, 20, 21
Forging load, 19, 21
Fractional factorial, 137, 138, 142, 159, 184, 185
Fractional factorial - 2^{7-3}, 175
Fractional factorial design - 2^{7-3}_{IV}, 45, 46, 47, 218
Fractional factorial design - 2^{k-1}, 39, 40, 169
Fractional factorial design - 2^{k-2}, 39, 43
Fractional factorial design - 2^{k-3}, 39, 43
Fractional factorial design - 2^{k-p}, 39, 43, 44
Fractional factorial design - 2^{2-1}, 39, 68
Frequency, 7, 9, 11, 12, 24, 209, 210, 235, 236
    272, 273, 275, 280
Friction welding experiment, 237, 242, 243, 245
    272, 273, 274

G

Generalised linear model, 137, 159, 160, 171
    193, 297, 298
Generalised non linear model, 232
Generalised non linear model for k = 2, 232,
    241, 263
Geometric representation of a face centred
    central composite design, 241
Geometry of a 2 factor central composite
    design, 236, 239, 241
Geometry of a 3 factor central composite
    design, 231, 240, 242
Geometry of the 2^2 factorial design, 29, 30
Geometry of the 2^3 factorial design, 33, 36

Geometry of the 3^2 factorial design, 231, 233, 235
Geometry of the 3^3 factorial design, 231, 233, 241
Global maximum, 285, 286

H

Heat transfer coefficient, 22, 148, 226
Heater bands, 24
Heating time, 26
High strength steel, 36, 109, 128, 130
Higher order interactions, 59, 70, 71, 109, 122,
    143, 189, 193, 194, 195
Hold down time, 26

I

Identity matrix, 67
Injection moulding experiment, 24,
    184, 190, 196, 204, 222
Injection speed, 204, 205
Inner array, 159, 160, 161, 184, 185, 189
Interaction, 139, 140, 142, 143, 149, 163
Interaction location effect, 58, 59, 68, 78, 227, 228
Interaction table, 73, 74
Isothermal incubation time, 17, 46, 101

L

Least squares procedure, 64, 66, 67, 85, 91,
    122, 173, 192, 255, 278
Level, 4, 12, 14, 30, 31, 33, 39, 51, 57, 59, 211, 215
Linear, 5, 6, 8, 23, 29, 43, 50, 231
Linear effect, 160, 231, 255, 256
Linear graphs, 50, 74
Linear one factor model, 256
LINEST, 67, 86, 265, 280, 281
Location effect, 57, 59, 257–259, 262
Location experimental design, 8
Long tailed distributions, 165

M

Main dispersion effect, 138
Main location effect, 57, 58, 68, 69, 76, 110, 195, 222
Mean on target, 11, 12, 165, 166, 195, 221, 298, 299
Mean response, 9, 138, 141, 143, 162, 164, 243,
    261, 304
Mean square error, 301, 304, 312
Minimising process variability, 137, 138, 170, 223
Minimum variance, 215, 218, 299
Mixed level designs, 263

N

Nickel, 4, 18, 128
Niobium, 18, 128
Noise, 109, 137, 139, 140, 142, 184, 212, 214
Noise–design factor interaction, 139
Noise factor, 5, 9, 139, 141, 143, 260, 298
Non linear, 6, 8, 14, 38, 167, 231, 255, 260
Non linear effect, 231, 255
Normal distribution, 147, 209, 297

O

Oil quench temperature, 26
Optimal design generators, 171
Orthogonal designs, 14, 45, 188
Outer array, 159, 160, 161, 184, 188, 207

P

Path of steepest ascent,
   285, 286, 287, 290, 291, 295
PerMIA, 137, 160, 166, 167, 169, 170, 174, 297, 302
Pipe shrinkage, 24, 223
Polynomial, 255, 273, 276, 277
Powder compacts, 18, 21
Powder size, 19, 171, 220, 222
Power of the mean model, 167
Prediction error, 137, 142, 143, 144, 146, 213
   215, 260, 264
Preform, 18, 21, 22, 171
Pressure, 24, 25, 26, 204
Principle fraction, 45, 46, 96
Printing defects, 26, 301, 312
Printing process, 26, 297, 301
Probability plot, 126, 128, 132, 133
Process optimization, 3, 4, 6, 13, 53, 209, 212,
   214, 221, 285, 297
Process variability, 6, 7, 8, 9, 11, 139, 159,
   209, 211, 212, 213, 215, 216, 219
Proof stress, 21
Pure dispersion effect, 167, 169, 170, 172, 175,
   176, 179, 180, 181, 259

Q

Quadratic, 14, 255, 257, 258, 259, 273, 278
Quadratic one factor model, 257, 258
Quality characteristics, 4, 6, 9, 57, 255
Quench rate, 17, 214

R

Random order, 32, 35
Range, 29, 31, 33
Replicated Design, 116, 120
Replication, 38, 66
Resolution, 185, 188, 195, 207
Response, 137, 138
Response surface, 137, 141
Robust design, 3, 9, 13
Rotatable design, 238

S

Scatter, 122, 131, 133
Second order interaction, 122, 149
Second order response surface model, 255, 260
   261, 263, 264
Sequential experimentation, 288
Signal factor, 166, 170, 211, 212, 297, 299
Signal to noise ratio, 167, 188, 191
Significance level, 124, 277
Simplified first order response surface model,
   133, 142, 143, 204
Simplified response surface model, 146, 149, 220
Simplified second order response surface
   model, 149
Sintering time, 221, 222
S-N_T, 160, 166
Solver, 212, 216, 217, 218, 223, 224, 301, 304,
   309, 313
Spherical design, 240
Spray pressure, 312
Spring free height, 25, 50, 51, 73, 74
Standard deviation
of an estimated effect, 109, 112
of the prediction errors, 123, 124
of the response, 162
Standard deviation of parameters in a second
order response surface model, 265
Standard order form, 29, 30, 31, 33, 36, 46, 64
   65, 67, 74, 264, 266, 267, 269, 271
Summary statistics, 160, 166, 168, 297
Surface area, 22, 148, 156, 226

T

t test, 237
Target, 7, 8, 9, 12, 57, 80, 85, 107, 135

Tensile strength, 118, 120, 133, 143
Three way diagram, 273
Titanium 21
Tolerance band, 17, 139
Traditional experimentation, 14, 30, 77
Transfer time, 26
Transformations independent of the mean, 165–167
Transmitted variation, 141
Transpose, 265, 273
Truck suspension system, 25
Two step optimization, 9, 12, 212
Two way diagram, 205

U

Unreplicated design, 36, 125, 146, 160, 231, 260, 269

V

Variance in response, 164, 173, 297
Variance of the prediction error, 215, 223
Variance stabilising transformation, 167–169, 300, 302, 303
Virtual experiment, 22, 24, 148

W

Weld amplitude, 24
Weld frequency, 24

Y

Yates procedure, 121, 130, 149, 150, 164, 169, 174, 175, 176, 189, 195

Z

Z table, 131
Z value, 126, 131, 133